PREPARING PROJECTS FOR SITE CONSTRUCTION

Smart Practices for Profitable Land Development

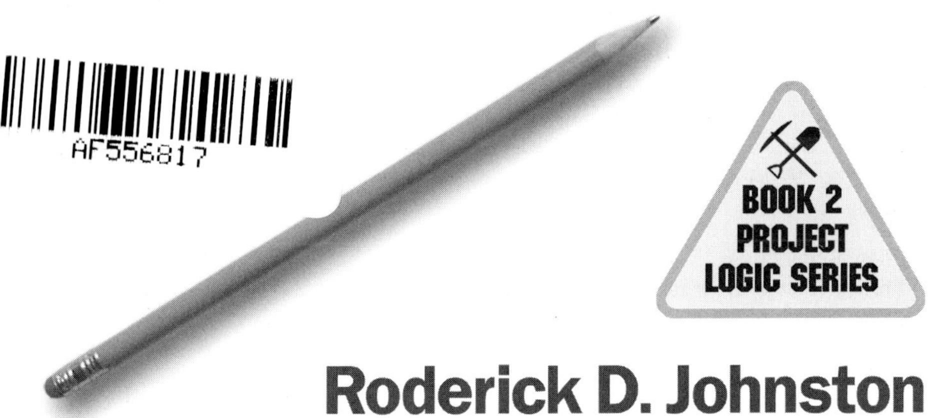

Roderick D. Johnston

Trans Mountain Publishing
Venture Scope Inc.

Fall City, Washington

Library of Congress Control Number: 2003098216

ISBN: 0971987211

This book is dedicated to my wonderful mother, Joanne

Rod Johnston, Venture Scope Inc.,
P.O. Box 1089, Fall City, Washington 98024-1089

Order this book online at Transmtn.com

Johnston, Roderick D.
Preparing projects for site construction / Roderick D. Johnston.
p. cm. -- (Project logic series ; bk. 2)
Includes index.
LCCN 2003098216
ISBN 0971987211

1. Construction industry--Management. 2. Building--Planning. 3. Project management. I. Title.

TH438.J59 2004 690'.068
QBI03-200918

TABLE OF CONTENTS

Illustrations

FOREWORD

In real estate development, much of what we know and believe is based on our experience accumulated over years of undertaking different projects. This is a great way to learn, but, unfortunately, the tuition can be very expensive. Have you ever said, "I wish I knew then, what I know now"? That's what this book is all about: providing information from years of experience so that the reader can *avoid* problems rather than have to learn from them.

For many types of projects, classes can offer most of what is needed. Vertical construction, for example, is quite predictable and has few of the unknown variables that exist in land development. After all, the bulk of the construction risk disappears when a project comes out of the ground. That is why the site-work phase of land development usually carries a greater contingency than the construction phase.

Fortunately, site work is usually a small part of a project since most buildings are built in urban areas, or built where utilities are available and the land has been previously cleared and prepared. But, the scope of the project increases when there are large parcels of raw land, or sites that have been passed over due to challenges. This is especially true today because in most areas the good land has already been developed. While the principles in this book can be applied to developing relatively small and easy projects to large-scale master planned communities, those in the industry contemplating or currently engaged in tough terrain or difficult projects will find it must-have reading.

Accumulated from over two decades of working with road location and engineering, site construction, and land development, Rod Johnston has assembled his wisdom and learning and written this hands-on guide to land development. It is clear that he has been in the trenches dealing with issues that owners usually don't understand and hope they will never have to address themselves. Some of his ideas may "ruffle the feathers" of consultants, contractors, and regulators, but they are all geared to one goal: helping an owner build a better project with less cost, time, and hassle. As a result, I expect this book will also stimulate dialogue among the many disciplines involved in site development. If so, we will all benefit.

This book is valuable not only for the project manager in charge of site development and preparation, but also for any owner who must assemble a team and provide overall leadership on a project. It offers insight, tips, and analysis that would be hard, if not impossible, to learn in the classroom. It raises important questions for an owner to consider at the very early stage of development and explains why early attention is so critical. Even for commonly encountered issues, it provides new perspectives. For example, it explains clearly why an area such as geotechnical work is "more art than science," involving subjective judgment and balancing of risks. The wealth of experience, insight, and advice in these pages can help anyone in site development and should make a lasting contribution to the real estate industry.

—Judd Kirk
President, Port Blakely Communities

PREFACE

In 1964, the state of New York hosted an extravaganza in Queens called the New York World's Fair. For school kids growing up in Uniondale, Long Island, a fair typically meant field trips: extended and exciting bus rides, an extraordinary lunch, some pocket money, and the main event. School trips into the city were highly prized outings, but for me, this one was special. Little did I realize how this fair would set the direction for how I would someday make a living and thus create this book series.

As with most global expositions, the New York World's Fair advertised futuristic living seasoned with a heavy dose of space exploration and gadgetry. It was simply a monster event. Cloaked in chrome, computers, and colors of endless neon, it beckoned—no—it dared us to dream of a better world more advanced and exciting than anything we could hope to imagine, much less absorb.

All the big boys showed up. Ford had a great ride. Sinclair Oil's dinosaur was there. Johnson & Johnson, Coca Cola, and General Electric all had terrific interactive displays. The list goes on, but for me, it was the General Motors Futurama tour that did it. Strapped in cars, we drove through a succession of exhibits, each growing in magnitude, describing how future growth would occur worldwide and I loved every second of it. It was my first taste of large-scale land development. There was rugged terrain, logging, road building and transportation networks that actually worked—and cities—cities were being developed everywhere imaginable from the most remote jungles and deserts to the bottom of the sea. Every step of the process was portrayed in sequence as we progressed from scene to scene. By the time, I had traveled that glorious tour repeatedly, a funny thing happened. Without realizing it, I was slowly grafting myself into the exhibit. I had, in essence, mapped my own personal Futurama. My head had been drilled, seeded, and backfilled with enough information to understand in basic terms how land is developed. It was at this fair that I also unknowingly began my journey towards authoring this book series with the intent to examine and explain the depths of site construction and land development.

Preparing Projects for Site Construction focuses on the narrow, inner workings of land development and preparation for structures as it applies to the site-work phase of construction. It's loaded with immediately usable information and poignant observations that apply to both public and private sector work. Though general in nature and not specific to any project, company, individual, or agency, this book is a beacon for those seeking more knowledge, savvy advice, and proven techniques for confronting and overcoming challenges related to land development and site construction.

Plainly, this book is written for owners from an owner's perspective. However, consultants, financiers, contractors, anyone associated with developing land will profit by studying its content. Nuggets of wisdom wait to be mined from an in-depth study of project related issues and points of contention seldom discussed beyond the confines of closed corporate meetings. All the reader has to do is contemplate, evaluate, and apply.

While reading and studying this book, perhaps you'll discover ideas and perceptions that you thought were only unique to yourself, your situation, or your project. Guess what—they aren't. Others may find the techniques and clarity refreshing and useful enough to save their project, increase profitability, or perhaps even help secure their job. Those that haven't yet broken ground stand to benefit the most from this book because they're blessed with time to set things up right. Pick up this book, read it, and study it. Prepare and enable yourself to handle today's problems, kick-off tomorrow's project, and get a handle on that grand thief of construction profits called site-work. Because—there's no forgiveness in dirt.

—Rod Johnston
Fall City, Washington

ACKNOWLEDGMENTS

I want to thank the following friends and professionals who took the time to review and to comment on part of or this entire book:

Jim Barborinas, Urban Forestry Services Inc.
John Barborinas, Bureau of Land Management
David Blake, Calthorpe Associates
Mike Bookie, Pachena Light Consulting
Jim Brisbine, Icycle Creek Engineers
Carl Cangie, Concept Engineering
Gerrie Degross, Degross Aerial Mapping
Stephen Dennis, retired President of Quadrant Homes
Ed Eilenfeld, Grasan
Pete Gonzalez, David Evans & Associates
Dave Hill, Concept Engineering
Judd Kirk, Port Blakely Communities
Ron Mertz, Williams Gas Company
Claudia Nelson, Port Blakely Communities
Dave Nelson, Dave Nelson & Associates
Dave Nemens, Huckell Wiemen Associates
Tom Nielsen, RCI Construction
Bill Pangburn, Materials Processing Inc.
Dick Prentke, Perkins Coie LLP
Brian Russell, URS Corporation
Victor Sostar, First Sterling Financial, Inc.
Bob Stokke, D.R. Horton
Bill Way, The Watershed Company
Curt Wittreich, CH2M Hill
John Zipper, Zipper Zemen Associates
Tina Talbot for her help in making a good book—a great one.

And, of course, my wife and family for everything else.

INTRODUCTION

It's an inescapable fact that an owner's chances for profitability and meeting schedule are dependent on success during site-work. This dependency is particularly stark for projects involving the development of raw land, mass grading and road building, and large-scale in-fill construction.

This book sets the table for laying out land development projects correctly from inception to the point short of choosing contractors and beginning work. The next phase, actual construction on a prepared site, is covered in book three of the Project Logic Series. Although anyone involved in construction and land development will gain insight and knowledge from reading this book, *Preparing Projects for Site Construction* has been specifically targeted for use by project owners and managers experienced and active in the art of developing land.

As is the case throughout the Project Logic Series, this book is designed for quick reference of valuable information. While construction and land development naturally attract problems, this book is about solutions—distilled, cohesive, and pertinent information that is immediately usable—because, when the pressure is on, owners need problems resolved.

What is site-work?

Some call it dirt work and others refer to it as site preparation. In simple terms, site-work can be defined as the phase of construction required to shape, change, or prepare land for a higher level of use. It also happens to be the single phase of work that most managers, consultants, and owners know the least about. Examples of activities within the sphere of site-work planning, design, and engineering may include but are not be limited to:

- Sensitive area analysis
- Wetland studies
- Wildlife studies
- Planning and layout
- Processing permits
- Off-site utility design
- Storm pond analysis and design
- Road location and design
- Streetlight design
- Traffic studies and design
- Landscape design
- Park planning and design
- Architectural work
- Utility facility engineering

Site construction may include, or be integral to any of the following cost centers:

- Surveying
- Timber cruising
- Logging
- Clearing and grubbing
- Chipping
- Demolition
- Dewatering
- Pile work
- Retaining walls
- Grading
- Utility installation
- Establishing access
- Installing hardscape
- Storm ponds and vaults
- Landscaping
- Reservoirs and pump stations
- Sewer lift stations
- Bridges

Before advancing, peruse the above list again. Why is site work information related to many of these subjects so difficult to obtain? There are a few reasons.

There is nothing glamorous about site-work. People don't visit construction sites and gloat over piles of stripped topsoil while casually rolling champagne around in crystal goblets. It doesn't have quite the same gravitas as, say, high-rise construction or erecting a sports stadium. Site work is dusty, muddy, and dirty. It's also the most incalculable and potentially diverse phase of construction, which may explain why few schools concentrate on teaching its inner workings. Site work, afterall, is a culture. An enlightenment learned through trial and error and, all too often, gross failure. If someone has never delighted in managing a grading operation or directed an army of trucks, scrapers, and loaders going everywhere, or has never tried to contain muddy runoff during near-flood conditions, or has never wrangled with a contractor over turbidity readings, then how can they preach about what to avoid, and how to avoid it? In short, if would be difficult.

Meanwhile, college students and professionals-in-training aren't gaining field experience as they used to. During the summer, it was once common to see students of engineering and architecture learning the mechanics of their profession by performing work such as laying pipe, pulling a survey chain, or shoveling concrete into forms. Without doing, it's not only hard to imagine how it's done; it's even tougher to direct someone else on how to do it.

Lastly, site-work is to land development what improvising is to music. With structures, everything is man-designed and man-made. It's prescripted. You can visualize and even mathematically conquer a structure without ever leaving an office. This isn't the case with site-work. Land development requires competing with nature, weather, and topography. It also requires interacting with specialized and sometimes arcane practices involving logging, explosives, pile work, and rock trenching. Things they didn't teach you at the university you get to learn in the field and most of the time, it's on your dime.

Throughout this book, expect to be mentored. Simple and plain truths are laid out in a manner to help prevent owners from wrestling with or being consumed by site-work. And, make no mistake about it, site-work can and will easily consume those who are unprepared. It's multifaceted and, to many in the industry, it remains the toughest facet of land development to master.

Terminology used in this book

For purposes of writing this book, many terms and labels have been used that are interchangeable and equal in meaning. Here are some examples:

1. Site work, site construction, and land development mean the same thing.

2. Parcel or lot can be any sized piece of land ready for structures construction.

3. Permit agency equates to public agency, the agency, jurisdiction, regulator, government, or municipality.

4. Builder, buyer, or client is an entity that obtains or purchases a parcel or lots from the project owner or other builder for purposes of constructing something.

5. Unless specified as a building or structures contractor, the word contractor denotes a site-work contractor or other contractor involved in preparing land for buildings.

6. Project owner equates to owner, developer, seller or you, the reader. Depending how it is used in context, it can also apply to landowner, homebuilder, or retail and commercial builder.

There's one more thing—this book is about saving project owners money. While there are many ways to plan for and perform site work, there exist only a few routes to handle land development in the most efficient, logical, and least expensive way without sacrificing quality. It's along those routes that we'll travel in *Preparing Projects for Site Construction.*

Chapter 1

CHOOSING CONSULTANTS

If you're free to choose your consultants without restraint, consider yourself lucky. Owners who surround themselves with the right people take a long step towards succeeding in land development. And since few decisions effect site-work more than choice of consultants, the size of the consulting firm, its relative experience, and its strength of personnel should all influence which companies or individuals are right for a project. Project owners bound by policy, low-bid, or precedent may not enjoy as much flexibility when contracting with consultants, however methods still exist to help narrow the list of potential firms and evaluate their roles relative to site-work.

Before advancing, two points need to be stressed. First, the right consultant for one project may be the wrong consultant for another. Depending upon the nature, jurisdiction, and complexity of various projects, consultants should be chosen based on their expertise, versatility, and price as applied to specific situations or tasks. Consultants who are able to jump right in and make an immediate positive impact are highly desirable.

Second, site-work design transcends structures construction. A structure may represent an owner's desires, chosen configuration, or architectural type. In contrast, owners have much less control over site design since land improvements must serve whatever is being built. For example, it's more critical that grading and peripheral infrastructure meet minimum rules for drainage and street geometry than be aesthetic. With site design, functionality rules and everything else comes in second. Once a site is developed, foundations require a certain level of grading and utility work regardless of product type. Where two projects or product-types diverge with different characteristics or architectural demands, the site-work serving both sites rarely changes proportionately. Schools require flat playgrounds in the same regard that retail or commercial development requires flat parking.

Profitable developers know that site design isn't as dependent on product as product is dependent on what a site will efficiently yield. Although owners have fewer alternatives when it comes to site design, predictability *in* design does generate benefits because site designers don't have to be specialists in any one architectural or product type. This allows owners greater freedom when choosing a site designer or engineer than when choosing architectural or builder services.

One last point is that successful projects maintain balance between in-house staff and consultants. Project control begins with controlling team members working under contract. This may mean augmenting in-house staff if resources or expertise is lacking. And, in-house staff should compliment and *not duplicate* the strength of your consultants. While many projects could profit from infusing insular management with a good dose of new and experienced blood, the key is balance. Recognizing and avoiding redundancy helps control soft costs and optimize team performance.

Hiring according to price

There are many ways to judge or evaluate consultants. Rules, regulations, and statutes inherent in public sector work may mandate that some consultants be hired according to lowest bid price. Surprisingly, some project owners in the private sector also operate similarly—they choose their consultants based purely on price. The practice of overvaluing initial price when choosing consultants may be the first true mistake that many project owners succumb to when setting up a project.

Stop! I know what you're thinking. Price *is* important but so are many other factors when it comes to using consultants. In many respects, the chemistry between you and the consultants may ultimately be the most important element to consider. All the dollars saved in using the cheapest outfit in town won't come close to compensating for a lack of quality, honesty, or performance in providing services that you may or may not even receive. That's not to say that you can't have it both ways. Low cost and good chemistry may exist in one neat package—but don't count on it.

A consultant's ability to work seamlessly with permit agencies also defines good chemistry. If you'll forever be at town hall trying to get permits released due to seemingly unworkable difficulties between your public agency and consultant, look elsewhere for consulting services. By then hopefully, it won't be too late to change horses.

One way to tackle this issue is to inquire first with your permit agency and find out which firms they've worked with before. With whom do they share history? Which projects went particularly well from their perspective? Dare to ask them who they like. They might not answer directly but it's worth trying. Also, inquire as to which consultants are already familiar with your agency's design standards, personnel, and permit procedures. A good match here is a good beginning.

Who do you need?

Most site construction relies on a relatively small number of consultants. Refer to *Figure 1-1.* This illustration tracks the usual mix of consultants applicable to site-work beginning with surveying and ending with sales. *Figure 1-1* illustrates

Site-Work Design and Activity Flow

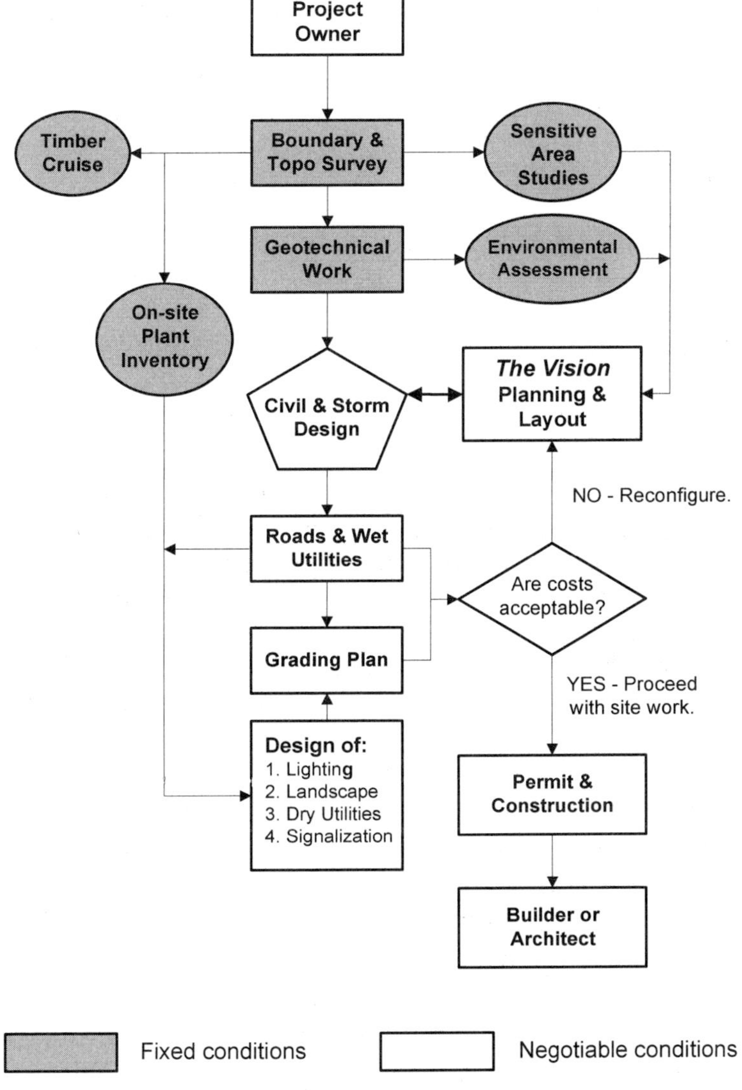

Figure 1-1

how planning decisions can be tested in terms of overall project cost. It also shows fixed project conditions represented by shaded activities and flexible conditions represented by white activities. Owners can plan, design, and engineer freely within white activities as depicted by the decision loop. Also, review *Figure 3-3,* which portrays the general use of consultants throughout the development process.

The most likely consultants who you may need during site-work include:

- Architects
- Attorneys
- Civil engineers
- Construction managers
- Geotechnical engineers
- Landscape architects
- Planners
- Surveyors

Other specialists may also be required including:

- Arborist
- Geologist
- Hydrologist
- Park designer
- Structural engineer
- Streetlight designer
- Wetland ecologist
- Wildlife biologist

While most of the above consultants have defined roles, owners may opt for one consultant to perform overlapping technical services or to provide project management. Project owners choosing to proceed with minimal in-house staff have the option of using a consultant under contract or hiring a new and independent consultant to assist in processing and coordinating work. Owners realizing too late that they need assistance in coordinating other consultants or in orchestrating project activity stand to lose in the way of schedule, cost, and opportunity.

One-size-fits-all consulting tends to diminish with increased slope. To understand why, peruse the following list which offers a few of the many ways that steeper terrain can complicate land development while at the same time, taxing engineering and design. Increased topography or slope breeds:

- Fewer options for access
- Challenging streetscapes
- Greater demands on piping and utilities
- Urgent gravity-driven erosion control issues
- More complicated and variable grading plans
- Greater variability in predicting earthwork volumes
- An increased dependence on slopes and retaining walls
- Higher incidence of encountering bedrock or unstable ground

- View potential from above and line-of-sight issues from below
- Potential for dealing with overly steep and undevelopable ground
- Difficulties in collecting, controlling, and safely detaining storm water
- Greater moisture, higher winds, and likelihood of temperature extremes
- Problems gaining vertical separation with roadways and pedestrian friendly walkways

Steep terrain can act as a great equalizer when deciding on consultants because those accustomed to handling simple designs on flat ground may not feel comfortable laying out and designing the same vision on steeper ground. That's okay as long as owners know beforehand which firms are experienced and efficient at producing cost-effective and marketable product on steep ground.

While firms may have individuals capable of contending with increased and broken topography, what are the chances that an owner will get them assigned full-time to his project? Taking it one step further, will totally occupying a firm's strongest or most valuable player come at a premium price? Or, is the firm willing to dedicate such an individual to a project that's considered small and unprestigious? Owners need to find out.

Once underway, problems with steeper terrain can ramp up when an owner or seller is asked to accommodate requests or demands posed by lot buyers, homebuilders, and civic entities. Layout, design modifications, and reconstruction on steeper ground all tend to be expensive. The right consultant can ease this process by being ahead on initial design or by at least creating a solution that's acceptable to the seller and all other parties. Owners must do their homework when choosing the right site-design consultants especially as terrain steepens.

What universal traits should you look for in any site-design consultant?

Firms are comprised of individuals each possessing his own character, personality, and abilities. Over time, many individuals forfeit this individuality and instead, acquiesce to whatever culture their firm embraces. Intentionally broadcasting their image or culture is priority for those firms that possess a formula for success. Other firms may inherit an image because of mistakes or misfortune caused in part by that firm's culture. Successful or not, consulting firms are perceived, branded, and judged according to their way of doing business based on their culture. What's left is a firm's reputation. Owners with projects already underway care little about street reputation. They've already formulated their opinion and will either rehire their existing consultant or switch partners.

Hiring a new consultant leaves owners little choice but to scrutinize potential firms or individuals based on reputation, interviews, and by inspecting their work. Inspecting their work both in the field and on paper should help owners decide whether a firm can produce a desired product. Since reputations aren't always accurate or correct, the burden falls on owners to separate truth from fiction. One way to cut through hearsay and anecdotal evidence is to openly table rumors and concerns with various people representing the firm in question. How do they respond? Does their attitude or position reinforce a culture that could support rumors as advertised or, might this be the right firm to use? Owners need to craft questions that pertain to their own needs and project, however, when the interviews end, it's important to understand how consultants rate in terms of:

- Objectivity
- Availability
- Competence
- Owner orientation
- Industry knowledge
- Honesty and integrity
- Genuine interest in the project
- Understanding the owner's goals
- Talent, interest, and willingness to solve problems
- Ability to inspect work through project completion
- Inclination to resolve complexities with an eagerness to please
- Determination and familiarity with shepherding work through the system
- Resources, experience, and expertise needed to excel on your type of project

Professional criteria to consider when choosing consultants may include:

1. References—Check their bank, client, and professional references.
2. Contractor opinion—Who does your site-work contractor recommend? Why?
3. Is the firm proficient in both Critical Path (CPM) and Gantt charting? Can they provide a credible and usable project schedule that dovetails with the plans?
4. Permit agency preferences—Check with your permit agency. If they won't commit to preferences, what firms do they use for their work and what firms do they maintain on roster?
5. Independence—It can be wise policy to avoid consultants who double as contractors or that have a vested interest in a contracting company, supplier, rental company, or a rival consulting firm.

6. Track record and credentials—Where have they been and what have they built? Is it still standing, still alive, or has it been long abandoned and boarded up? What is their history on safety and lawsuits?
7. Value engineering—Does the consultant overtly advocate value engineering (VE) and have a VE program in place? Can owners review examples of VE studies and resultant savings attributable to prior VE work?
8. Teamwork and style—Is the firm a team player or does it operate from a pedestal? Obviously, you want team players who are proactive, eager to please, and that don't routinely return your calls from a mobile phone shortly after 5:35 pm.
9. Long-term value—Don't place too much importance on hiring a firm specifically to solve immediate or short-term problems. You can accomplish quick fixes with a purchase order or a short form agreement. A request for proposal (RFP) should be used only when hiring consultants for the long haul.
10. Peer review—How will you know if the consultant that you're considering is the right one for your project? By doing a lot of land development or asking others. What do your peers say about them? Learning from the mistakes of others is good but aligning with their success is better.
11. Office location—Where are they? How close is their office to your project? Don't underestimate the potential cost and hassle of dealing with long-distance consultants. Video conferencing is great when it's working but nothing beats being able to have a spur of the moment on-site meeting when it's needed.
12. Reliable expertise—If an owner is consistently meeting with one of the firm's top individuals, ask if you can expect to have this person dedicated, or at least heavily involved, in your project. You don't want to be sold a company based on meeting their best people only to end up with a new college graduate or someone serving an apprenticeship.
13. Proposals—Utilize, prepare, and make decisions according to what you learn from requests for information (RFIs) and RFPs. Is the consultant's presentation convincing or, is the presentation all fluff with lots of PowerPoint and keyboard clicking? Are you getting real information and substance? Can you believe what you are being told or are you being sold?
14. Work product—Are their plans complete and easy to comprehend? What you see in the beginning is probably the best plan-set that they've ever created, so carefully evaluate their approach to detail and design. If you like the plans, can you afford to pay for the same quality of work? Or, are you apt to get "private plans" at a "public plan-set" price?
15. Consultant by default—Conflict-of-interest between an owner's

competing projects or between jobs within a large project may lead owners to choose the wrong consultant by default. Is may not be wise to use the same consultant on concurrent projects because "that's the way it's always been done," and it might not make sense to change consultants on a large single project merely for the sake of diversity. Beyond convenience, there should always be reasons for using a particular consultant.

16. Prestige—High-profile developers and municipalities may want a well-known consultant's name to be associated with their project. If a "name" consultant can add real or perceived value to a project—that's great. However, if a consultant has earned a name based on anything other than a successful history in site-work, then that's not okay. The wrong consultant leading site-work can be disastrous for the budget and schedule. Everything changes with site-work. It's the wrong place for schooling.

17. Horsepower—How much capacity in people and production can they muster? Are they slow now and looking for work, or do they have what it takes to deliver under any conditions and in any market? Ask yourself, what resources *could* your project require if the market heats up, the permits arrive on time, or the weather cooperates longer than usual. What about capital assets? Some consultants offer access to powerful and expensive office resources (and capable operators) negating the need for owners to purchase the same services independently.

18. Cost—As discussed prior, too many consultants (and contractors for that matter) are chosen on the basis of their advertised start-up pricing. Judging consulting firms solely on the basis of start-up pricing isn't only foolish, it can be suicidal to budgets. Owners unfortunately don't find this out until it's too late. We'll discuss start-up pricing and what it takes to minimize exposure to this phenomenon later, but, for now, suffice it to say that cost and pricing are important but rate a distant second to *value*. Although it comes disguised in many forms, value manifests itself through total lower cost, quicker delivery, and less—measurably less—mental trauma and potential legal exposure. Value is a by-product of fitting the right consultants to your project and in maintaining good chemistry between those consultants and owners.

19. Full-service capacity—Should owners consider a full-service firm capable of supplying architectural, engineering, planning, and construction management? Some owners are well-served by using consulting firms that supply an array of general and broad expertise, however, this leverage tends to disappear with site-work. Although standardized, site construction is specialized work and not normally

the forte of structures people. A full-service or a- risk consultant with a bent towards planning and architecture may, as a function of keeping the project moving, pay minimal heed to site-work, geotechnical exploration, and earthmoving. If schedule at any cost is acceptable to an owner, this arrangement may work fine. However, if an owner's primary intent is maximizing profit and minimizing cost, a site-work specialist will probably perform more efficiently in preparing a project for structures.

RFIs and RFPs

Assuming that you've already qualified potential firms to work with, having them prepare and submit a package of information and a proposal helps owners in a number of ways. For one, it helps owners decide whether there's a fit between them, their project, and the prospective consultant. These tools also force an owner to think about project needs and in doing so, commit those thoughts to paper. As an owner, what specifics are you looking for in a consultant and what do you want the consultant to accomplish for you?

RFIs and RFPs allow owners to get a feel for firms interested in their project while getting a first-hand look at the style, offerings, and responsiveness of various consultants. Receiving an actual proposal or RFP allows owners to evaluate a firm's vision or interpretation of their project including a scope of work. They also enable owners to review a consultant's proposed pricing, schedule, and approach to problem solving and design. As a product to itself, an RFP represents a degree of commitment by the consultant. Owners can improve their chances of enjoying a successful RFP campaign by:

1. Knowing beforehand what they want.
2. Being ready to begin work and commit funding.
3. Quickly turning the preferred proposal into a working and complete contract for services.
4. Allowing adequate in-between time for mailing, phone calls, requests for more information, field trips, and evaluating each consultant's response.
5. Beginning early and giving themselves time for preparing a sharp and succinct proposal that clearly states their goals and mission to potential consultants
6. Considering hiring an independent third-party consultant to assist in reviewing the proposals. A third-party consultant can share their impressions concerning each proposal and offer insight into what they believe each firm can realistically deliver.
7. Keeping those who will manage the consultant *in* the evaluation and

decision-making process. Give all in-house or hired management enough time to interview all firms, evaluate proposals, and offer input based upon what they've learned through the proposals.

8. Being forthright with all potential consultants. Clearly state that prospective consultants aren't to deviate from the questions, content, or list of information as requested by the owner. It's okay for the consultants to add information, brochures, and the like, but owners need to review all consultants with an even hand. This happens when owners dictate the rules and format.

Although every project is unique, common themes abound when requesting RFP's for land development services. Here are some questions that owners should ask of potential consultants during the RFP process and follow up interviews:

1. What challenges do you foresee?
2. What key milestones do you recognize?
3. What appeals to you about this project?
4. Do you anticipate using any subcontractors?
5. Do you foresee any potential conflicts of interest?
6. Describe how you would provide public outreach?
7. What is the consultant's relationship with the permit agency?
8. Describe similar projects that your company has been part of.
9. Do you believe that the owner's goals are attainable as stated?
10. Describe your project responsibilities and define a scope of work.
11. Can you breakdown your cost for services? Where is the money going?
12. What direction is your company taking over the next six months to five years?
13. Will you be able to provide a consistent level of service throughout the project?
14. How many full-time and part-time employees does your proposal consider using?
15. What advice would you give the owner to better improve profitability and schedule?
16. Is the consultant's contract acceptable or should you use your own contract for services?
17. What do you interpret as being the greatest risks to beginning and ending this project? Why?

18. Is the consultant, planner, or architect set on using a set team or are they willing to work with owner-chosen consultants such as the geotechnical and civil engineer or landscape architect?

Finally, ask yourself if the consultant's proposal is overkill or negatively-skewed relative to your vision for the project. Note: Owners of difficult or one-of-a-kind projects may find it more productive to allow consultants free reign to offer RFPs as they see fit. Consultants who are encouraged to develop and propose alternatives early in the process can cover new ground, provide a counter-perspective, and help refine the owner's vision for the project.

Who should complete your site-work plans?

Civil engineers are well suited for developing grading plans. They have the know-how, office equipment, and software required to dissect raw topography, transform it into a road network and grading scheme, and calculate earthmoving volumes. Civil engineers can handle change orders and quickly redesign aspects of the grading plan.

But most project owners want more than a generic road layout and grading plan. They thirst for designs that save money and provide maximum value. Saving money may equate to losing native stripped topsoil on-site or creating a balanced grading plan that results in moving a minimal amount of structural cut material the shortest distance possible. An example of maximizing value can be overexcavating a hole, perhaps in a park or other future unstructural area, for purposes of mining quality material for sale and then turning the hole into a short-term revenue-producing commercial dumpsite.

Owners need consultants capable of squeezing as much value out of a piece of property as the property will yield. The problem is, where should they look? The civil engineer or construction consultant is most inclined to attempt saving money through creative design-balancing earthwork but they're not always capable or experienced enough to smell opportunity. Architects, planners, and other consultants untrained in site construction or lacking exposure to land development rarely know enough to make a difference. Contractors can be great sources of wisdom depending upon their ability, contract status, and relationship with the owner.

Surveying and layout is another element for project owners to consider. Whoever does the civil engineering should be responsible for surveying and laying out their work. Why? Because there's a critical interplay and dependence between civil engineering, surveying, and construction staking. These three disciplines rely so heavily on each other that many civil engineers employ their own in-house survey teams. We'll discuss surveying and engineering in more detail in the chapter titled "Rules of Measurement and Control."

What about using landscape architects to develop grading plans?

Engineered grading plans are considered "primary templates" upon which other entities such as the power company, an architect, or park designer can base their design. Firms that have completed their design can export it back to the originator for incorporation into a more complete set of plans and documents. In the case of a base grading plan, it must be accurate and produced in both scale and appearance suitable for planning, engineering, and design work. If the content or accuracy of an original consultant's work is questionable or useless to other consultants, it's probably not being produced by the right consultant. Although landscape architects can propose grading concepts that a civil engineer can work with, in most cases, the landscape architect isn't the best choice for designing an overall site grading plan.

Should producing any grading plans be included in the landscape architect's scope of work? Sure—it can work when the landscape architect remains the lead consultant on a landscape-oriented project, or when he is developing a plan for his own use such as in pocket parks, planter areas, and small landscape amenities. The probability is low that other consultants will rely on the landscape architect's plans as a primary template for their own use. Surveyors or the civil engineer will still have to digest a landscape architect's plans when calculating coordinates for construction staking.

From a management or technical standpoint, earthwork-related change orders, surveying, and volume calculations are not strong points for most landscape architects. In addition, landscape architects may not be prepared to visualize and understand issues such as erosion control, utility requirements, pavement design, specific builder and product needs, or how the current marketing plan may affect grading. This type of work simply isn't their forte.

Last but not least, grading-plan designers must consider earthmoving patterns, equipment, and techniques. Smart grading plans account for each phase of a project's cut-to-fill sequence. The more earth that's being moved and the longer the duration for moving it, the more impact earthmoving techniques should have on grading-plan design. Consider these points when evaluating whether or not a landscape architect should complete your project's grading plans.

What about an architect?

The same argument can be applied to architects. Although some architects understand the relationship between topography and product type, many architects view site-work as a necessary evil—an ugly affair that's in the way of something more beautiful. This is understandable since some architects aren't equipped to deal with site-work, let alone design or value-engineer a grading plan where the end product may have little or no consequence to structures layout and design. But that's exactly what must happen, especially on challenging terrain. Structures design must accommodate whatever the ground will give. This means thinking about the ground independent of architectural design.

Being structures people, architects generally subcontract out overall grading design. This in itself can create problems for owners because an architect, or any consultant not sensitive to site-work, may not allocate sufficient lead time needed for a civil engineer to design grading, access, and utility alternatives. Many architects also specialize in designing products such as residential, multifamily, commercial, or government structures. So, owners have to know what they're building early on in order to commit to the right architect. Multiuse developments may require the services of many architects thus making it impractical to choose one for overall site layout and design. Ultimately, architects are expected to deliver a product hinged to a master grading plan designed by the firm performing road and utility design, and that's usually the civil engineer.

On-Site Consultants

As a buyer or builder, it makes sense to use the seller's consultants that may already be established on-site. This is particularly true with surveyors that have already established base control, recognize boundaries, and are familiar with local datum. Existing geotechnical engineers that understand the project's subsurface conditions and soil characteristics and civil engineers already possessing base maps and an understanding of permit agency requirements are worth pursuing for your own use.

Outsourcing

Using consultants that outsource routine detailing, computer-aided design, architectural, and engineering work out of your area, or out of the country, may be possible in land development. Since wages and the cost of doing business are lower in some regions of the world, outsourcing is popular for reducing cost and because work can proceed theoretically, around the clock. While such positives are undeniable, land developers must proceed with caution because unlike the design of fast-food restaurants, site design tends to be custom work. It's rarely routine. For instance, issues can arise if a distant agent is unable to:

- be available for crisis collaboration;
- juggle on-going earthwork balancing;
- perform pre-design field reconnaissance;
- understand local site construction methods;
- specify or correctly-size non-metric materials;

- factor-in the influence of local weather patterns;
- handle on-the-spot, design-oriented change orders;
- visualize existing ground and vegetative conditions;
- relate to variable subsurface soil and groundwater conditions;
- comprehend the site construction schedule and work sequence;
- properly translate municipal requirements (or not confuse your design standards with those from another agency or project) and;
- work continuously and cohesively on a multi-year project.

Sloppy, incomplete, and unsophisticated work will reflect negatively on you and your company. It's also dangerous for owners to proceed with unrealistic, inaccurate, or porous "boilerplate" site construction documents and contracts developed by distant people who have limited knowledge of site construction. Here, the legal exposure is just too great. In addition, owners that have invested heavily in planning and creating unique site layouts may not want their "site template" in the hands of someone that can replicate it and offer it to multiple clients around the globe. Especially if those clients also double as your competitors. While it's possible to overcome some of these items, the benefits and cost-saved by outsourcing site design must out-weigh the risk. Site design professionals aren't easily interchangeable.

Who should manage your site construction?

Owners inexperienced in site construction will want a construction manager who doubles as a navigation aid: a beacon of wisdom during dark times, and an active manager who works to keep the project on course and in the black. Owners mildly experienced in managing site construction may hire a construction manager to act merely as a psychological handrail. In any situation, good construction managers are a rare breed. Good *site* construction managers are even tougher to find.

Project owners with in-house site-construction management or who are undertaking a project too small to justify the added expense of hiring outside management can move on to the next topic. Otherwise, owners should consider hiring an agent or firm for managing site construction. *Figure 1-2* diagrams three routes available to owners when choosing site construction management. Item one passes all responsibility for site construction management onto a third party, while item three results in an owner assuming partial responsibility for managing site-work in league with a third party. Item two links to having in-house management. Here's a list of possible candidates to consider for managing site construction:

Alternatives for Owners Choosing
Site-Construction Management

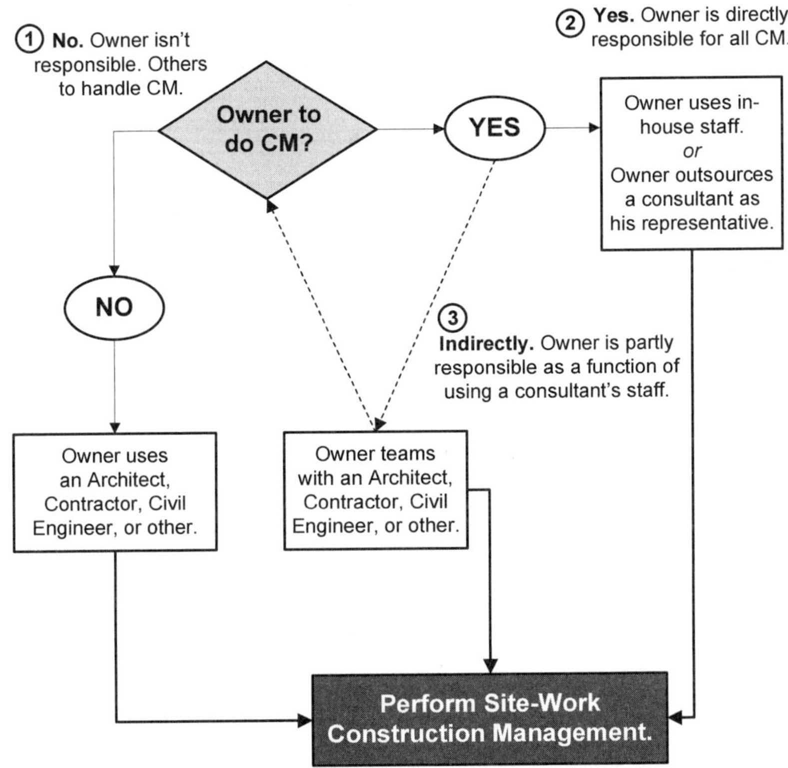

① Owner has minimal control over site work. Construction management is handled for a fee.

② Owner has maximum control over site work. Construction management is handled in-house.

③ Triangle of questionable responsibility. Construction management is shared for a fee.

Figure 1-2

1. An architect
2. An at-risk agent
3. A testing laboratory
4. A general contractor
5. A civil engineering firm
6. Yourself or in-house team
7. A construction management firm
8. A full-service firm comprised of any combination of the above

Each of the above disciplines brings unique strengths to the table, yet few consultants are equally strong in all services that they provide. Can a consultant perform construction management and other design or planning services with equal quality and without conflict of interest? Will choosing only a few services from a firm that offers many result in having that firm siphon key individuals from your project? Could your project become less important to that consultant? What service is the consultant's strong point?

Individual versus corporate performance

Construction management firms are driven by many factors but, for most, profit, year-round employment, and professional pride rank highest. The first two, profit and tenure, are easily measured and remain within an owner's sphere of influence. Pride is more difficult to gauge because it's complex and involves individuals, personalities, and reputations. It also relates to performance or lack thereof.

Consulting firms are often judged according to the performance of one or more individuals. Some know them as Rainmakers. Is this unfair? Can it be grossly inaccurate? Perhaps, however, there's no doubt that every firm has its breadwinners. Certain select individuals who bring in the work, enhance repeat business, and who do things right. Forget the firm—these are the individuals that a project owner should want for his project. Who an owner gets individually is usually more valuable than the name of the firm that individual represents. The best individual from the worst firm can very easily out perform the worst individual from the best firm. And the same holds true for hiring contractors. It's who you get that counts.

What owners really want

In the end, owners want leadership and management that are relevant, consistent, and able to meet project needs. They should abhor hiring management that's easily swayed, distracted, or prone to poor decision-making. Project affairs must be conducted with precision and clarity, free from influence by association or employment. Site-work managers must remain ahead of the contractor and not allow the contractor the freedom to dictate the terms of project control because, given the chance, many contractors will. The following critique examines the benefits and drawbacks of using various disciplines for managing site construction.

Should you consider using a construction management (CM) firm to manage site-work?

The prerequisite for managing site-work is experience in site-work. As practiced as a firm may purport to be in overall construction management, there's a clear line of distinction between the skills required to manage horizontal construction and those required to manage vertical construction. Managing structures is a totally different game from managing site-work. Don't be distracted by the wonderful design, scheduling, and planning services that some construction-management firms offer. While important in other ways, contemplating these accessories can distract owners in the hunt for strong site-work management skills. Be wary of consultants or contractors who downplay or trivialize site-work on even moderately challenging projects. If something can go wrong, it usually happens while *they're* on watch during the site-work phase of construction.

The benefits of using a construction management firm to manage site-work can include:

1. Flexibility—CMs need only be on-site as required for reducing cost.
2. Connections—Most CMs network with other specialists and have access to a variety of problem-solving sources.
3. Experience—Most CMs are equipped to handle a wide spectrum of work. Owners can benefit from broad experience and seasoning.
4. Creativity—An experienced CM used to full potential may provide creative and unique solutions to problems and offer cost-cutting strategies.
5. Specialization—Owners who find a CM specializing in what the project requires win by fulfilling a need. They also avoid inheriting a CM by default or hiring the wrong CM out of desperation.
6. Objectivity—CMs can offer impartial reviews and critical counterpoint to issues where, otherwise, an owner's emotional involvement or reluctance to abandon a position may work against the project.
7. Consistency—CMs can offer proven techniques for managing work. Successful construction management firms develop systems for paper flow, documentation, and communication, which are ready for immediate implementation.
8. Focus—In comparison to general contractors and other multidisciplined consultants, CMs don't have to protect other on-site

profit centers. Independent CMs are less likely to get embroiled in conflicts of interest, to fall prey to skewed judgment, or to make decisions that may compromise managing construction. As an extension of staff, a CM can operate clear and unencumbered by outside influence.

9. Counsel—Owner-oriented CMs are more likely to question all activity on-site. In contrast, a general contractor may question or criticize other contractors on-site but who will question the general's activities? Or, an architect may question the planner but who's going to be looking over the architect's shoulder? Empowered CMs are better positioned to strategize and interact on behalf of an owner. The presence of a strong CM can also neutralize adversarial relationships that may flourish between an architect, engineer, or contractor. For this reason alone, owners should consider using an independent CM on design-build projects.
10. A clean *adios*—When site-work is completed and the CM is retired, it's unlikely that an owner will continue to receive ongoing billings for work that is plainly finished. When the site-work and building CMs are one and the same, owners are vulnerable to receiving uniform and regular billings from clearing through occupancy regardless of what's being performed on-site.

The disadvantages of using a construction management firm to manage site-work can include:

1. Over commitment—CMs are not always available when you need them so it's crucial that the CM agrees to be available until seasonal shut down or completion of work.
2. One-size-fits all—CMs who offer services from land purchase through managing tenants may not possess enough expertise to handle both site construction and site-work contractors. Few firms can handle it all but, when they can, they come at a premium price.
3. Changing partners—Stability and continuity are important. However, once a project develops an aire of predictability, CM firms may move their better people to newer and fresher assignments. It's important that owners insist on working with the same CM people throughout a project.
4. Cost—CMs are expensive. Many charge fees based on a daily minimum. Owners should continually monitor their CM's activities relative to budget constraints and management. Establishing gross not-to-exceed thresholds can be a useful cost-cutting device and baseline for evaluating a CM's value.

5. On board too late—When owners procrastinate and hire a CM after construction is well underway, they accept the risk of the CM's never "catching up with the project". By the time an owner realizes that his CM is chasing more problems than he's solving or that the CM habitually dodges sticky or chronic problems, it may be too late to correct.
6. Limitations—CMs should contract directly with the owner. When the CM is not under contract, other consultants or contractors may try to circumvent the CM and petition the owner directly on issues. This can result in a gumbo of expert opinion seasoned with a driving need to be heard, to fulfill an agenda, or to unduly maximize someone else's bottom line. Owners must support their CM or else risk having the CM lose control and be less effective in managing the project.
7. Cookie-cutter management—It's easy to manage in ways that are comfortable or familiar, but owners don't need cookbook style management. CMs must adjust to varying conditions and not overly rely on stale or misguided solutions. This means changing tack on a project-to-project basis and reacting to situations with fresh ideas and a strong field presence. Electronic wiz kids and computerized accounting systems are great but nothing beats implementing systems and management schemes tailored to saving money and reducing problems.
8. Level of effectiveness—Finding the right CM for the right job isn't automatic. Claiming to be a consultant doesn't mean being able to handle managing an owner's site work. Generally speaking, CMs must be experienced, educated, and technically competent in whatever type of construction that the owner will be performing. They must be skilled in questioning, persuading, and directing a multitude of other consultants and contractors while maintaining everyone's respect and cooperation. For purposes of this book, that only comes with having a deep and complete understanding of site construction.

Also, understand that a professional engineering (PE) license needn't be a prerequisite for managing site construction. Owners led to believe that an engineering license should be a basic requirement for managing site work may end up hiring the wrong people for the wrong reasons. Good engineering doesn't necessarily guarantee effective site-work management and vice versa. Engineering and management require different skill sets. In fact, managers who rely on a license or certificate to gain respect or to validate their leadership style can be counter-productive—and expensive. It's also worth noting licensed engineers are far more abundant than seasoned site construction managers. If an owner expects an individual to perform both engineering and construction management, that owner will discover that there aren't enough hours in a day to perform either job adequately, which leads us to the next topic. That is, what should an owner consider when contemplating hiring a civil engineer, an architect, or a general building contractor to manage site construction?

When should you choose using a civil engineering firm to manage site-work?
The civil engineer's basic task is design. Design can include anything from developing grading and erosion control plans to performing utility line calculations and road engineering. Some civil engineering firms may incorporate in-house surveying, and larger firms often offer planning, architectural, and project management services. But, can a civil engineer successfully manage the construction of what they've designed without conflict of interest? Unless the civil engineer and owner have a special and unique relationship, meeting both conditions may be a tough order. On the other hand, when design is set and meeting budget and making schedule becomes the primary concern, a civil engineer can fit the role for managing site construction.

The benefits of using a civil engineering firm to manage site-work can include:

1. Designer's edge—Designers usually procure permits and approvals. They should know the conditions and specifications—cold.
2. Specialization—Owners may elect to use a civil engineer to only manage construction of specialty items such as pump stations, reservoirs, pile and shoring work, and small public works structures. This can free the owner's team to manage other on-site activity.
3. Understanding—One-stop service can simplify complicated projects. Nobody should understand a project more than the designer, so, a designer responsible for site-work management can be highly effective as long as the designer understands the mechanics of site construction.
4. Convenience—Hiring a civil to perform multiple tasks results in fewer people in meetings and should pay off in lower overhead and management costs. Plans and information get distributed more quickly to team members. If the civil is supplying survey crews, owners enjoy the benefits of streamlined communication.
5. Tracking—In the same manner that it's easier for owners to track mistakes in design, in documentation, and in surveying when they are performed by the same firm, *removing* a CM, an architect, or other consultant from the loop makes it easier for owners to identify and track shortcomings in construction management.
6. Efficiency—Transmittals, change-order processes, and communications are quick and clean. Civil engineers and surveyors in the same firm constantly communicate. This allows contractors a direct line of

contact with the office engineers. Owners save time and money and enjoy consolidated and simplified billings and accounting.

7. Presence—As the CM, the civil engineer would be always available for meetings, field management, and office coordination. Emergency field problems can get immediate attention and issues with other departments can be handled in person. Being on-site allows civil engineers to resolve problems before they evolve into change orders.

The disadvantages of using a civil engineering firm to manage site-work can include:

1. Objectivity—Can the civil separate design from management?
2. Professional liability—Is the engineer insured to perform construction observation services on projects of his own design? Are owners covered?
3. Incompatibility—Many engineers don't understand entrepreneurship and how it links to profitability. Good managers do more than herd people and process; they inspire everyone around them to achieve higher goals that result in lowered costs, in meeting or accelerating schedule, and most importantly, enhancing team trust.
4. Confidentiality—Can the engineering firm be trusted with delicate cost and sales data? Will a firm admitted to an owner's "inner circle" remain ethical and loyal, or use proprietary information as a pivot point for picking up a competitor's work? Can an owner be comfortable sharing budget and balance-sheet information with an engineering firm that's absorbing a continual flow of the projects soft costs?
5. Narrow focus—Give most civil engineers a problem, draw a fence around it, and they'll solve it. They're very good at point-specific problem solving, however, scheduling site-work, inspiring dirt contractors and other specialty subcontractors, and being able to create solutions in the field may not be their forte. And that's understandable considering the number of years that one has to invest in site-work in order to learn how-to construct something without spending too much money.
6. Overdependence—It may be unwise for an owner to keep all his eggs in one basket. For example, will the owner's schedule and bottom line suffer if an existing firm is asked to leave the project, to compete for future work, or to accept a reduced scope of work? More importantly, would your project survive? Once an outside firm establishes an inordinate amount of on-site authority, billings can also increase substantially leaving owners with little recourse. Firms drunk with power can become hypersensitive or obstinate when questioned and may resist control.

7. Ability—Hybrid civil engineering firms may perform one service better than another. On your project, which one will it be, better engineering or better construction management? And, if the firm touts construction management as a service, do they really understand site-work as it pertains to land development, or are they better suited for municipal work or vertical construction? What about overload? An engineering firm faced with a surge in demand for design work may pull experienced people from your project and replace them with new and inexperienced personnel.

When should you consider using an architectural firm to manage site-work?

Like engineers, architects offer a well-defined, and predictable scope of work. Many architectural shops offer a full array of services in addition to design work that may include, but not be limited to, planning, streetscape design, park design, and project management. The key word to focus on here is project management.

There is difference between project management and overall construction management in the same spirit that overall construction management differs from site-construction management. Being primarily structures people, most architects aren't as equipped to manage site construction.

Evaluating pay estimates, passing along other consultant's change orders or requests for information, and conducting meetings are important but as an owner, aren't you looking for more? Don't you want seasoned management that will think creatively, continually pursue value engineering, and perform optimal control over activities, such as erosion control, pond construction, earthwork, and utilities? Ultimately, a second question arises concerning the architect's (or any consultant's) willingness to be client-centric. Can the architect separate his genius from his responsibilities—meaning, is he able to accept fault? In private land development maximum profit and minimum cost requires mastering the art of site-work and questioning everything. Just going through the motions doesn't cut it.

The benefits of using an architectural firm to manage site-work can include:

1. Image—The right architect can attract leaseholders, advertisers, awards, and prestige.
2. Versatility—Having access to many in-house services and disciplines can result in overall quality project management. Large firms offer ample resources and manpower.
3. Simplicity—A one-stop service means fewer people and less paperwork. Owners may be able to focus on other issues leaving the bulk of project management and design coordination to one entity.

4. Global perspective—Architects are structures-conscious. They should know why something in the field has to transpire in order to bring a project into vertical construction as quickly as possible.
5. Tradition and experience—Owners use architects because they like their product. Or, they may like the look or function that a certain architect provides. When styles, attitude, and vision blend harmoniously, comfortable long-term relationships develop. Architects who provide value and predictable pricing set themselves up for repeat business.
6. Vision—An architectural firm should be able to envision the owner's desires, to mix those desires with exciting and innovative design, and to produce a product acceptable to everyone including the permit agency. The benefits of having one firm coordinate everything from engineering to construction, through lot layout and home design, are compelling.
7. Presence—Constant presence in one form or another is desirable. Meetings are easier to schedule and field-to-office coordination becomes more efficient. Architects who are constantly on-site get plugged into current events allowing them to quickly change direction in planning, layout, and product. Overseeing site-work allows architects to resolve problems before they emerge into change orders, thus minimizing impact to cost and schedule.

The disadvantages of using an architectural firm to manage site-work can include:

1. Objectivity (See civil engineer above.)
2. Professional liability—(See civil engineer above.)
3. Incompatibility—(See civil engineer above.)
4. Confidentiality—(See civil engineer above.)
5. Overdependence—(See civil engineer above.)
6. Ability—Is the architect better at building design or leading site-work contractors? Can an owner afford to begin work without knowing?
7. Wrong priorities—Any efficiencies gained by having one firm handle an entire project can evaporate if the firm in charge doesn't have its priorities straight. Site work emphasizes horizontal construction. It cannot be treated as an appetizer for vertical.
8. Focus—An architect's primary focus is on designing structures. Like most consultants, architects gravitate to portions of the project where they feel most comfortable, where they can easily establish control, and where their understanding is strongest. Familiarity with design and

structures does not translate to familiarity with managing site-work.

9. Cost—An architect is expensive and their hourly scale can vary widely. Predicting soft costs for architectural services isn't easy, especially if the architect is providing both design- and project-management services. Increasing an architect's scope of work further away from his areas of strength may result in greater costs, including on-the-job schooling.
10. Effectiveness—An owner is always at risk of hiring an architect who doesn't understand the geotechnical report, the civil engineer's work, or what's involved with site construction. Add a site-work contractor who may have had previous difficulties with the architect or that senses the architect's lack of construction knowledge, and that contractor's respect and desire to cooperate may dissolve on the spot. Right or wrong, projects cannot survive under serious duress. An architectural firm that triggers the above reactions will have a hard time questioning, persuading, and directing site construction. From there, it's downhill for owners.

When should you consider using a general building contractor to manage site-work?

Considering a general building contractor to manage project-wide site-work seems like a brilliant idea. Afterall, contractors are builders and buildings require site-work, so it only stands to reason that those who wear hardhats understand site construction. But do they *really* understand the nuances of clearing and logging, dirt, utilities, and erosion control? A closer look reveals that building contractors focus on where the money is and, for them, it's usually in the structure, not in site-work. Many building contractors consider site-work a nuisance and, for some, it's a potential subcontracting nightmare.

Still, circumstances may make the general contractor a worthwhile candidate for managing overall site construction. For instance, is the general contractor scheduled to build a structure or series of structures on lots concurrent with perimeter grading, utility, or road work? If so, a general building contractor may be the right choice to manage everything from clearing to gutters because it's reasonable to assume that they'll find solutions if site-work threatens to impact the schedule, his subcontractors, or his profit. But, it's *not* a given. Knowledge in structures doesn't equate to knowledge in site-work, meaning that a building contractor may be susceptible to difficulties working dirt, preparing subgrade for paving, or coping with weather. Which brings us back to the question, is a general building contractor the owner's best choice when it comes to managing off-site and on-site site construction? Not necessarily.

The benefits of using a general building contractor to manage site-work can include:

1. Convenience—Using a general contractor already mobilized or expected soon on-site makes tactical sense.
2. Comfort—An owner may feel more secure using a general contractor expected to possess the expertise and experience needed to handle peripheral site-work.
3. Planting seeds—Using a general contractor for both contract and fee-management work may enhance trust equating to long-term positive relationships for both parties.
4. Image—Using a general contractor who has a favorable or easily identifiable name can promote a strong project statement. It can suggest that the owner is serious and willing to invest in capable representation.
5. Expertise—General contractors offer a wealth of varied experience, however, they're usually strongest where they stand the greatest chance of making a profit—in vertical construction. Familiarity in structures doesn't automatically translate into being able to control project site-work—but it helps.
6. Vision—A general contractor should comprehend land development as it unfolds. This includes knowing his role and the role that other contractors play in bringing a project home to completion. Although not guaranteed, general contractors who exhibit such understanding can work out well managing site-work for project owners.
7. Resources—General contractors lacking the equipment or the personnel needed to handle emergencies or change orders usually know how to quickly mobilize resources needed to complete work. Project owners may also find a general's on-site staff, office equipment, and temporary housing convenient for use by consultants and the project team.
8. Carrots and leverage—A general contractor who is going to later perform work for an owner, but hasn't yet obligated his site construction to any firm, can hang that carrot in front of potential site-work subcontractors as leverage in getting them to perform better for an owner. Many things can motivate site-work contractors, but two things seem to carry more weight than others. One is being able to stay on-site indefinitely and the other is getting a free hat.

The disadvantages of using a general building contractor to manage site-work can include:

1. Gun smoke—General contractors who get into shoot-outs with inspectors and agencies on their own projects will probably carry negative residue throughout the site. Heavy inspections and absolute unyielding compliance with the specifications may follow them everywhere.
2. Friction—For whatever reason, whether resentment, personality conflicts, unfamiliarity, disrespect or outright competitiveness, some contractors just don't get along. Putting one contractor in control of another contractor's bottom line, work regime, and professional pride can be risky.
3. Limitations—A general contractor will, in good faith, recommend solutions or perform services limited by his ability, company experience, or philosophy. An owner with multiple varying site-work challenges may end up handicapped when he relies solely upon a general contractor for advise in spite of that contractor's limited knowledge of site work.
4. Cost—Using a general contractor can be expensive since some of them will only manage site-work based on a fixed percentage of gross project dollars. Since fixed fees can be difficult to evaluate in terms of performance, this type of agreement usually only works for owners interested in one objective. That is, getting to project completion regardless of cost. And, fixed fees aren't immune from increasing with revised schedules, change orders, or increased responsibilities. Unless agreed to beforehand, the likelihood of an owner receiving a rebate on fixed-fee work resulting from a decreased scope of work is remote.
5. Purpose—Since general contractors are usually on-site for the sole purpose of making a profit on structures, they may consider anything that doesn't maximize or influence profit to be secondary in importance. Many general contractors perform fee management as a service only. In a worst-case scenario, they may not increase on-site staff to handle the owner's site-work management. Instead, they may expect on-site staff to perform both contract and fee-management work resulting in the owner paying twice for diluted services. Some call this "double-dipping". Whatever the name, it can be a raw deal and a true disservice to owners.
6. Inexperience—A general contractor who is on-site to build a structure will staff his crew with structures people. If the owner needs storm utility expertise or creative ideas for better balancing earthwork, the chances are slim that he will get the best advice from people versed in roofing or HVAC. Many generals understandably want to keep their top personnel guarding profits tied to contract work. Worse yet, once the general's contract work is under control or the profit is sealed, he may send his best people to greener pastures, to new challenges, or to the office for job estimating. This leaves owners with—you guessed it—neophyte managers.

7. The Boomerang—If the owner's relationship with the general sours over anything, such as job conditions, violations, change orders, poor plans and specifications, personality conflict, or poor circumstances, the first territory covered by the general will be his own contract work. In general contracting, profit is king. Attention that should go to managing other site-work might inadvertently suffer. Conversely, the risk assumed by an owner in having "all his eggs in one basket" is extremely high. As problems are brought to the table, either party may be tempted to use one project function as leverage against the other. In the end, it can be self-defeating.
8. Conflicting schedules—General contractors assuming site-management responsibility may have to decree when to limit access or deliveries, when to disrupt traffic, or which contractor gets to use specific lay-down areas. When the schedules of other contractors conflict with the general's operation, the general must act objectively or else problems will develop. The general may have to place the schedule of other consultants ahead of his own. Schedule problems of a different nature can also occur if the site-work contractor's start-to-finish schedule doesn't correlate with the owner's overall project schedule. In this situation, the general acting as the construction manager, may have to closeout the job and pursue miscellaneous site-construction punch-list items. Does the general want this responsibility? More importantly, does the owner want the general to assume an increased scope of construction towards the end of the project?

   Note: If an owner chooses to have a general contractor also manage overall site-work, it's best to place the general under a separate contract for construction-management services. This technically and legally separates the general's structures' work from that of managing the owner's site construction.

At-risk, site-construction management

Owners can also contract their site-work management to a consulting firm or agent who in turn, teams *with* the general contractor. In this scenario, the consultant's role is similar to that of a general contractor. The consultant assumes risk for performance and schedule, takes responsibility for controlling work, and plays a leading role in initializing change orders and contract modifications. At-risk agents can assume an overpowering role that may be too influential for many owners to handle. Instead of being in complete alliance with an owner, a CM assuming risk has his own interests to consider and those interests may not always align with an owner's desires, vision, or budget. Owners who employ at-risk or third-party management should have a backup plan should they wish later to change managers or revert to using in-house people.

Self-managed site-work

Many project owners choose to manage their own site-work. Owners confident and qualified to do it alone will save money. Owners who aren't so sure should have professionals waiting in the wings in case their in-house staff becomes overwhelmed or delaminates. As an option to tackling site-construction management head-on, owners may instead adopt the role of project manager. It takes considerable time and ability to keep a team of specialists on schedule and relevant to each other. Acting like the hub of a wheel, overall project management keeps owners connected to planning, design, permitting, engineering, construction, sales, builder issues, and property management. Each discipline needs coordination and, more importantly, leadership. Owners who use a third-party site-work management firm have to rely on that firm to plug into all project functions, to do what it takes to make the project succeed, and to stay ahead of problems.

Speed is the key

Before proceeding on to the subject of managing site work, here's a nugget of wisdom. The key to successful and cost-effective site-work is speed. The faster that the construction manager figures it out and executes, the higher will be the probability of meeting schedule and beating budget.

You see, every manager eventually succeeds in completing a project's site-work. Getting the work done isn't the challenge. Getting the work done quickly and with as little staff as possible is the objective. It's like the cat-in-the tree analogy. When asked if a cat stuck high in a tall tree would ever find it's way down, the fireman replied, "I've yet to find a cat skeleton in a tree." Well, you'd be hard-pressed to find a project abandoned, fenced off, or infinitely stuck in limbo because nobody could figure out how to accomplish the site-work phase of construction. Everybody, no matter how green and inexperienced, eventually figures it out. They may petition every contractor and consultant in the county for advice, or vacillate forever trying to make decisions, but they'll eventually get it done. The kicker is, at what cost to the owner's budget and schedule will it take to complete site-work so that lot sales and vertical construction can begin?

Setting parameters

Owners who use in-house site-construction management likely will grant their own people maximum power and leverage. However, owners who employ third-party site-work management should first decide how to structure their CM agreements and ask themselves the following questions:

- What degree of power and control will the CM possess?
- What types of issues does the owner want to know about?

- Will the owner provide office resources and personnel to the CM?
- To what extent does the owner want to be included in decision-making?
- How much latitude does the CM firm have in shaping its role and limits of authority?
- What maximum-dollar threshold is the owner willing to allow the CM to operate under before requiring that the CM petition the owner for approval?

Highlight items to be aware of when writing consulting contracts

Refer to *Figure 1-3.* This chart breaks down the primary and more common ways to structure consultant agreements. Of course, there are permutations on these alternatives, but the idea remains the same. Once an owner decides to either bid or negotiate work, the next step is to determine how the owner will accept consultant billings. Is time-and-material or hourly backup supporting progress payments going to be required? If so, it must be well detailed and in a format useful for measuring the consultant's performance against billing.

Most consulting contracts adhere to a set structure. Here is some boilerplate information to consider for CM agreements.

- Life of the contract
- Consultant fee structure
- Terms of the agreement
- Terms for confidentiality
- Consultant responsibility
- Taxes and audit procedures
- Framework for dispute resolution
- Define what is considered deliverables
- Definition of common goals and objectives
- Provisions excluding third-party beneficiaries
- Terms, procedures, and backup required for billings
- Owner's window for fulfilling approved pay requests
- Contingency clause for changing the scope of services
- Liability, indemnity, and business insurance requirements
- Who the consultant and owner will each assign as lead people
- Stipulations for protecting and recovering data in case of an emergency
- A list of key employees including their phone numbers and e-mail addresses
- Clearly defined, fixed scope of work including a detailed list of expected services
- The option for the owner to electronically access the consultant's billing system at anytime

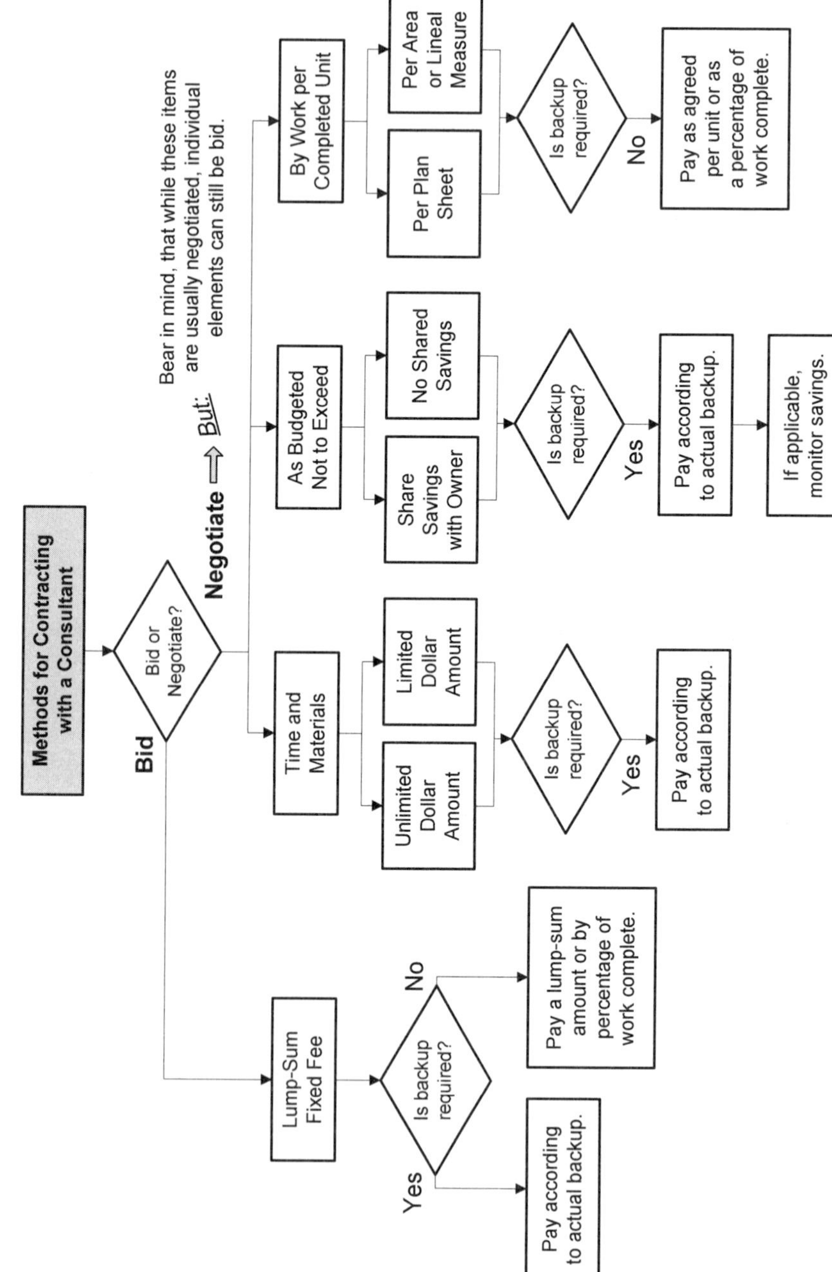

Consultant Contract Alternatives
Methods for Contracting with a Consultant
Bid or Negotiate?
Bid
Negotiate
But:
Bear in mind, that while these items are usually negotiated, individual elements can still be bid.
Lump-Sum Fixed Fee
Is backup required?
Yes
Pay according to actual backup.
No
Pay a lump-sum amount or by percentage of work complete.
Time and Materials
Unlimited Dollar Amount
Limited Dollar Amount
Is backup required?
Yes
Pay according to actual backup.
As Budgeted Not to Exceed
Share Savings with Owner
No Shared Savings
Is backup required?
Yes
Pay according to actual backup.
If applicable, monitor savings.
By Work per Completed Unit
Per Plan Sheet
Per Area or Lineal Measure
Is backup required?
No
Pay as agreed per unit or as a percentage of work complete.

Figure 1-3

More detail and color can be added to consultant contracts by considering the following topics as they apply:

- Subconsulting—If allowing a consultant to subcontract out work to other consulting firms is a concern, add language that prohibits, controls, or severely discourages such action.
- Surprise exit strategy—Under what conditions can an owner terminate the consultant contract? Will a consultant who decides to leave remain obligated to bring a replacement up to speed and, if so, who will pay for this service? Place time limits on everything.
- Transfer of documentation—Specify how and when to submit plans, reports, and documentation to the owner. Include language specifying that unless sent to the owner or owner's agents, the consultant can only transmit plans, reports, and documentation after receiving the owner's permission.
- Clarify personnel commitment—Which personnel is the owner getting and for how long? The owner's goal should be to retain the consultant's best people for as long as possible. Consultants who appear uneasy about committing quality people for extended periods of time may agree to establishing milestones that, once reached, allow the consultant to pull and replace individuals.
- Training—Do you expect the consultant to mentor or teach in-house staff? Are the people in your firm or on your project open to learning, or will retaining a management consultant end up costing more than it's worth? Be honest with yourself. If a consultant is able to help steer your team towards a better and more profitable project, then lead others to that vision and include it in the consultant's scope of work.
- Get all costs upfront—Cost proposals from consultants should be complete and categorized in as much detail as possible. Examples of consultant activities that should be clearly priced and described in hours include planning, design, field monitoring, inspections, processing change orders, handling requests for information, conducting meetings, completing as-builts, travel expenses, subcontract services, and office allowances. An initial hourly breakdown per task helps owners later when monitoring performance to cost ratios.
- Control ancient billings—Define how far back a consultant can go in billing the owner. Leave nothing to memory. On all projects, there is a line behind which nobody remembers anything and, short of third-party invoices for reimbursement, consultants when pushed, may not be able to positively demonstrate or convince anyone that what they are charging is legitimate. I know, it's tough to swallow—but it happens. To control catch-up billings, owners need to identify a point of closure on anything billable. This also forces a consultant to better manage his own hours, to-date charges, and pass-through billings.

- Who pays for the consultant's mistakes? Schedule impact and costs related to redesign, repermitting, reconstruction, restaking, being pushed into wet and unseasonal weather, extra meetings, and reverberations felt by other consultants are just a few ways that a consultant's mistakes can cost an owner. Now, we all make errors and consultants are no different regardless of what they state in their brochure. However, when an unusually high number of misdesigns, screw-ups, or poor decisions occur, short of firing the consultant, an owner has to do something to right the situation, protect himself, and guard schedule and budget. Because a point exists beyond which a consultant's blunders can't be swept under a rug, the contract should include a mechanism that enables an owner to place blame and pursue damages. But beware—owners who pursue this strategy must remain clear of providing misdirection, interfering, or poor process management that could self-incriminate or weaken their case.
- Ownership of intellectual property—We know who pays the bills but do we know who ultimately owns and controls all the paperwork and electronic copy produced? Studies, plans, data, specifications, documents, and all records should stay under the control and ownership of the owner. In addition to maintaining control, owners and other consultants avert having to continually pay a third-party consultant to search, make copies, and transmit paperwork or electronic files in order to keep a project functioning. Access to such vital information allows owners to provide information on-demand to team members. Another benefit to having owners possess duplicate information is safety through redundancy. If a dispute between the owner and a consultant that has been paid to create or hold information arises, that consultant cannot gain leverage by freezing access to information already held in duplicate by the owner. Maintaining in-house files of all intellectual property entails having ample data disk storage, electronic backup systems, and office space for storage cabinets, plan racks, and boxes of historic documents.
- Completion of work and run-on billings—Be clear, detailed, and exact about what constitutes the end of a consultant's scope of work including any limitations on invoicing owners. Unfortunately, some consultants have an unwritten policy of billing well after their work is complete. Unless challenged, they may continue billing until they match an original budget number. Now, the chance that a consultant's cumulative time and material billings invoiced at regular intervals will exactly match a rough-order-of-magnitude gross budget estimated months earlier is statistically remote, particularly when the budget estimate is five figures or more. When this situation arises, it brings into question the validity of the invoice and the intent of the consultant. Rather than go there, protect yourself by crafting contracts that clearly differentiate not-to-exceed amounts with savings *from* lump

sum fixed pricing. Include measures that allow an owner to request billing updates at any point in time. Also ensure receiving a percent-complete-to-date commitment with each consultant pay request. By coordinating with contractors and getting a feel for work complete to date, owners can trace a consultant's activity on-site, in the laboratory, or in attending project-related meetings. Clearly state that rough-budget estimates as submitted by consultants early on are only that—rough-budget estimates—and not benchmarks upon which to base billings.

- Out-of-scope services—We've already covered the point that consultant contracts must sufficiently detail services that the consultant is providing. However, complicated projects may also mandate listing what the consultant *isn't* providing. For instance, during negotiations, there may be warm and tender conversations concerning items important to the owner, such as value engineering, constructability, triple-reference-point control staking, slope staking, generation of road and slope profiles, road-location design according to the data contained in the geotechnical reports, or pavement alternative analysis. The problem is that well-intended conversations can be deceptive. The owner thinks that the consultant wouldn't mention these services unless they were included in the consultant's budget, whereas the consultant thinks that the owner is acknowledging the benefit in spending additional money for such services. In this example, both parties begin by grossly misunderstanding each other. Ultimately, the consultant performs work and bills a surprised and unhappy owner, or an owner anticipates receiving services that never materialize. In this predicament, both parties may be well intentioned but lack of follow-up and documentation results in miscommunication and potential impact to the owner. The solution is for owners to ensure that all desired services are included and delineated in the consultant's contract. When it's possible that a consultant will be called upon to provide extra services, include mechanisms that allow the consultant's scope of work to easily expand or diminish.

Errors and Omission insurance

Consulting in general is largely an unregulated business. Reputation and experience are some of the many ways that owners evaluate management consultants. The game changes when consultants rely on licensing, or certification to perform work. Consultants who design, engineer, or develop plans or construction standards based upon and backed by a license or board of certification for such work should be required to carry errors and omissions (E&O) insurance.

E&O insurance and unlicensed consultants

Beyond licensed professionals, requiring E&O insurance of other less formal consultants may be futile and result in unnecessary costs, overly conservative decision-making, and a decrease in your pool of potential consultants. Here's why:

1. Insisting that all consultants carry E&O Insurance can result in damping innovation.
2. Owners who require unlicensed consultants to obtain E&O insurance can expect to spend more time and process more paperwork transmitted for both defensive and offensive purposes.
3. If a consultant's work isn't regulated or backed by a standard, a board of oversight, or a license, he may be E&O uninsurable and, therefore, unable to purchase E&O insurance.
4. Insurance costs and related office expenses get passed onto owners. Do owners want to pay for insurance that may not be absolutely necessary or may be difficult, if not impossible, to collect against?
5. Consulting work performed having no precedent, no accepted standard for performance, or no benchmark to measure against is difficult to gauge. For instance, in tracing a problem, how would an owner know if a project manager omitted something or if a sales-and-marketing consultant was in error? In error of what? Omitted from whose standards or criteria? How does an owner objectively and fairly measure subjective performance?
6. As an owner, you cannot have it both ways. You can't manage, affect, or impact a non-licensed consultant and not bear joint responsibility if things don't work out. As mentioned above, calculations and commitments to engineering, design, and surveying are traceable, quantifiable, and objective. Conversely, a plan or strategy for managing affairs is subjective. Success, however measured, is dependent upon many soft variables including the responsiveness of consultants and contractors, whether or not an owner follows a prescription or advice, markets, buyer's whim, uncooperative municipal authorities, and unforeseeable circumstances.
7. Having E&O insurance doesn't automatically qualify consultants, licensed or not, as easy targets if something in site-work that they're involved with fails. Too many factors including, but not limited to, weather, soil changes, water, inaccurate base topography, poor design, and mismanagement can negatively affect site-work. So, if pursuing an E&O payout from a licensed consultant isn't guaranteed, then how does it make sense to pursue damages from non-licensed consultants for something more complex involving perhaps many other subjective factors? Unless it's easy to identify a mistake or there's no question who's at fault, you have to ask yourself if an E&O claim is worth pursuing.
8. Lastly, an owner should be convinced that his in-house management team is clean and without blame before leveling charges at consultants.

Unfortunately, land development cannot be bound into a neat and tidy risk-free package. Nothing short of perhaps an expensive, project-wide blanket insurance policy, may blunt the effects of poor planning, clouded thinking, and inept decision-making and even then—someone's head may roll. Like the weather, some things just aren't insurable.

E&O insurance and geotechs

No doubt, geotechnical engineers, like any licensed or certified consultant should carry E&O insurance. Realize, however, that, depending on the issue, there is a stark difference between requiring a geotech to carry E&O insurance and executing a successful claim against the same geotech.

Why? Some geotechnical practices, such as laboratory work, compaction testing, approving fill material, and designing retaining walls, are usually black and white exercises. Culpability can be traced and responsibility is clear. Trickier items involve judging a geotech's work based upon the completeness of subsurface exploration in regions of variable and unpredictable soils. Examples of everyday and subjective calls can include recommendations for water infiltration, identifying the limits of unstable soils and slide areas, soil typing, proof rolling, and subsurface water mapping. An owner who ignores a geotech's recommendation for lab testing or increased exploration such as test pits, borings, or geophysical surface surveys (surface wave penetration), may bear a significant degree of responsibility if related problems later occur.

Other gray matters pertaining to geotechs and E&O insurance include the effects of consultants or contractors who forge ahead with design or construction without geotechnical recommendations.

This should mean two things to owners. First, spend enough money on exploration to make a geotechnical consultant's E&O insurance relevant. Be candid with your geotech and understand the degree of exploration required to develop a reasonable probability for accurate ground mapping. Get the geotech's recommendation in writing and go one-step further by inquiring whether the geotech can live with less. If the answer is no, approve what's recommended and concentrate on performing the work as cheaply and efficiently as possible. This action reduces the likelihood of geotechnical problems typically resulting in a better project. Second, make sure that all design consultants have and are abiding by the geotechnical engineer's reports and studies during contract and change of conditions work.

Consultant billings

How a consulting agreement is structured goes beyond the owner's preference for style. It also defines the ease and options available for controlling consultants. And nobody appreciates this more than the people who review and approve consultant billings. Again, refer to *Figure 1-3.* The consultant contract alternatives shown can also be thought of as ways to better manage consultant billings.

Here's a list of possible contractual payment terms to consider when working with consultants:

- Lump-sum payment or fixed monthly billings
- Fee based upon a percentage of gross construction cost
- Open-ended time-and-material or hourly billings based on an agreed fee schedule
- Fee arrangements based upon finished units, such as per plan-sheet count, per square area, or per lineal footage of utilities
- Value or performance-based agreements that result in consultants and owners sharing revenue or cost savings as well as imposing fees due to not meeting schedule, missing milestones, or exceeding budget. This approach works only after both parties agree on how results will be measured and who will do the measuring. Moving from cost management to performance and quality evaluation demands that goals and objectives be perfectly understood by both parties prior to beginning work. It also helps to avoid using these types of agreements on work that's prone to conglomerate decision-making or that may be at the mercy of uncontrollable factors such as poor weather, jurisdictional instability, or unpredictable permitting. Evaluating a consultant's performance requires being able to clearly assess value, merit, or efficiencies based solely on that consultant's work void of outside influence. Also—since it's typically not possible to construct something twice in order to measure savings one way or another, establish an accurate and detailed benchmark cost estimate from which to judge performance.

A common example of performance based consulting is hiring a value engineering team where the potential savings identified by the team should clearly offset the cost of performing the study. Agreements can also be based on contingency pricing or "variable fees" where consultants may perform work for partial payment or without guarantee of any payment. Depending upon the degree of savings or value produced, the consultant may collect all, part of, or none of their fee. Conversely, some consultants can collect a windfall if the incentives are set high enough based on performance. Here are some examples of how a consultant's performance may be objectively measured.

- ✓ Winning an award
- ✓ Shortened construction schedules
- ✓ Quicker than expected permit issuances
- ✓ Lower than projected consultant billings
- ✓ Improved quality without additional cost
- ✓ Additional quantity without additional cost
- ✓ Overall reduction in time it takes for occupancy
- ✓ Lowering costs through value engineering or creative management
- ✓ Windfall profits resulting from exceptional design, value engineering, and creative problem solving

In site-work, numerous variables can make any equation relating construction cost to consultant fees irrelevant. Charging consulting fees by the square footage or as a percent of total construction cost can be cumbersome, prone to error, and isn't recommended in site-work. On smaller jobs, owners may encounter disproportionately higher fees when consultants charge by percentage of total construction cost.

Fee arrangements that are based upon units can result in owners receiving a product skewed towards whatever defines payment. For instance, paying consultants per plan sheet may result in receiving a continual and inordinate number of plan sheets, or sheets laden with redundant specifications and extraneous detailing. Paying consultants by the lineal foot of utilities may result longer and more utility lines than are absolutely needed. Similarly, owners wishing to trim costs may deceive themselves by cutting out plan sheets that on the surface appear irrelevant, but that are useful to field personnel in ways that a disconnected owner wouldn't understand. The same saving-pennies–to-spend-dimes rational applies to cutting back on utilities or other construction units in order to save design costs.

A strategy that works well with lump-sum, fixed-cost, and periodic not-to-exceed billings is to tie payment to construction milestones. This strategy puts a brake on runaway billings. Milestones should be easily identifiable and wholly within the consultant's control. Although milestones vary according to each consultant, some examples may include a time frame for submitting plans for approval, completion of a study or analysis for storm water or traffic impact, transmittal of complete and final as-builts, and final plat submittal. Consultants should be able to tag and to substantiate dollar amounts to milestones. Linking payment to milestones works to keep everyone attentive to the process and provides a platform for measuring progress. It can also foster a keen interest in attending schedule and coordination meetings.

To minimize heavy change-order pricing, owners pursuing design work on a lump-sum basis should only proceed if site planning is fixed, complete, and unlikely to change. Change orders, incomplete plans, and reduced quality are prevalent problems when contracting with design consultants on a lump-sum basis. Skip ahead to *Figure 12-1*. Substitute the words, "lump sum" for the word "cheap" and compare the results for getting plans lump-sum and right, and lump-sum and quick. Hourly or time-and-material pricing is usually a better way to go when design is incomplete or at-risk of changing. When combined with not-to-exceed thresholds, hourly pricing can be more predictable and result in less change orders. Lump sum works best when the consultant's work is all-inclusive and the likelihood of changes or modifications is small.

Chapter 2

DESIGN-BUILD

Projects typically follow a set and linear pattern of events culminating in construction. Moreover, plans are developed, budgets are estimated, work is bid or negotiated, and construction begins. This sequence of events is commonly referred to as the *traditional project delivery system.* Consultants and contractors working in the traditional project delivery format operate separately from each other and contribute individual expertise and services as needed.

Design-build differs from the traditional project delivery system in that it pairs upfront, and usually under one contract, a single entity typically comprised of an architect/engineer (AE) and a contractor. Other consultants, a client or buyer, or the owner's staff may also be part of the team.

As a single-source entity, design-build teams are intended to collaborate, massage, and design a project, from its beginning through its completion. Gone, at least in theory, are individually inspired elements and solo decision-making. Also absent is the progression of consecutive activities where one activity can't proceed until the prior activity has passed. Instead, contractors are free to work in stages before final design and detailing are complete and without having fully approved sets of working drawings. Design-build is a preferred style of management for many in the construction industry.

People usually equate design-build with structures rather than with site work. Why? One reason is that structures can be built *independently* of most anything. Once the foundation is in and the roof is in place, building a structure tracks continuously along an uninterrupted *calendar* schedule. Only man-made problems, such as design errors or lack of materials prevent a structure from being completed. For purposes of understanding design-build as it applies to site-work, the key words to remember are *independent* and *calendar.*

The limitations of design-build as they relate to site-work

Calendar

In terms of calendar, site-work is performed seasonally on an intermittent basis, or continuously on a year-round basis. For most project owners, site-work has a season. The analogy I like to use is that intermittent site work resembles a lump of chewing gum on a bicycle tire. It comes around every so often on the clock and thumps you. Then it's gone until it comes around again next year at the same time and same place.

Year-round site work performed regardless of weather or season lends itself more to public works projects where adhering to schedule or reducing public disruption can trump the cost of construction. Not to imply that cost isn't important but, unless there is sufficient slack in the schedule or unless permit restrictions or technical reasons interfere, public works projects generally continue through thick and thin until they're complete. Understandably, the cost can be great.

In private-sector work, owners may be obligated to continuous and unseasonal site-work for any number of reasons including meeting market conditions, satisfying conditions of a sale, or regaining schedule. Large-scale developments adjust regularly to whatever makes sense in maintaining or increasing profitability. Sadly, there are times in site-work when a manager must balance moving too fast with idling long enough to allow his contractors time to perform change order work. For instance, as fine grading progresses in order to meet a paving deadline, modifications to in-street utility stubs, and re-alignment of driveway aprons may also be transpiring in the same location in order to satisfy terms of a newly negotiated purchase and sale agreement for adjacent lots. In the name of cost control, these layers of activities must happen when the "gum on the tire hits the street"—during the site construction season.

As a rule, profit-oriented project owners don't perform site-work out of season or in violation of budget and schedule unless:

1. Permits or special conditions are set to expire
2. They have a limited window of opportunity to obtain financing
3. Markets and profit dictate accelerating the site-construction schedule
4. It makes sense to work into inclement weather in order to complete that year's work
5. Delaying site-work may result in disrupting civic services, schools, or poses an ongoing hardship or safety concern to the public
6. It coincides with low-cost construction (That is normally when erosion potential is lowest, when it's economical to move dirt, or when it's advantageous to perform concrete flatwork and paving.)

Owners conscious of controlling costs have to respect the calendar and its impact on site-work.

Independence

Using structures as an example, let's consider variables affecting the construction of a building. Structures are created in someone's mind, are born in an office, and are built as designed. They're a known quantity. During construction, design problems and conflicts are regularly solved on paper and impacts can usually be traced to human error or mismanagement. Vertical construction is unlikely to be affected by unpredictable factors such as weather, poor dirt, or subsurface conditions. This is because consultants such as architects typically specify design-build for a project *after* the foundation has been designed. And, site design is a prerequisite for drafting a foundation plan. Without having to contend with grading, utilities, and infrastructure, building contractors are able to control their environment.

This, of course, isn't the case with site-work. Site work doesn't happen in a bubble. It doesn't happen independent of unknown variables or take place under controlled conditions. How do you "roof" a site so that it remains free of water and protected from weather? If were able to do so, how would you contend with the full spectrum of belowground soil possibilities perhaps seen only once before by a wooly mammoth? And then there are issues with wetlands, turbidity, erosion, and contaminated soils. Plainly, site-work is susceptible to many factors beyond an owner's control regardless of management style. This lack of independence can undermine the purpose and intent of design-build when applied to site-work.

When the design-build concept works for owners

Exceptions exist where design-build can prove invaluable in land development. For instance, it can be a used as a tool to maintain or to accelerate site-work. Some owners simply require site construction to proceed faster than allowed by the traditional way of doing business. It's also an effective tool for kick-starting construction and getting site-work underway without complete or approved plans. *Figure 2-1* illustrates how designs can be stunted allowing construction to begin without fully developed sets of plans.

Here are a few more reasons to consider incorporating design-build into the site-work phase of land development:

- Negotiated work. Owners who create design-build teams are more likely to negotiate pricing. This saves time and expenses required for bidding.
- Increased lead-time. When the contractor is part of the team, long-lead items, fabrications, and shop drawings can be processed and ordered quickly.
- A chance to construct with the weather. With or without plans, owners

want to build during favorable weather. Design-build is a vehicle that allows site-work to begin in season without permits or completed working drawings.

- Quicker reaction to changes than with traditional project delivery. If an owner obligated to complete lot construction becomes stalled, say, in waiting for final lot grading plans from a builder, design-build will allow that owner to react quickly once it is clear what the builder wants.
- One-of-a-kind jobs. Design-build works well for unique and "haven't tried it before" type jobs where the contractor is free to do whatever it takes to complete work. Here, design is committed to paper after construction is complete in the form of as-built drawings.
- Increased scrutiny of documents. Owners of challenging projects may feel better knowing that more people are involved in completing work. More eyes reviewing the construction documents can result in concentrated levels of value engineering and constructive analysis.

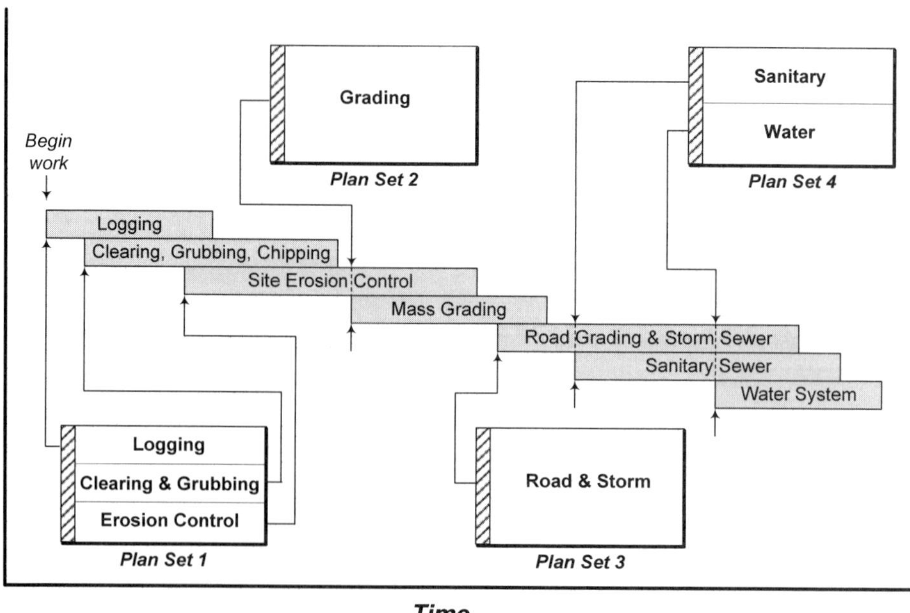

Figure 2-1

- Nimble as needed. Individual items that may drag or become locked in dispute under the traditional delivery system can remain in motion as the design-build team shifts to other issues or alternative phases of work. The civil engineer may be preparing road and storm plans while the contractor is working off of incomplete clearing and grubbing plans. Refer to *Figure 2-1.*
- Less exposure to claims and disputes. Design-build teams assume greater risk and responsibility than those under individual contracts do. So, it is in the team's best interest to work out differences and to remain cohesive. A design-build team under one contract is less apt to file a lawsuit against the owner. The downside for owners is that averting risk never comes cheap; they pay for it one way or another.
- Contractor input. Design-build shifts more responsibility to the contractor allowing them to provide balance, know how, and critical counter-point to design and planning issues earlier in the process. The same holds true for site-work subcontractors and site-work specialists. The trick has always been getting contractor input early without obligating owners. Design-build allows contractors and their subcontractors to get involved early enough to make a difference.
- As previously mentioned, design-build can act as a vehicle for beginning work if an owner doesn't possess complete sets of plans. Instead, individual design elements can be isolated and approved piece-meal as smaller plan sets or even as single plan sheets thereby allowing fieldwork to proceed. As layers of permitted construction drawings unfold in sequence, construction advances steadily towards completion. This approach also benefits owners who start late in the season without complete plan sets. *Figure 2-1* portrays, this "divide and conquer" strategy, which allows the owners to break ground and the construction to progress.

On a side note, be aware that bonding capacity and tolerance-for-risk may define who leads the design-build team. In most situations, that entity is the contractor.

When the design-build concept works against owners

The leverage gained through design-build can evaporate when owners have to contend with issues beyond their control. For instance, there can be severe consequences to fast tracking work without first understanding seasonal weather patterns, soil conditions, and topographic inconsistencies. Another example may include not being able to procure dry utility crews from local service providers as scheduled or not having permits for off-site improvements or sensitive-areas work. Design-build or not, in each case owners are powerless. Some municipalities may even allow an aggressive owner operating under a design-build umbrella to install something unapproved only to come back later and require that the work to be re-done or installed elsewhere. The point is this—irresponsible management and being too quick on the draw can lead an owner head long into cost and schedule disaster. Certain facets of managing land development, namely, site-work planning, cannot be ignored or compromised. This particularly applies to design-build.

Project owners should also evaluate if the design-build process is worth the effort and expense. In the same amount time that it can take to design, permit, and build an individual structure on a site already graded, a land development project may languish while waiting for a political determination, a permit to impact wetlands, or acceptance of a storm-water discharge application. Again, lack of control and not being *independent* bears hard on site-work. Oddly enough, what makes the design-build process so attractive in saving time may defeat owners who prefer a slow methodical management style that is steeped in evaluating all alternatives and preserves maximum flexibility. Here are some potential issues associated with using design-build as a vehicle for accelerating site construction.

- The clock gains greater importance—Time for meetings and proper coordination become more critical for design-build team members.
- Lack of site-work experience—Owners need site-work oriented consultants and contractors who have established a culture for success with fast track design-build projects.
- Quality control—When permit agencies fail to require quality control, owners should ask consultants and contractors to extend warranty and maintenance clauses to cover quickly hatched design-build work.
- Design cost—With design-build, design and engineering costs can be open-ended. Team members may ask for a retainer, request to bill a percentage of construction cost, or insist on being able to bill "whatever-it-takes" charges based upon time and materials or hourly rates.
- Greater odds of increased construction costs—Depending upon topography and its challenges, site construction cost is likely to increase with design-build. Downtime, time-and-material billings, and expenses related to inadequate resources can exacerbate uncertainties inherent in design-build.
- Exaggerated tolerance for error—Examples of site-work design elements that, under the best conditions, leave little room for error include wall design, utility location, pipeline gravity flow, road grade, fire flow versus elevation restrictions, and shallow culvert design. Fast-

track design and engineering can stress these and similar elements that already push the envelope or limit of design.

- The unknown—The word design-build may suggest that a project is in danger of proceeding without adequate base information. Short of the boundary and topography survey, no other base information can be more critical than sufficient and properly interpreted geotechnical data. Owners cannot let quick processing come at the expense of shoddy or non-existent geotechnical exploration.
- Owner's role—What will the owner's role and responsibility be? For example, can the owner challenge the design-build team's choice of subcontractors or specialty consultants? Will the owner impose penalties for missing schedule? If the owner accelerates schedule, what affect will that have on the site-construction contract and the terms of liability? Clarify these issues when kicking off a design-build program.
- Budget formulation—Land development cannot be estimated exclusively using two-dimensional measure whether lineal or square. So, without nearly complete plans, budgets are difficult to develop and even more difficult to estimate—especially while on the fly. Owners wanting budgets void of complete plans may have to accept "range estimates," which are addressed in Chapter 13 on cost estimating.
- Schedule uncertainty—Site-construction schedules already contingent or uncontrollable can be further strained by design-build. For instance, design-build may prohibit taking the time to draft multiple grading plans and compare construction costs of each alternative. Permit approval can drag if agency demands fluctuate, are inconclusive, or produce hanging variances, or if *related* approvals linger en route from other public agencies.
- Permit restrictions—Increased awareness of erosion control, environmental protection, and general liability has left many public permit agencies reluctant to permit any work that could backfire on inspectors. Unlike structures, site work can expose neighboring properties, surface water, and soil to increased risk. Potential gaps in engineering may also concern permit authorities making them reluctant to allow design-build site-work to proceed.
- Stereotypes die hard—Some architects may find it difficult to trust contractors. They may still feel that it's up to them to design everything after which it belongs to the contractor. Architects and contractors who have been at odds before exacerbate the problem. Likewise, contractors may harbor suspicions, be painfully overt concerning poor design, or complain about an architect's shortcomings. This may not go over big

if errors or inadequacies are openly debated in the owner's presence. Some contractors may not respect an architect's opinion on site-construction sequencing and methods.

- Insurance and liability—Will your consultants and contractors remain fully insured and protected while performing in a design-build capacity? Will firms be offering unlicensed design and construction services? Can each firm handle increased exposure to potential claims of negligence? Check the status and licensing requirements for design-build entities in your state or domain. Inquire as to whether or not your design-build team will operate and be licensed as a separate legal entity. The process usually begins by first seeking counsel from a construction attorney before committing to a design-build contract.
- Life after—Design-build works for structures construction and for smaller site-work jobs that can be built without interruption. It may not provide owners with the same leverage on large scale, multiyear land development projects where the cycle of site-work repeats itself, season after season. Choosing to change design, to pursue another product type, or to operate differently on a multiyear project may prompt an owner to also change personnel on the design-build team or even revert to the traditional process of doing business. How does an owner reconcile with a design-build team used earlier but which now seems a bit too expensive or out of style? Some project owners may find the cost, the upheaval, and the turbulence created by wholesale team changes to be unforgiving and for most, their margins may not allow it.

Performance-based work

Performance-based permitting is another form of design-build. With performance-based permitting, the site contractor proceeds with only a concept, with an incomplete design, or with complete but unapproved drawings. Work happens under the direct supervision of a licensed engineer or designer and can only progress in stages as work is inspected, approved, and documented. Performance-based permitting may be, in itself, the purest form of design-build. However, this method only works with public agencies that trust the owner, contractor, and design team. It is best suited for work items that are both low in risk and are limited in size and scope.

Chapter 3

EARLY LOGISTICS

All too often, owners mistakenly overlook golden opportunities available only during the early phases of a project. That's unfortunate because it's during this period that owners enjoy maximum flexibility to set the stage for cost effective and efficient site construction. Many factors influence the proper preparation of site-work but they all, in one way or another, relate to:

1. Type of project
2. Execution
3. Available funds
4. Ease of gaining approvals

Of the above items, the last two have a life of their own and don't directly relate to optimizing site construction. However, the first two items, type of project and execution, offer owners a wide palette of opportunities for properly preparing for site construction.

Type of project is defined by constants and constraints such as product type, topographical relief, and project size. Type of project includes items that an owner must react to and that aren't easily manipulated or open to negotiation. Where as type of project can be largely defined as inelastic or fixed, execution is its antithesis in that it's supple and adaptable. In simple terms, execution is the manner and methods used by an individual or team to oversee and manage a project.

Style, vision, knowledge, experience, and innate ability are basic elements of execution. Collectively, they influence the degree in which a project succeeds of fails. The more that one understands the mechanics of preparing land for structures, the higher the odds that one will successfully execute a profitable land development project. In a strange way, recognizing one's fear and vulnerability during site-work acts as a lubricant—an incentive that compels one to focus on smart strategies and preventative practices leveled at averting cost overruns, schedule disruptions, and logistical disasters. Fear of failure forces people to seek solutions. Many develop survival skills. Managers who go through this trying and sometimes unpleasant experience also learn how to tap into a mother load of opportunity only available during site-work.

Once they're there, ideas on how to effectively save and make money during site-work emerge like clowns pouring out of a circus jalopy. Just when the car appears empty, another one comes out, followed by two more, and then, yes, a midget with a monkey in his hat. Yikes—it doesn't end and neither do profitable opportunities on projects that are properly set up for site construction.

Type of project and land requirements

Before purchasing land, it's smart to consider how you're going to build on it. What in blazes are you getting into? You may already know what type of product motivates you and your team. Are you building a school or an industrial park? Residential housing or big-box retail? Do you need roads and parking? You already know that parking isn't a hit on hillsides and that residential views are tough to get on flat land. What about potable water? Do you have enough volume and pressure to build without requiring off-site facilities? These types of questions should season any decisions related to how and where a project owner chooses land.

It's here that we'll briefly separate from planners and architects leaving them to wrangle with rice paper crayon drawings of ghostly figures holding hands while streetscaping alongside red 1968 Pontiacs. Our goal is not to dissect visual design, finance, and permitting as much as it is to benefit *owners* with techniques to save them money, to make schedule, and to ease site construction.

In the beginning—without planners

Before you can employ planners, you have to start with land—but how do you choose land? If you already possess property—hold on—we'll get to you later. But for now, let's discuss how land can be evaluated with an eye trained towards site-work.

Flat land is the easiest land to develop but when it's in tight supply, many owners are forced to construct on steep terrain in, sometimes, tough-to-get-to places. As a result, development is becoming more expensive, feasibility is frequently challenged, and battles rage over maintaining profitability. Some project owners are competing in an environment where growth is severely restricted, if not wholly prohibited. In general, complex, topographically challenged, and technically innovative projects are prone to:

1. Lack utilities
2. Limit product
3. Attract resistance
4. Challenge design
5. Stress construction schedules
6. Increase the timeframe for planning and design

7. Narrow choices for consultants and homebuilders
8. Increase public agency risk expressed through onerous permitting, specifications, and agreements
9. Breed newfangled concepts and amenities that people with nothing to lose want to try at an owner's expense

Predesign indicators

Before committing to developing on any project, use every conceivable tool to help figure out its mysteries. Here are some guidelines to consider on major issues when moving to carve raw nature into something useful. With consultants or qualified staff, evaluate:

- Sunlight—What is the site aspect? Where does the sun sit relative to the site?

  Site work ramification: Wherever the sun's rays strike, their heat and light help to prolong the workday, prevent freezing conditions, and to retard snow retention. Dirt dries out quicker and is workable longer when exposed to southern and western aspect. This warmer and drier aspect results in diminished site-work costs and allows for a longer construction season. On northerly and easterly facing slopes, the opposite is true—these slopes are wetter and cooler, and they receive less sunlight.

- Access—What is the political and financial cost of establishing access and how far into the public road network will infrastructure have to be improved? Before first occupancy, reliable access will have to be established.

  Site work ramification: Access governs how grading and utility installation proceeds. Poorly positioned access is costly if it doesn't serve hauling, grading, and utilities. As a result, owners may have to invest prematurely in expensive road work just to begin construction. Front-loading excessive and premature capital expenditures into roads and utilities can financially cripple a project or force profitability further into the future. Poor access can be somewhat blunted by manipulating project build-out before committing to a development schedule and proforma. For example, in order to avoid generating huge cut volumes by constructing an entry road through a hillside and into a project where land hasn't yet been prepared to absorb fill material, it may make sense to first develop flatter ground if it can be accessed with cheaper temporary infrastructure. Even though there may be market demand for the type of product planned in areas of significant grading, the up-front cost of moving large volumes of cut material to ground not yet cleared and stripped may be prohibitive not to mention other likely issues involving schedule, logistics, or permits. Don't under-estimate the impact of poorly situated access to or through a project.

- Timber—Read the forest, plant life, and grasses. Depending upon your region, some tree species can indicate standing water, seasonal wetness, potential fire hazard, and generally—trouble. Large stumps may indicate that windy conditions probably aren't going to be an issue. Conversely, lots of blow-down suggests heavy wind, wetness, or perhaps a serious land-moving issue. Avoid entitlement or permit conditions that require saving native trees on land naturally prone to losing trees or that harbor species difficult to save. See the chapter "A Word About Native Trees" for more information.

  Site work ramification: Trees and plants that indicate water may also be telling you something about the soil. That is, it may be fine-grained and holding water. Or, trees may be indicating shallow soils above hardpan or bedrock where plant life easily finds subsurface water. Larger trees and stumps can mean deeper soils more suitable for grading and structural fills.

- Soils—What type of soils are you inheriting? Are they fine-grained, dusty, and slick or are they course, sandy, and rocky?

  Site work ramification: In general, fine-grained soils, such as silts ands clay, are tough to work, are moisture sensitive, and have a notorious reputation for shortening the construction season. Soils that hold rather than pass water can impact every phase of site-work. They also tend to erode quicker and to have a greater affect on water quality than do course-grained soils. Saturated and unworkable soils impact schedule and generate greater quantities of unusable soils. Conversely, course-grained soils are workable under broader weather conditions. Unless they lack binder (fines), are poorly graded (pure beach sand), or are too dry to shape and compact, coarse-grained soils are preferred for construction. Topsoil depth is also important. Most projects can utilize a fixed amount of raw native topsoil, but surplus topsoil that is unable to be wasted on-site or to be sold will have to be exported to an off-site dump. If infiltration or septic systems are designed for the site, percolation, mounding analysis, or other tests will have to be performed on non-organic native soils in order to establish design parameters.

- Terrain—How steep is the ground? Is it broken and undulated, or are slopes continuous and flat? Is the terrain too steep to support the right mix of product without inordinate grading and loss of land to slopes or sensitive areas? At what elevation is the project beginning and ending?

Site work ramification: The likelihood of longer work schedules, higher costs, and tougher construction increases proportionately with a steeper terrain. Producing viable lots and parcels becomes more challenging. Steep terrain resists development resulting in the need for deeper cuts, higher fills, increased dirt volumes, and in a greater probability that retaining systems will be necessary. Slope also limits product resulting in fewer alternatives for development. With elevation gain, soils tend to get shallower, risk of encountering bedrock increases, and weather worsens. Microclimates created in extreme or mountainous country, when valley to higher elevation gradients combine with draws, ridges, and plateaus, can produce rapid air movements, erratic temperatures, heavy precipitation, and fog. Slopes exceeding the on-site soil's natural angle of repose are vulnerable to high erosion potential from running water. Cross drainage issues between neighboring properties can blossom with increased slope. Not all contractors or engineers handle steep terrain as well as they handle flatter ground. Steep terrain may also force civils to engineer roads and lots to the maximum-allowed grades, encumbering the design and construction process with minimal room for error. Enough of an elevation difference can impact domestic water pressure significantly, thereby increasing design and construction costs. However, increased slope can also make the design and construction of gravity wet-utility design easier.

- Drainage courses—Running water reveals much about a site. Comparing the color of running water before and after a rain can indicate a site's soil erosion potential. Equally important is scour. What do the side walls of a drainage course consist of? Look for rock. How deep does the surface soil appear? Are shallow tree roots sticking out into the drainage course or are they hidden? Abundant shallow roots can suggest the presence of thin soil or bedrock, or can reveal that the land is rich in trees and plants that thrive in moisture—and are finding it.

Site work ramification: Granted, water is beautiful, and as an amenity, it's tough to beat, but drainage courses can mean significant loss of developable land to buffers and setback requirements. Drainage courses can also indicate a need for bridges and culverts. Some drainages are treated as linear zones of intolerance where discharge of construction runoff is prohibited. Site-work costs increase when the presence of water restricts access, requires crossings, or heightens emphasis on controlling runoff and erosion.

- Sensitive areas—What type of wildlife habitat are you dealing with? Does the land hold wetland- and water-dependent wildlife, abundant upland birds, and maybe a few beavers? Check with a wildlife biologist before committing to anything. Does your agency consider slopes too steep to build on as sensitive areas? What about archeological or mining features?

  Site work ramification: Early in the planning process, locate, mark, and fix all sensitive areas as control points. When mitigating habitat or wetlands, locate and reserve suitable areas as soon as possible. Above all, avoid scrambling to locate mitigation land at the end of a project. The odds of finding land suitable, for say a wetland, decline as a project approaches completion. Once suitable wetland mitigation areas are known, they're set aside, or they receive low impact treatment, or they are graded out completely and re-established as a wetland. For example, an owner may curtail logging or eradicating

Ability to Influence a Project Diminishes Over Time

Considering a Hypothetical Project Model

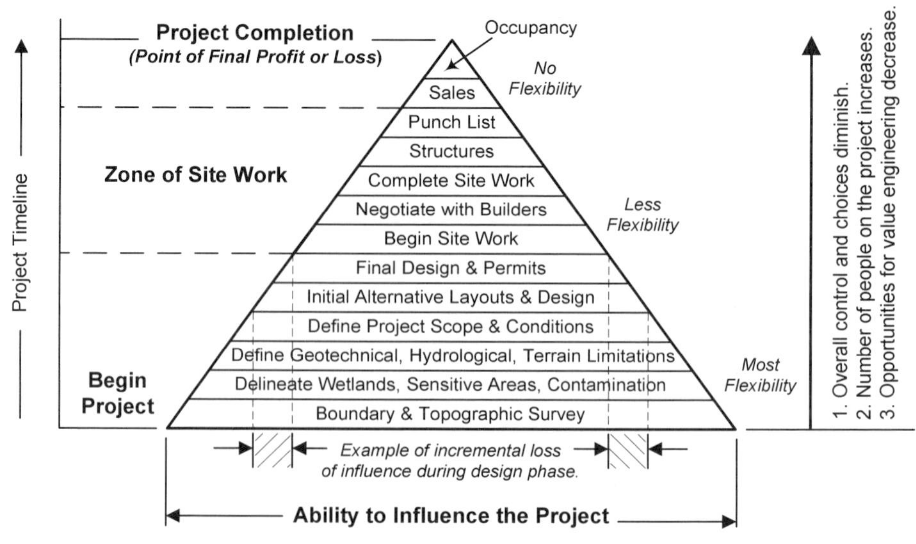

Figure 3-1

native plants in areas slated for future wetlands. Or, there may be requirements to leave dead wildlife snags and islands of native plantings only in areas planned for grading and filling with import hydric soils. How does leaving ground natural work, you wonder, when surrounding land is being intensely graded and developed? Unless you can dedicate acres of open space, easily match grades, and guarantee proper drainage, it usually doesn't. Not all created wetlands work.

If an owner finds it feasible to develop a property, the next step is to corral, measure, and devise a strategy for conquering it. Before moving on, refer to the *Figure 3-1* triangle. This chart illustrates that an owner's ability to influence a project rapidly decays over time. Once site construction begins, an owner's opportunities for value engineering and alternative strategies will have diminished considerably.

Executing early in the process

Part of the process in conceiving a project includes deciding what's going to be built. Initial ideas and parameters, such as targeted number of units, total building square footage, or number of acres to be developed, flow into design and construction cost estimating. From here, owners need guidance and it usually comes in the form of in-house staff and hired help such as contractors and consultants. Their task is to fit and test the owner's vision to the land. How much is the project going to cost? How much revenue is it going to produce? How long is it going to take? Once these questions are answered, an owner can take the steps necessary to address design, permits, and site-work. What's critical? Where is there exposure to unknowns and extraordinary costs? Where does the value in developing a particular product-mix break even with other development alternatives or sales considerations? The answers often are contingent on site construction.

If he is going to influence a project, a site-construction consultant must be invited to the party early enough to make a difference. While some owners ponder if they'll include a site-construction specialist as part of the initial team, others know that they need the help but are uncertain as to when they'll bring a site-construction professional on board. The remaining owners aren't aware of the benefits of hiring a site-construction advisor. Unfortunately, the wrong decision here can handicap a land development project before design begins.

Provided that an owner has or brings a good site-construction "mind" onto the team, the next step involves lining out geotechnical exploration to evaluate the soil's character, variability, and structurability. Geotechnical information is one of the three key elements necessary for controlling site construction. These three key elements are:

1. Attaining complete geotechnical work
2. Proceeding with complete plans and documentation
3. Building a detailed and comprehensive schedule of values and bid structure

Each of the above subjects will be covered in ensuing chapters, but, for now, it's important to understand that within them exists the lions-share of opportunity to control costs and effectively manage the site-work phase of construction. Other opportunities for lowering costs, meeting schedule, and maximizing profit derive in one way or another from these three elements. Once an owner has a handle on these key elements, a course can be charted that:

1. Minimizes cost
2. Maximizes profit
3. Ensures project control
4. Delivers a marketable product
5. Minimizes environmental exposure

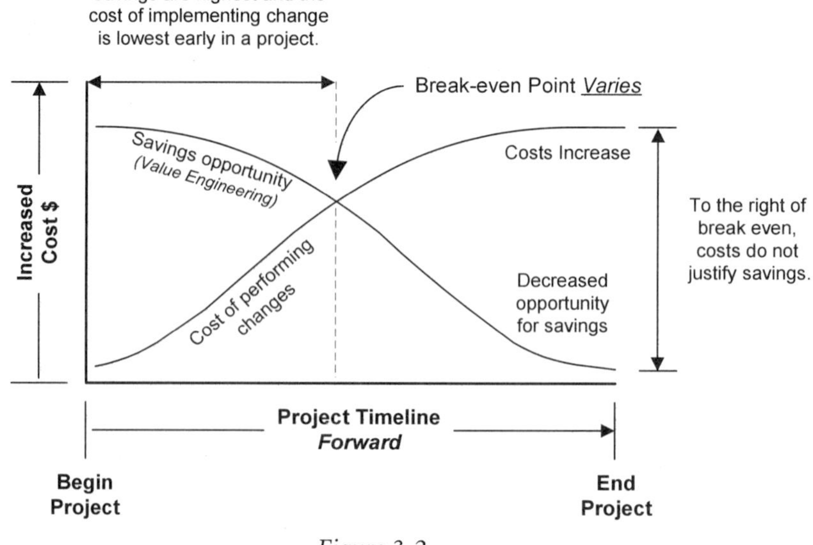

Figure 3-2

6. Meets or beats schedule with no added premium
7. Adheres to a regimented and chronological plan for development

Influencing a project versus time

By now, it should be clear that the greatest opportunity to influence a project occurs at the point of inception and decays rapidly as the project matures. Refer to *Figure 3-2.* This graph shows that the opportunity for saving money as a function of performing changes diminishes over time. The lesson portrayed on *Figure 3-2* is that the ability to salvage or improve a project's internal rate of return or net profit declines as that project reaches the break-even point where the cost-of-performing-changes curve intersects with the opportunity-for-savings curve. To the left of break even, savings can be economically attained with changes. To the right of break even, the return doesn't justify performing changes; in fact, the cost differential worsens as a project advances in time. Every project has a gross-cost–to-benefit curve derived from an array of subcurves germane to each individual phase of design and construction. The cumulation of sub-activity curves defines the composite project curve.

Spending to save on site-work

Depending upon a project's location, geology, and what's being constructed, successful owners exhibit one common trait—they invest early. Ways to invest early include:

- Commissioning adequate topographical mapping and surveys.
- Contacting local well drillers to find out where water tables exist.
- Honoring grading and its potential impact on profitability and schedule.
- Collecting sufficient geotechnical data through boring and numerous shallow test-pits.
- Bringing specialists into the process early to allow ample time for studying and designing critical project components.
- Identifying, scheduling, and laying out global elements that either drive the project or hold a position upon which the project is contingent.
- Knowing your project's microclimate. They research regional and local weather station data and arrange to obtain this data on a regular basis.
- Identifying storm planning control points early. They pursue basic storm-system layout and design while value engineering the system as the project progresses.
- Deciding on and remaining true to a project scope. Though not always possible on mixed use, multiyear projects or master planned communities, the sooner that a project's scope is fixed, the better.

- Scheduling adequate potholing to ensure that existing on-site and off-site utilities have been thoroughly inventoried and correctly mapped. (Don't rely on municipal construction drawings or as-builts as the sole template for utility design.)
- Commissioning an existing condition survey (American Land Title Association survey, or ALTA survey) while working to verify the existence of known or suspect items such as utilities, septic fields, underground tanks, possible contaminated sites, and encroachments.
- Preserving and monitoring pre-existing culverts during storm events. Successful owners use existing culverts as a tool to develop baseline runoff and flow calculations instead of only relying upon new calculations and hydrograph interpretation. They expect their consultants to measure upstream high-water marks as part of the basis when sizing culvert replacements. As a defensive tool, they understand that knowing the historic working capacity, hydraulic radius, and date of installation of pre-existing culverts may come in handy later if an owner wishes to challenge regulations and the sizing or overdesign of storm ponds, conveyance systems, and culverts.

In the beginning—with planners

Preparing any project for site construction begins by first quantifying the land. Measurements by surveyors and civil engineers and data collection by geotechnical engineers define property three-dimensionally. Once information, such as vertical and horizontal data, existing land features, soil logs, and sensitive area delineations, is known, topographic, open space, and soils base plans can be developed. Although they may not know final product type and density, owners should possess a general idea of market needs and of what builders are looking for. And, they must have a sense of what *they want* to build. At this juncture, it's time to add design consultants to the process.

Figure 3-3 illustrates the roles that various consultants play through typical site construction as defined by major milestones. While reviewing *Figure 3-3,* note that it's worth retaining team-oriented, capable, loyal, and economical consultants through periods of inactivity. Besides the benefits of maintaining continuity in communication and work, changing consultants is expensive. On large multiyear projects, changing consultants through periods of inactivity can mean sacrificing knowledge and history, dealing with gaps in the learning curve, having to create new paper and documentation, and possibly getting a firm that's no better or cheaper than the former.

Referring again to *Figure 3-3,* note the dark horizontal line denoting that many tasks can and should be completed on-site before drawing a planner,

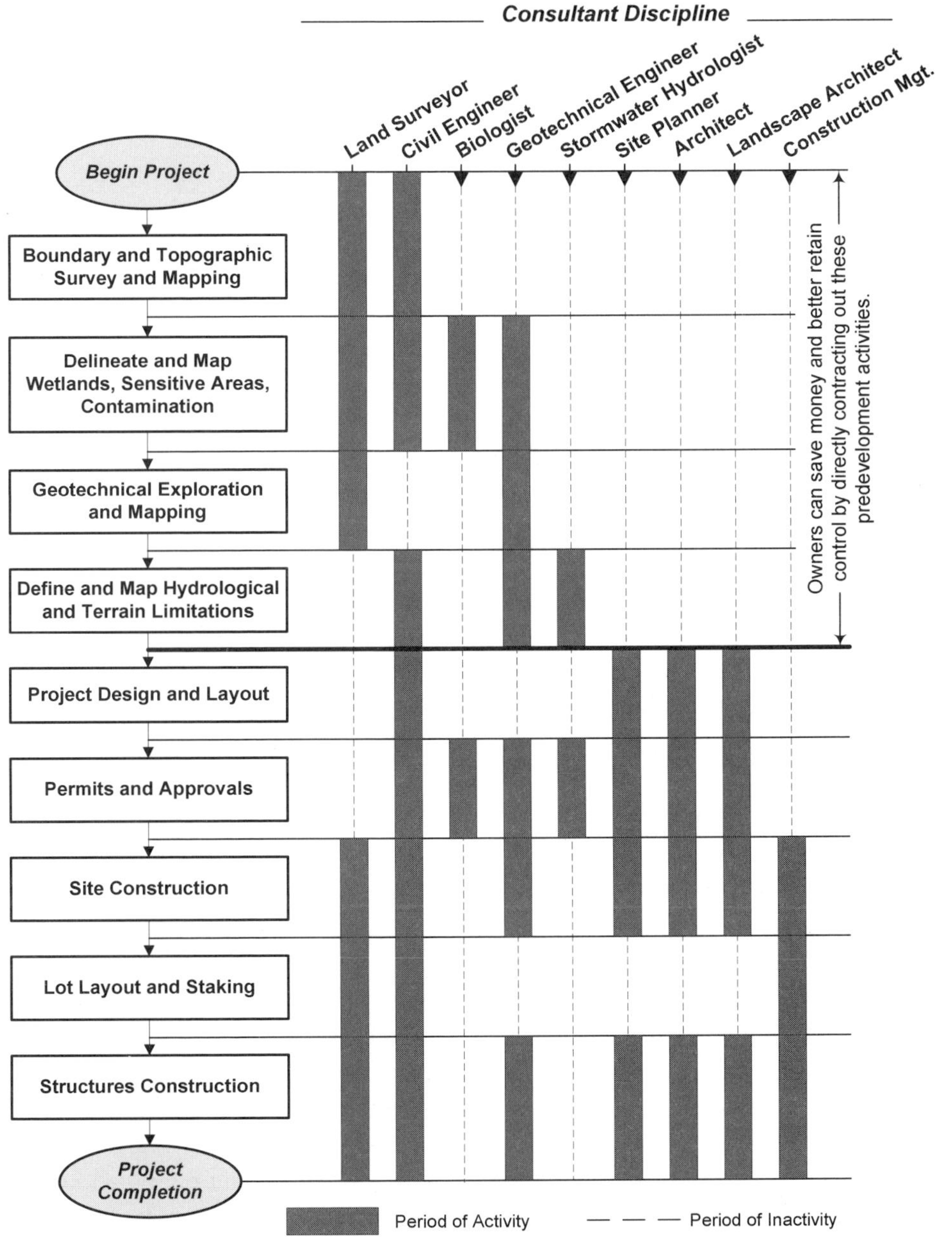
General Use of Consultants
Throughout Project Timeline
Consultant Discipline
Land Surveyor
Civil Engineer
Biologist
Geotechnical Engineer
Stormwater Hydrologist
Site Planner
Architect
Landscape Architect
Construction Mgt.
Begin Project
Boundary and Topographic Survey and Mapping
Delineate and Map Wetlands, Sensitive Areas, Contamination
Geotechnical Exploration and Mapping
Define and Map Hydrological and Terrain Limitations
Project Design and Layout
Permits and Approvals
Site Construction
Lot Layout and Staking
Structures Construction
Project Completion
Owners can save money and better retain control by directly contracting out these predevelopment activities.
Period of Activity
Period of Inactivity

Figure 3-3

architect, or landscape architect into the project. It's not a mistake to immediately go to an architect or planner to discuss possibilities, however, it can be a cardinal error to stay there too long allowing the wrong consultant to control the process forward, especially when you don't have to. The cheapest and most expedient path for owners to take is to handle the early tasks of site-work planning themselves, namely those activities surrounding the first four boxes depicted in *Figure 3-3.* By handling boundary and topographical surveying, sensitive area delineation, geotechnical work, and studies pertaining to hydrological and terrain limitations, owners save on overhead expenses and discuss first-hand strategies and requirements with those professionals who will actually perform the work. While this approach will place owners well ahead of the curve, it is crucial to remember is that this work must be completed to the fullest extent in *preparation for* architects and planners. It doesn't have to be *performed by* them. Instead of allowing the wrong consultant to handle important predevelopment work or risk having them subcontract with other consultants possibly according to bare-bones bid or shear-budget allocation, instead, redirect any saved overhead expenses into gaining better and more complete base information about terrain, water, soils, and sensitive areas. Spend more where it counts. This allows owners to retain greater process control of the schedule, soft cost expenses, and future site construction.

Moving on to planners, as capable as many of them are, they still need to understand, to digest, and to respect a project's base topographic parameters and control points. They must know what the ground will give before attempting to fit a master scheme or vision onto it, and, this should happen before an owner spends one dime on design time. On a smaller parcel or lot-by-lot scale, the same requirements apply to architects.

So, what am I suggesting here? I am stressing that on sloped terrain; one must develop a series of rough preliminary road-access and grading-plan alternatives *before* summoning a planner to the table. Grading alternatives will vary with percent slope, lot size, view potential, road layout, slopes versus walls, or what ever else makes sense on a particular piece of property. In a primeval way, the point of this exercise is to determine what the land will give. Although a planner could perform this analysis, it may be preferable to use a civil engineer that is only concerned with slope, gravity utilities, minimal grading, and establishing maximum easy access. Planners can always benefit from more information.

If there exists only a few ways to access and to grade a particular project, this work will prove invaluable. But—by no means should an owner allow this exercise to limit or to compromise a planner's creativity. On the contrary, encourage planners to flex their creative muscle when tackling alternative grading plans and layouts including giving them the freedom to alter anything in accordance with their conceptual designs. From this exercise, planners gain a known and quantified platform from which to germinate ideas. Preplanning layout and evaluation also works as a reality check by preventing months of high-cost, trial-and-error meetings and by eliminating design based on concepts that are otherwise unworkable on a given slope and terrain.

This leads to another question worth asking which is, "who controls the grading plan". The civil engineer is usually best equipped to assume the lead role when developing grading plans, especially as terrain steepens. This doesn't prohibit the planner or the architect from performing changes or further refining grading schemes. However, civil engineers can produce accurate and widely usable base maps and grading plans and usually possess the software needed for calculating earthwork volumes. We'll explore this issue later in the book.

Here are some guidelines for planners developing concepts for steep ground.

- Aspect and landscaping—Since aspect affects temperature, moisture regime, natural lighting, and wind, it has a huge bearing on the survivability of certain plantings, trees, and shrubs. Owners shouldn't spend a dime on landscape design until they know that their landscape architect and planner have carefully analyzed planting stock and its probability for on-site survival.
- Dirt volumes—Mass excavation and earthmoving numbers can be sobering. From a cost-and-schedule perspective, it's imperative to know earthwork volumes before investing too much time in planning and design. A grading plan with costly earthwork volumes, or that offers minimal or no value, is the wrong template for a planner to work from. Planners must remain careful not to change the preferred grading plan enough to compound earthwork quantities and create grade separation issues.
- Stay flexible—On large projects, a planner's design solidifies as adjacent infrastructure, grading, and product type becomes clear and set. The further out that a planner designs from known or fixed infrastructure, the more flexible that planner's design must be to accommodate potential changes. However, if base topography is accurate, planners and designers who understand slope can suggest grading schemes and spot elevations from which civil engineers can develop road alignments and ultimately, grading plans.
- Topography—Planners must begin with and stay within the limits of what the ground will give. A productive strategy is to have the civil engineer serve to the planner a platter of possible road and grading alternatives that accommodates a wide range of product types *before*

the planner begins conceptual work. This helps to guarantee that the planner proceeds with precise topography and accurate horizontal base control. From this point, the planner and engineer are free to collaborate on creative design solutions in pursuit of a final grading plan.

- Road location—Planners must remain flexible until road design, utilities, and vertical-ground separation are defined. Since planning and road layout commonly happen *concurrently*, planners should coordinate their layout changes with the road designer before offering the owner any final planning concepts. Why? Because the planner's changes may result in unworkable road grades, the need for walls, or in the loss of net buildable area thereby resulting in useless meetings, lost time, and wasted money. When planners and engineers work together, other possible road layouts and grading schemes may emerge. Final planning concepts, though, should follow road layout.
- Other priorities—Planners must honor other influences and priorities before becoming too set on any particular design, concept, or direction. By the same sword, planning shouldn't be allowed to ignore or to proceed in spite of other preordained issues such as:
    - special municipal design elements and minimum parameters;
    - the desires of pre-existing builders and community or neighboring residents;
    - the need to consider potentially unglamorous, but sometimes required, affordable housing; and,
    - rerouting or changing the alignment of roads already designed close to maximum grade, minimum curvature, or that possess sight-distance issues.

What if the property is flat resulting in minimal earthwork and easily placed roads?

In situations where road grades and locations aren't an issue, or where cut-and-fill volumes are low or, at least, going to be consistent regardless of how a property is developed, encourage planners to proceed earlier and allow them a greater hand in determining final layout. When site-work can be accomplished with little risk to cost or schedule or when an owner is unlikely to receive a design or concept that is unconstructable, half-baked, or grossly expensive to build, planners or architects can lead engineering provided that they're also able to coordinate and to control surveying and layout.

Many projects—one headache

Complex projects can demand differing sets of skills and understanding. For instance, master planned communities are best approached with a team experienced in a multitude of product types. Retail and commercial park development requires people who understand the interplay between various retailers, building layout, and parking. People who understand affordable housing, trail circulation, and service-enriched amenities are typically equipped to handle assisted living or active adult communities. Most large projects require individuals who design storm facilities and utility systems. Size also matters. It may be unwise to rely on individuals with limited backgrounds to control large-scale development in the same manner that they would be trusted to handle smaller jobs. Again—the cost of tuition is just too high.

Few owners retain enough qualified people in-house to handle everything that comes down the pike, especially when it comes down the pike all at once. So, owners of large complex projects benefit when versatile and shrewd individuals, who are capable of juggling multiple subtasks, anchor their team. Here's a breakdown of possible smaller projects (subprojects) organized into subgroups within a hypothetical large complex development.

PRODUCT— SUBGROUP A	INFRASTRUCTURE— SUBGROUP B	BUSINESS— SUBGROUP C
Single-family	Fiber optics	Joint ventures
Multifamily	Water supply	Special conditions
Retail	Storm facilities	Adjacent land issues
Commercial	Roads & utilities	Negotiating with buyers
Affordable housing	Off-site improvements	Development agreements

Figure 3-4 graphically represents these three subgroups and their interrelationship with each other. Notice that the subgroups contain subprojects or aspects of land development all of which are integral to the success of a large mixed-use, multiyear project. Subprojects within subgroups are innocuous—merely a means to an end. It isn't until they overlap and link with other subprojects that action occurs towards producing an outcome. Managing a project with many subgroups requires owners first, to truly understand the challenge, and, second, to foresee what upfront analysis or events must occur in order to maintain, as scheduled, the future outcomes sought by an owner.

Owners concentrating on current subprojects generally understand that future subprojects will have to be dealt with, but they may be blind to how future subprojects are linked, let alone, to how they relate to today's subprojects. For example, permits for multifamily work may be contingent on an undesigned segment of looped-fire access or wet utilities. Commercial or school construction may trigger permits for off-site road improvements, traffic analysis, or satisfying the requirements of a development agreement. Negotiations with buyers interested in a village or retail center may generate the need for additional storm-pond design and construction by the owner or seller. It can become a tangled web.

Projects managed by owners who are blind to dependencies between subprojects, generally lack sufficient planning. When unexpected links between subprojects materialize, unprepared owners end up scrambling to locate consultants capable of quickly administering the work. This commonly results in details falling through the cracks, higher stress levels, and greater reliance on permit authorities and contractors to "come through" in pulling off a task. Cost is usually secondary in importance to keeping things going resulting in a satchel of unexpected expenses.

At any given time, subprojects and, thus, subgroups may be:

- delayed by each other;
- connected to each other;
- dependent on each other;
- interrelated as groups or individual topics;
- stacked, sequenced, or strung-out in set order; or,
- any combination of the above.

One way for owners to handle large, complex land-development projects is to envision the whole pageant from clearing to occupancy, and then parse out a project into relevant subgroups. This helps owners to get their arms around issues and to begin to understand interrelationships and dependencies that they'll eventually have to manage. It also requires them to pay supreme attention to detail. They must continually inquire as to which studies, design work, or negotiations should be currently underway in preparation for future site construction. The *Figure 3-4* subgroup model provides an excellent starting point for team "brainstorming" sessions and a base for building a more comprehensive project schedule.

Since owners are privileged only to know where they stand today based on yesterday's decisions and the consequences of their chosen style of management, what should you, as a project owner do now in preparation for tomorrow's desired outcome?

Try de-focusing off of residential

On mixed-use projects, conventional thinking typically encourages residential construction to precede retail, commercial, and industrial development. The thought is that services follow residential sales. Instead of investing only in the minimum amount of global infrastructure needed to support and to permit

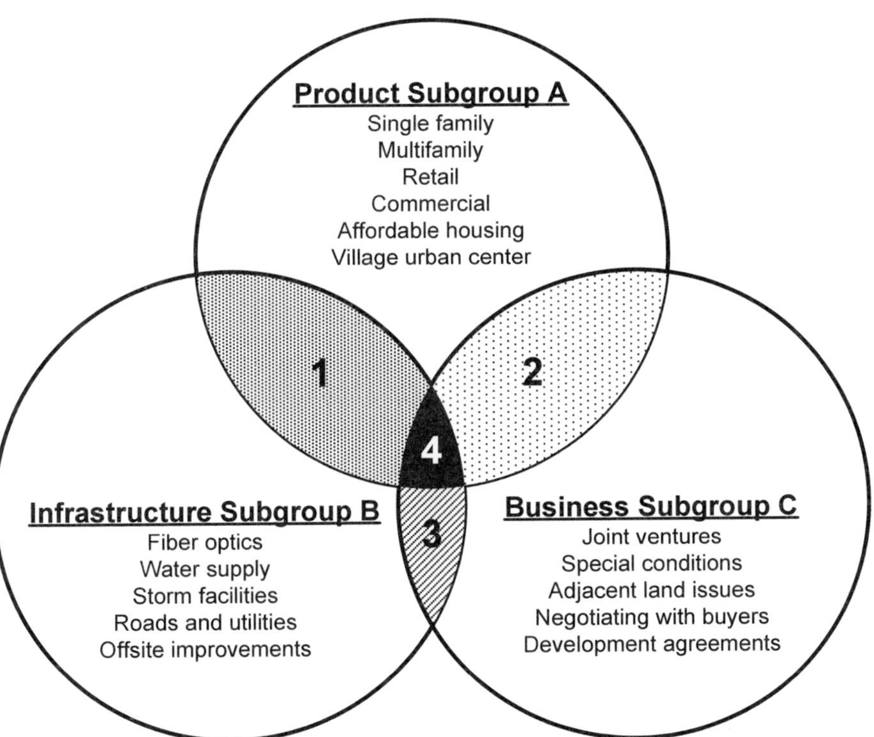

Examples of outcomes where dependencies between subgroups overlap:

1: Grading, retaining structures, project layout, amenities and parks, permits

2: Sales, product mix and limitations, ERU restrictions, custom lot work

3: Entitlement, land giveaways, servicing and accessing adjacent properties

4: Success or failure, profit and loss

Figure 3-4

residential development, owners should consider doing the opposite. Consider building *just* enough global infrastructure to support multiple uses. Then, approach planning with the freedom and the flexibility to develop other project sectors concurrent with residential development.

The other option of adding capacity and infrastructure as immediate demand dictates is cumbersome and more expensive. While offering a grocery store, convenience shops, or medical facilities may improve a project's profile and attract greater numbers of varied buyers, it can also trigger the need to upgrade utilities, to expand storm ponds, and to lengthen roadways. Installing infrastructure in readiness for future varied product isn't necessarily a negative because such work can usually be completed cheaper during the early stages of a project. The trick is being ahead of the process, and knowing how to build just enough infrastructure, but no more than absolutely necessary, to meet incremental demand.

So then, where does an owner begin? Since each development sector demands varying utilities and infrastructure, first rough out the project's year-by-year build-out schedule and ignore potential changes in market conditions, builder pressure, or perceived ease of cash flow. These items must be temporarily blocked out in order to objectively isolate off-site and on-site global infrastructure, utility, and storm-pond requirements that are needed to support each *individual* product sector. Next, estimate the cost to build out each individual sector, both singly and then incrementally, together with other sectors.

This provides owners a base from which to plan minimal but necessary site-work required to support a "premeasured" range of mixed development. This analysis doesn't have to be complicated and may be intuitive for those who understand the project. Ultimately, the goal is to optimize costs and to be flexible enough to supply builders and buyers with ground ready for any combination of product type or structures. Project owners who focus on this approach won't be guessing at costs for global infrastructure, and they'll be less likely to invest unwisely in planning, permits, and construction for insufficient, overbuilt, or unnecessary infrastructure.

Other reasons exist for defocusing off of residential. For one, the sooner that retail and commercial can be established on-site, the better. Mixed-use feeds off of both but leaves owners with a chicken-and-egg syndrome to contend with. For instance, grocery store developers usually require a minimum critical-mass of housing before moving in, while restaurant developers traditionally resist building until the final hour. Obviously, they both want people living on-site as do project owners or sellers who are marketing residential sales with promises of soon-to-be retail, neighborhood shopping, and clustered services. Who's buying or leasing first; the grocery store, the single homebuyer, or the commercial builder? From a site-construction perspective, it doesn't matter provided that the owner or the seller has the designs, and optimally the permits, to construct access and stub utilities to whatever lots or parcels are in demand. Residential may lead, but mixed-use quickly follows. The same strategy holds true for active-adult development communities where social amenities, such as a clubhouse, sport courts, and walking trails, should be in-place or well under construction concurrent to initial sales.

And try refocusing on affordable housing

Project owners need to focus early on the tougher and more volatile aspects of land development including affordable housing. Inclusionary zoning that results in requiring land and construction of affordable housing by nature can force owners to stray from the basic principles that guide profitable land development.

For many owners, affordable housing may not demand the attention it should in part because profit margins can be limited or non-existent. Other issues that may entail questionable funding, elusive programs, and policies that shift with elections can plague affordable housing. Advancing with a land use plan that remains adaptable to market conditions or that allows entitlement properties including affordable housing to shift according to how a project develops is preferred however for some owners, such conditions may encourage ignoring certain product types and amenities. Moreover, if acted upon too late, affordable housing can collide with an accumulation of other project-ending issues including budget overruns, disappointing cash flow, unanticipated loss of developable area, and having to locate such housing closer to other incompatible product. For strategic reasons, it's wise for owners to plan affordable housing on par with other sectors of development. The same holds true when budgeting grading, utility, and infrastructure expenses. Even though tradeoffs involving the cost of structures construction and potential income can place affordable housing into a separate and more tenuous economic category, site preparation costs for all product types on similar terrain within a project should be consistent commensurate with grading objectives, building footprints, and foundations.

Because affordable housing is a specialized product usually pursued by a limited number of builders or nonprofit organizations, owners need to procure their builders early. Lucky project owners may get to negotiate work with a small pool of affordable builders. Unlucky owners may end up desperate to find a single affordable housing contractor, particularly when markets are tight. Whether owners are lucky or unlucky, both cases test profitability and almost guarantee zero leverage for owners.

Affordable housing that is deeded away *before* street frontage is designed can lead to site-work problems. If the deeded property sits idle and utility stub or ingress and egress locations remain a mystery while the balance of the project gets built, it can get even worse. Driveway aprons, utility stubs, street lighting, and landscaping, to name a few, may end up going in anywhere. In some cases, road standards dictate splay distances and dimensions for dry utilities but that still doesn't guarantee placing infrastructure and utilities in the correct locations needed to serve future development of the parcel. Until a municipality assumes the right-of-way, project owners or sellers may be on the hook to return later in order to cut open the roadway, install additional stubs, or modify curbs and planters. On privately owned roads, having the seller return later to modify infrastructure is almost a certainty. Some owners may choose to leave streetscapes undeveloped adjacent to idle vacant property, but this course of action does nothing to enhance marketing or to support finished product deeper into a project.

One way to handle issues concerning vacant parcels is to insist that the project owner or seller controls utility and streetscape design throughout the project. Homebuilders and buyers of land adjacent to proposed infrastructure would get the chance to review and comment on the seller's road and utility plans before the seller procures permits to begin site-work. This protects sellers from having to tear out and replace curbs, sidewalks, landscaping, and aprons at the seller's expense should a buyer or builder later want to change something. Sellers then have the option of installing utility stubs and aprons to lots or parcels that remain on the market. Of course, the likelihood that services and access will end up in locations that work perfectly for any buyer or builder is remote. In this scenario, owners typically return and re-install whatever they have to in order to make a sale. Handle affordable housing and other "wildcard" properties early enough in a project to allow time for planning, budgeting, and scheduling.

In-fill

Not all site work involves developing raw virgin land. Redevelopment of existing properties is becoming more common as society strives to make better use of urban land. Better use can imply any of the following:

- Increasing density
- Environmental cleanup
- Increasing surrounding property values
- Developing under-utilized or vacant land
- Replacing a blight with something more appealing

- Gentrification (investing in aged and rundown urban areas)
- Changing product type to better serve a district or neighborhood
- Concentrating growth to align with existing transportation systems
- Adding or centering a development around an amenity, such as a transit stop

Where growth management controls or a short supply of raw buildable land discourage or prohibit development, in-fill is a solution. In terms of site work, in-fill poses a different set of issues for owners. It typically begins with the need to assemble a series of properties into one contiguous and efficiently shaped piece of ground with enough mass to make redevelopment viable. Then, developers may be faced with an array of existing conditions that need to be dealt with including:

- Unusable corners and low spots
- Unclassified and suspicious fills
- Possible ground and water contamination
- Encroachments or unstable walls on property lines
- Buried debris, old foundations and roads, and dumpsites
- New wetlands resulting from poor drainage, neglect, or non-use of a property
- Asbestos, or aboveground containers or equipment containing hazardous materials

Commercial contractors understand how to build in congested areas. Residential and mixed-use developers new to downtown construction may find themselves negotiating a different set of rules regarding:

1. Noise constraints
2. Continual traffic control
3. Specific and fixed haul routes
4. Limitations on worker parking and job shacks
5. Limited opportunity for material storage and staging
6. Limited areas on-site to store excavated or imported material
7. Requirements to improve or construct parking and transit facilities
8. Increased pressure to wash trucks and maintain surrounding streets
9. Limited area for cheap construction water treatment and detainment
10. Requirement to add non-revenue–producing amenities, parks, and street furniture
11. Requirement to add or upgrade wet- and dry-utility stubs to neigh-

boring properties

12. Requirement to repair, upgrade, or improve existing internal and external hardscape
13. Increased demands for security and night lighting due to the threat of ongoing crime
14. Requirement to upgrade, replace, or add to existing internal and external wet and dry utilities
15. Requirement to upgrade or add signals, pedestrian controls, or intelligent transportation (IT) devices
16. Requirement to install noise, safety, and sight-mitigation measures, such as expensive walls, fencing, and landscaping
17. Increased limitation on hours of construction or deliveries in areas approximate to residential housing, schools, or high-traffic areas

For the site-work project owner or contractor accustomed to working raw land, in-fill is usually significantly more expensive per square-unit area than developing raw green land.

Entitlements

Correct easement widths and locations, strategically located open-space giveaways, and smart preplatting are items that owners need to be conscious of if they intend to maximize profit, flexibility, and net buildable area. This means catching poor decision-making, unjust restrictions, and unworkable constraints *before* gaining entitlement to pursue building a project. After signing agreements, it may be too late for some changes. When changes to entitlement are possible, be smart about when to pursue them. The time for owners to seek more favorable terms or seek regulatory relief isn't when they've painted themselves into a corner due to inappropriate design, unrealistic engineering, or weak planning. A sure way to lose a public agency's respect is to disregard its specifications and requirements and return later asking for leniency when your back is against a wall.

Besides, trying to compensate for deficiencies in entitlement after the fact is costly, time consuming, and, sometimes, futile. Given a choice, why spend political capital, pay consultants to formulate a case, or go through contortions for something like additional buildable area or higher density when those issues would have been easier dealt with during entitlement? Procure fair and generous development guidelines during the entitlement phase—not after it.

Land giveaways

When possible, grant land giveaways that are physically off of and separate from the project. We've already discussed some of the problems associated with independently owned vacant land sitting idle and in the path of development. The example used was in regard to affordable housing, however, land giveaways for schools, pump stations, power substations, fire stations, and city facilities are no different. When placed within the boundaries of your project, these facilities can:

1. Retard sales
2. Impact views
3. Impact overall grading
4. Impede site construction
5. Impact traffic and parking
6. Place increased demands on utilities
7. Negatively affect surrounding land use
8. Contribute to storm treatment and detention requirements
9. Indefinitely freeze streetscape design and location of curb cuts, utility stubs, and driveway aprons
10. Completely disrupt the dynamics of existing utilities and infrastructure if the land is sold or shifted to other uses later
11. Become an eyesore and affect sales if they aren't designed, approved, and built concurrent with surrounding development and architectural guidelines

Before finalizing design and engineering, know how land giveaways can affect your project. An effective way to do this is to keep new or absentee owners informed. Be clear about critical road and utility elevations. Disclose how grades surrounding their land are going to change. Encourage those that own or control land giveaways to design accordingly. On steeper ground, design and layout of every parcel should work initially from the owner's overall grading plan. Avert miscommunication about grading issues that can affect adjacent plots of land and result in unwanted walls, encumbrances, and drainage issues. And, communicating this information in writing is important for project owners who want to avoid having a municipality or other benefactors later level charges that their land is unbuildable or that it is not what they were promised because of surrounding improvements.

Other forces can come into play when dealing with land giveaways. Parties that provide financing or that grant money for affordable housing may attempt to impose constraints on a project owner. Politicians control how land given away for public facilities is developed and that can change overnight at election time. Groups or trusts that inherit green space have a say in how trails connect, which can affect road location. These are just three examples of how a project owner can lose control when giving land away. For protection, clarify in writing undefined and subjective matters early on in order to minimize the need later to reinterpret, renegotiate, or redefine issues.

Non-revenue–producing land and layout

Although expensive to plan and construct, parks can add premium value to close or adjoining lots. Techniques for maximizing lot values adjacent to or in view of parks are no different from those used for maximizing values along waterfront. For instance, keep lot widths narrow allowing as many buyers as possible a view of or access to the amenity while locating larger lots a distance away. Smart design also calls for planning rectangular and linear parks that border as many lots as possible. Given fixed-lot size and density, numerous small parks will serve a greater number of lots than will one or two large parks while further increasing opportunities for pricing product higher. (Note: the exact opposite is true when laying out storm ponds. The cheapest storm-pond configuration usually correlates to a minimum number of ponds each having the shortest possible perimeter distance. Clearly, round is the most economical shape for storm ponds. *Figure 6-1* depicts shape and its relationship to area.)

Conversely, prime access from a few select lots does little for the masses, which makes providing adequate parking critical to planning. Newly constructed lots rarely abut public parks or lakes without a road or alley between them. Perimeter roads serve as access and provide room for public parking. Single-loaded roads or alleys with a park, lake, or other amenity on one side and lots and intersections on the other are expensive per unit to construct. For example, four single-loaded roads that encircle a square park means that four road sides are not going to produce as much revenue as the four outer sides that front lots. In effect, half of the surrounding infrastructure built around the park is not going to serve lots with access or utilities and therefore not produce measurable revenue. At best, the cost and revenue basis for constructing such roads is less than optimal. At worst, it's a total bust. Will lots contiguous to a park or waterfront increase in value enough to justify the inefficiencies of constructing more expensive infrastructure that supports only one side of the road? Meanwhile, double loading a road with lots that back onto a park or amenity effectively blocks public access. The solution may be to compromise by offering both single and double-loaded lots. Other factors beyond lot premium, including aesthetics, atmosphere, recreation, and meeting municipal demands to provide a community amenity, also season the cost-benefit structure of land giveaways such as parks.

Utilizing entitlement property to everyone's advantage

Increasing the value of land slated to go to a municipality, school, or institution before the date of sale or conveyance can be an effective strategy for owners. Here are a few ways that owners can profit from improving land within their project before transferring title.

1. Communication begins—Improving someone else's land forces people to communicate.
2. Enhanced schedule—The odds of installing utilities and infrastructure on schedule and within budget increase when an owner controls all land.
3. Retain contractor control—Owners can prolong project control by utilizing their own site-work contractor on parcels destined for sale or to be transferred to others.
4. Reduce project impact—The optimal time to remove trees, raise dust, make noise, and run trucks on buyer or seller land is during the early stages of a project before occupancy.
5. Improving land increases its value—Cleared and graded land is more valuable than raw property. Project owners can improve property in many ways, including adding or removing fill material or grading a property so that it drains properly.
6. The owner's contractor benefits too—Additional work awarded to an on-site contractor to perform third-party site work can reap positive results for all parties. For instance, an on-site contractor can fall back and work on property to be given away or sold while waiting for the project owner's permits to break loose. Property under construction and out of the way can be a great place for temporary offices, storing dirt and materials, and parking. Greater volume of work should reflect, or at least solidify, lower pricing for project owners.
7. Cost economies for new owners—Municipalities, schools, or institutions can benefit when project owners use on-site contractors to perform site work on land that will be theirs. How? They save by not having to bid or coordinate work. They can capitalize on the project owner's volume, on already bid or negotiated unit prices, and by using contractors already mobilized on-site. Savings can be even greater if non-union contractors perform work on land still under a project owner's private ownership and control. Sites already cleared and rough-graded allow designers to begin work immediately with known topography.

Alternatives for future utilities

When formulating development agreements for projects occurring on the cusp of urban areas where sanitary sewer, storm sewer, and water services are planned but not yet available, include language that allows the use of alternative on-site storm and septic systems should anticipated public systems fail to reach your project. On the flip side, projects approved for on-site water, storm, and sanitary disposal should include language in agreements that grant an owner the option to connect to newly installed public wet-utility systems should they ever reach a project's border. Having the option to connect to off-site utilities may save an owner later if on-site storm, water, and sanitary requirements become too burdensome, expensive, or prohibited over time.

Agree to infiltrate all or a portion of your storm water runoff only after performing a complete and sometimes expensive geotechnical subsurface analysis. On sloped terrain, this includes at minimum, identifying, mapping, and measuring the outlet elevations of all surrounding springs and wells located in close proximity to the project. The problem for most owners and jurisdictions is that they do not understand the level of study or the amount of money and time needed upfront to analyze whether a large property can safely and indefinitely infiltrate water. In some cases, owners and permit agencies may never know how a system will perform until it is built and operational. Project approvals based upon minimal and shallow geotechnical exploration that suggest infiltration will work can cost an owner later if it's discovered that water doesn't percolate in volumes as expected or that infiltrating water results in changing the subsurface water regime resulting in unstable ground, aquifer contamination, or the creation of wetlands and new springs. In addition to installing belowground water injection galleries, infiltration systems commonly require the construction of full-sized stormwater treatment and detention ponds or vaults and bypass pipes or armored channels for handling overflows. As a result, owners may find that providing infiltration may result in having to design, permit, and construct dual stormwater handling systems. In the end, infiltration can be expensive.

It's always good policy for owners to pursue back up alternatives should a development agreement become unworkable, counter-productive, or used as a tool for antagonism. Without solutions for handling storm, sanitary, and water, projects don't exist.

To grade or not to grade

Developable land destined for *sale* must, at some point, be minimally serviced with wet and dry utilities and vehicular access. Beyond that, grading is an option—not a necessity. For instance, an owner selling a lone in-fill parcel may do absolutely nothing except sign some papers to complete the transaction, leaving all construction improvements to a buyer. Conversely, owners intent on developing and parceling-out raw acreage have to make a decision. That is, whether or not to grade lots and parcels for sale or leave them raw. In most instances, only three options exist. Owners can log, clear, and grade or demolish everything and completely develop the site; they can provide only infrastructure leaving everything between sidewalks natural; or, they can establish a combination of infrastructure and wholesale land improvements while leaving some portions of the site undeveloped. Owners may have many reasons to not completely develop a project but, for most, it boils down to end-value. Is resale value going to increase enough to offset total development costs? Ignoring deflationary mechanisms on reasonably sized and smaller parcels that require logging, clearing, and grading, the answer is invariably, yes. No builder or buyer should be able to mobilize, survey lot boundaries, log, clear, install erosion control facilities, strip topsoil, perform geotechnical inspections, grade, and install walls cheaper than the owner performing overall project development. When it comes to completely improving parcels or leaving them raw, the owner or seller holds many advantages over a buyer or builder. For example, owners or sellers enjoy:

- Better pricing on timber logged in higher volumes
- Greater odds of installing gravity utilities and aprons in the right location
- A tremendous advantage in being able to export and import on-site material between parcels
- Less chance of having roads and peripheral grading conflict if they are constructed concurrently
- A non-interrupted build-out schedule. Everything happens as it should in preparing lots and parcels for product
- Less room for buyers to negotiate when land is fully prepared and thereby void of unknowns in topography, walls, construction, or dirt
- No late-project disruption. They avert having buyers perform total land development in the midst of residents and children, community traffic, and ongoing commerce
- Greater control over the site including *all* trucking and erosion control, working hours, dust and mud abatement, and protection of newly installed infrastructure, such as curbs and landscaping (Here, all that remains for a buyer or builder is fine grading and foundations.)
- Increased efficiencies when the same utility and earthwork contractor who is building infrastructure devotes more resources and time to handling site grading and interior piping (Increased scope should result in lower overall over-unit pricing, cheaper overhead, and predictable performance.)

Besides cost, other possible reasons for an owner to decide not to fully develop parcels may relate to:

- Limited capital
- Unsure markets
- High financing costs
- Restrictions or phased permitting
- A need to reduce or share liability
- Negotiations with a buyer or builder who is only interested in raw ungraded land
- A long-term strategy to minimize involvement or possibly to sell the project before build-out.
- Efforts to avoid generating excess volumes of cut material that cannot be lost anywhere on the project, or, to avoid prematurely turning vegetated areas into troublesome zones of unsold impervious ground with increased risk of erosion.
- A desire to capitalize only on a particular resource such as valuable timber, pit- or bank-run material, or rich topsoil, or to use the land temporarily as a location for a truck wash, HAZMAT fuel pad, stockpile area, dirt recycling operation, or a temporary siltation facility.
- Parcel size—Beyond a certain minimal critical area that fluctuates from project to project, it *can* make less financial sense for the seller to clear and grade parcels and lots as they increase in size relative to total project net developable area. Another consideration concerns balancing dirt-heavy parcels and lots with other plots of land that are in need of fill material.

In most cases, buyers who complete parcel and lot development concurrent with the seller's overall grading and construction operation should in the end, pay less for developable land and create for themselves, greater value.

Platting

Preliminary and final plat requirements vary according to jurisdiction. Regulations aside, here are simple guidelines that owners can follow to ensure efficient execution in platting:

- Entitlement—Obtain preliminary plat approval on an entire project when possible. This will ensure that requirements remain constant.
- Plat for density—Preliminary plat for maximum density. It's easier to modify a plat later by decreasing density than it us to try and increase density.
- Collaborate—Wherever possible, collaborate with an agency's planning department. This strategy will often yield quicker results with less chance for error.

- Phase platting—Design and submit large parcels when preliminary platting for multifamily, retail, or commercial product. Modify the preliminary plat or phase platting, as more detail becomes known.
- Plan well—Site planning that results in a fixed layout supports expeditious platting. Planners and engineers must know the required standards for roads, utilities, easements, and lot densities for the jurisdiction in which platting is to occur. Plat redesign is costly.
- Replatting can impact utility stubs and right-of-way planning—Separate and third-party designs (by entities such as dry utility companies, landscape designers, and street lighting consultants) based on a preliminary plat may have to be redesigned if the plat changes.
- Final plat—Make sure that builders in your area can, and want to, build and sell product on the shape and size of lots and parcels that you are platting. It's expensive to revise lot dimensions on paper and in the field, and then tear out and relocate aprons, driveways, utility stubs, landscaping, or street lighting in order to satisfy a buyer, or builder's requirements *after* you've recorded a final plat.

When time isn't your ally

Or, as my dad would say, "Strike while the iron is hot." Planning, wishes, and vision can give way to growing technical specifications and evolving municipal requirements. Also, given enough time, opportunities for approving and permitting unique money-saving ideas can vanish. Owners of incomplete or pending projects can become easy prey and leveraged into bearing a disproportionate cost-burden related to local population growth, traffic congestion, utility strain, public infrastructure, or political whim. The result is often higher fees, slower approvals, and, possibly, curtailed development. The answer is to plan, design, permit, and build as soon as possible, or else risk increased exposure to:

- Political changes
- Wider regulation
- Costs and inflation
- Market fluctuations
- Internal personnel changes
- Forced diversity of product
- Creeping construction standards
- Growing environmental restrictions
- Changes in public agency personnel
- New or re-packaged public opposition
- Wider and greater requirements to improve or add public infrastructure
- Increased exposure to neighbors determined to extort or gain owner-paid improvements to their property

- Increased public agency demand to develop non-revenue–producing type amenities—This may include assuming increased stewardship or community indebtedness for landscaping and trees, amenities and parks, noise abatement walls, stream enhancement, buffers, wildlife tunnels, parking, or surface water detention and treatment facilities.
- Time—project owners gain more time to think and make unnecessary changes.

Sadly, it's possible for all of the above to happen concurrently over the course of a project. In today's political and social environment, development and construction isn't getting easier. Conditions and circumstances can change a project in ways that an owner would never expect. "Striking while the iron is hot" reduces exposure to costs overruns, stress, and disruption.

Chapter 4

RULES OF MEASUREMENT AND CONTROL

Site work is based upon three basic sets of surveying and control measurements. Collectively, they include preconstruction surveying, construction staking and layout, and surveying as-builts. In this chapter, we will be concerned only with the first, preconstruction surveying and site measurement.

Horizontal measurement

Property considered for development should already be marked with corner monumentation and property lines. If survey markings aren't pinned, blazed, or flagged in the field, they should exist on record somewhere in a local courthouse or title office. Whether physically present or not, property corners and lines must be verified with all characteristics and features located and mapped so that owners know what they have. This information is also a prerequisite for permitting and development. Examples of land characteristics that should be surveyed and plotted can include but are not be limited to any of the following:

- Bridges
- Clearings
- Cliffs
- Culverts
- Exposed bedrock
- Fences
- Gates
- Gravel pits
- Groups of trees
- Holes and mine adits
- Large boulders
- Pipelines
- Roads
- Septic fields
- Structures
- Utility lines
- Surface water
- Wetlands

All property needs to be vertically measured resulting in a topography plan, or "topo" for short. Flat terrain with minimal relief can pose drainage issues resulting in the need to survey raw ground using a denser grid of, at minimum, one-foot contour intervals. As ground steepens, the contour interval increases commensurately with the level of accuracy required for design and engineering.

Topography plans attached to newly acquired property should always be check-surveyed in the field (also called "field truthed") to verify accuracy and completeness. In land development, accurate topographical plans are indispensable. If a project owner is operating without a topo plan or if the current topo plan is inadequate, corrupt, or in any way suspicious, then move to the square on the gameboard that says, "develop a new topography plan."

The field topography plan

Topography plans describe ground vertically according to elevations represented as contour lines. How topography plans are developed can affect the accuracy of base maps used for engineering and design, and this should interest owners. Why? Because an owner who understands the rudiments of land development is better equipped to ask questions, visualize issues, and make informed decisions.

Topography maps are typically created by using one of three methods, each dependent on how ground survey information is collected. In the first method, a survey crew shoots and records the coordinates of points describing fixed elevations corresponding with whatever contour interval is desired. Vertical elevation drives where shots are taken and then plotted. The resulting topography plan describes ground contours without the need to interpolate or average vertical survey information. Points of elevation are simply connected resulting in a contour line. Although chasing contours is usually more time consuming than taking random shots and interpolating data, it can make technical and economic sense to use this method when conditions allow surveyors to take multiple long-range shots from one setup over land that's clear of foliage, uniform, and predictable in slope.

A second and more common surveying method, called *random-radial,* involves collecting random shots but neither fixed horizontal nor vertical data drive the process. Instead, survey crews collect ground information wherever they can get access and gain clear lines of sight. This data is plotted as field-shot with one exception; the vertical component is interpolated to locate elevations in alignment with whatever contour interval is desired. The smaller the contour interval, the more accurate the plan. How data is interpolated and where contour lines are drawn is critical for accuracy. Nevertheless, data averaging leads to averages, and averages lead to variances that can later cause errors in design and quantity estimating. With this method, surveyors must take an adequate number of shots representing ground conditions.

Baseline and cross-section is a third method where surveyors shoot and measure points on the ground correlating with a fixed horizontal grid pattern. Nothing is located subjectively as all points measured in the field are predetermined according to intersecting lines. Separate from establishing initial baseline layout and control, elevation or altitude plays no part in determining how field information is gathered with methods two and three.

Once, while working in a remote and largely uncharted region of Alaska, we had to create our own grid pattern in order to establish control. This required taking shots true to the grid which involved negotiating everything from muskeg swamps and rock chasms to dealing with large angry mammals. The real test came when we hit an unexpected canyon and weren't able to negotiate the rock walls, which dropped hundreds of feet straight down into a dense forest of hemlock and spruce. Grid or not, the next stop for us was a floatplane ride back to camp.

While this is an extreme example, it's critical that surveyors remain true to the grid in spite of dense timber, brush, or tough terrain. They must also take diverse shots of a large enough sample size to make the effort worthwhile. Afterall, an entire project is only as good as the accuracy of base topography.

Topographic error

Mistakes attributed to erroneous base information, initial surveying, poorly averaged ground shot data, and misplotting of information can all produce an inaccurate topography plan. Survey mistakes can occur as a result of many variables ranging from sheer personal incompetence to ill-planned logistics when negotiating broken terrain or thick forest foliage and ground cover.

Surveyors were once free to clear, chop, and cut lines of sight to gain maximum range for taking shots. Provided issues related to curvature and refraction are avoided, longer shots result in more accurate geometry, fewer equipment set ups, and quicker data collection. Meanwhile, cleared paths facilitate ingress and egress for survey crews lugging equipment, stakes, and backpacks in and out of the brush. Today, surveyors may not be able to clear lines-of-sight due to land restrictions and environmental concerns making more shots necessary. Additionally, jogging and offsetting around individual trees further increasing the likelihood of mistakes and errors and the cost of the survey.

Broken ground can also contribute to inaccurate topography and survey mistakes. On one hand, surveyors must take ample shots necessary to collect complete and representative information while on the other hand; they must be practical and efficient in not overshooting an area. The art is being able to read terrain and knowing where to set up once to take a maximum number of representative shots before moving on.

Owners and topographic error

Here's an example of how price-bidding topographical survey work can harm owners. After bidding and awarding topographical survey work on property that is sloped and densely vegetated, the owner comes to realize that the topographic plan is off. Wanting to understand why, the owner may be think that the surveyors may not have prepared for tough conditions when they bid the job or perhaps, that they didn't consider the weather. Afterall, some surveying equipment doesn't perform well in rain, wind, and fog. Or, perhaps they didn't account for mud-swollen access roads. Didn't they plan on walking in a few miles early everyday? The bad news here is that bidders protect margins. They bid to get work and not necessarily produce the highest quality product. This may mean fewer shots taken, quicker mapping, and poor base maps. In the interim, years can pass before an owner actually tests topographic accuracy through design and construction. One mistake in topography can put the whole plan in question.

While these are just a few of the many issues that can negatively affect surveying, owners need to understand that complete and accurate raw land topography mapping takes time. As foliage thickens and ground steepens or becomes irregular, it becomes tougher to bid topographic work. Understandably, owners don't want to spend more than necessary on anything however, this is one cost center where they cannot afford to take chances. It has to be right. The more remote and challenging the terrain—the cheaper it is to pay for complete and accurate survey information once. Topographic plans having the accuracy of a downhill ski trail map put owners at risk everywhere including having to re-commission additional survey work later.

Contour Interval and Error

Another source for error involves the ground between contours. Minimize the interval between contours commensurate with the level of accuracy needed for design and construction. A two-foot contour interval indicates that two feet of vertical ground exists between contours however those two feet can vary significantly and certainly not be uniform. Refer to *Figure 4-1.* This graph provides an example of variance within contours. Where one contour line begins and another ends, we tend to think that a straight line of slope extends between them such as the solid base topography line shown in *Figure 4-1.* But as this diagram also illustrates in a dashed line, nothing may be further from the truth.

As important as the topography plan is in setting base datum for all design, layout, and construction, there still exist both economical and practical reasons to limit the number of survey shots taken when formulating a topography plan. The ground dictates the number of shots required to capture an image of its surface. As discussed earlier, consistent and uniform ground requires fewer shots than ground that is broken and variable. Again, refer to *Figure 4-1.*

Figure 4-1 shows a topographic plan and profile section displaying four contour lines (elevations 100 through 106). The straight solid line on the profile represents theoretical ground, as one would ascertain by reviewing the topographic plan above it. The dashed line on the profile represents actual ground conditions that have approximated according to survey shots taken in that area. Cross-hatching represents the difference (or error) between actual and theoretical. Since the difference falls *within* the contour lines, it's considered allowable error for a two-foot contour plan. In other words, there is nothing technically wrong with either the plan or the profile as detailed and the surveyors did their job correctly. However, a degree of error *still exists* within contour between the actual ground conditions and those represented by the topography plan.

Statistically, a two-foot contour interval could average one-foot plus or minus, cut or fill, in error. It's true that an owner may get lucky and possess a topo plan that mirrors the ground extremely well but it's also possible that an owner may fall into a topo plan with a large and consistent error to one side or another of the average expected error. An overage or shortage of dirt and probable budget impact can result.

Similarly, a five-foot contour interval could average two and a half-foot plus or minus error imposing higher inaccuracies and potentially, much more expensive penalties. Errors in contour can produce any number of topsoil and earthmoving volume busts such as:

- Fine-tuned and "balanced" sites may be in peril of not balancing. Owners may be forced to move more material on-site, make deeper fills, or export and import more material than budgeted.
- On sites where topsoil stripping must precede cut to fill, depending upon the depth of topsoil, the variances in dirt volumes can be wide and wildly unpredictable. Unpredictability in site-work usually leads to trouble.
- Actual on-site dirt volumes will change from those as estimated from the topography plan. Although a running plus or minus one-foot error will proportionately have less affect on a deep excavation than will a two and a half-foot plus or minus error, the changes in volume can be staggering. Consider that an additional foot of cut or fill over an acre equals 1,613 bank cubic yards or, 2,097 loose cubic yards (1.3 factor) or, perhaps 2,420 truck cubic yards (1.5 factor). At say, $7.00 per cubic yard, the penalty per acre could equal $11,291.00, $14,678.00, and $16,937.00 per acre, respectively.

Cranking it up a few notches let's assume that this overage exists uniformly over 30 acres. The same costs increase to $338,730.00, $440,340.00, and $508,110.00 respectively for cut or cut to fill. Going the other way, a deficit of fill material can be more expensive resulting in the need to import greater volumes of structural material. And, it only gets worse, much worse, when errors happen within a wider contour interval.

Now, fast forward to *Figure 9-1.* Being off vertically can also produce horizontal error. Holding a value for vertical error static, the flatter the ground, the worse the horizontal penalty. This phenomenon can be especially painful in city environs and on steeper in-fill sites.

But this isn't the end of the story. Throw in some survey error, engineering/ design error, or wet weather and budgets can shatter like fine china on a concrete floor. I'm confident that some readers have experienced this deadly convergence. In site work, it rarely gets worse than dealing with unexpected variances in excavation and just think—you haven't started structures yet.

Project owners facing this prospect on hard-bid jobs may find themselves in a predicament. For instance, a contractor bound to earthmoving numbers yet experiencing huge overruns not adjustable by unit price may formulate a case citing the owner as responsible for the overage. And, the contractor may be right but for the wrong reason. While the owner, designer, surveyor, and contractor bicker and nit-pick every possibility under the sun, all along, contour error may have been the culprit.

Proving contour to be the problem can be difficult. Since the ground is no longer intact, verifying error between base topo and actual ground may entail studying staking records. If existing staking is in place, it may explain some of the error. If stakes are missing, the next move is to study survey cut sheets however they can be wrong or insufficient when used to re-construct original ground conditions. Plus, other issues concerning design and survey calculations can also come into play. Instead of sleuthing after the fact to find answers to topo error, work to prevent topography problems by:

1. Field truthing topography shot from aerial surveys.
2. Randomly profiling challenging terrain and comparing the results with the topo plan.
3. Being wary of any contour plan that exceeds a 2-foot interval. Keep the contour interval short.
4. Inflating earthmoving volumes and being better prepared both schedule and budget-wise for dirt overruns.
5. Scheduling topography and land surveying in late fall through early spring when the leaves are off deciduous trees.
6. Considering using a small bulldozer to flatten thorny and obnoxious brush prior to surveying giving surveyors more room to see the ground and measure points.
7. Candidly speaking with the surveyor about the frequency of shots and if recommended, spending additional funds in broken terrain in order to gain complete coverage.

Variance within Contours

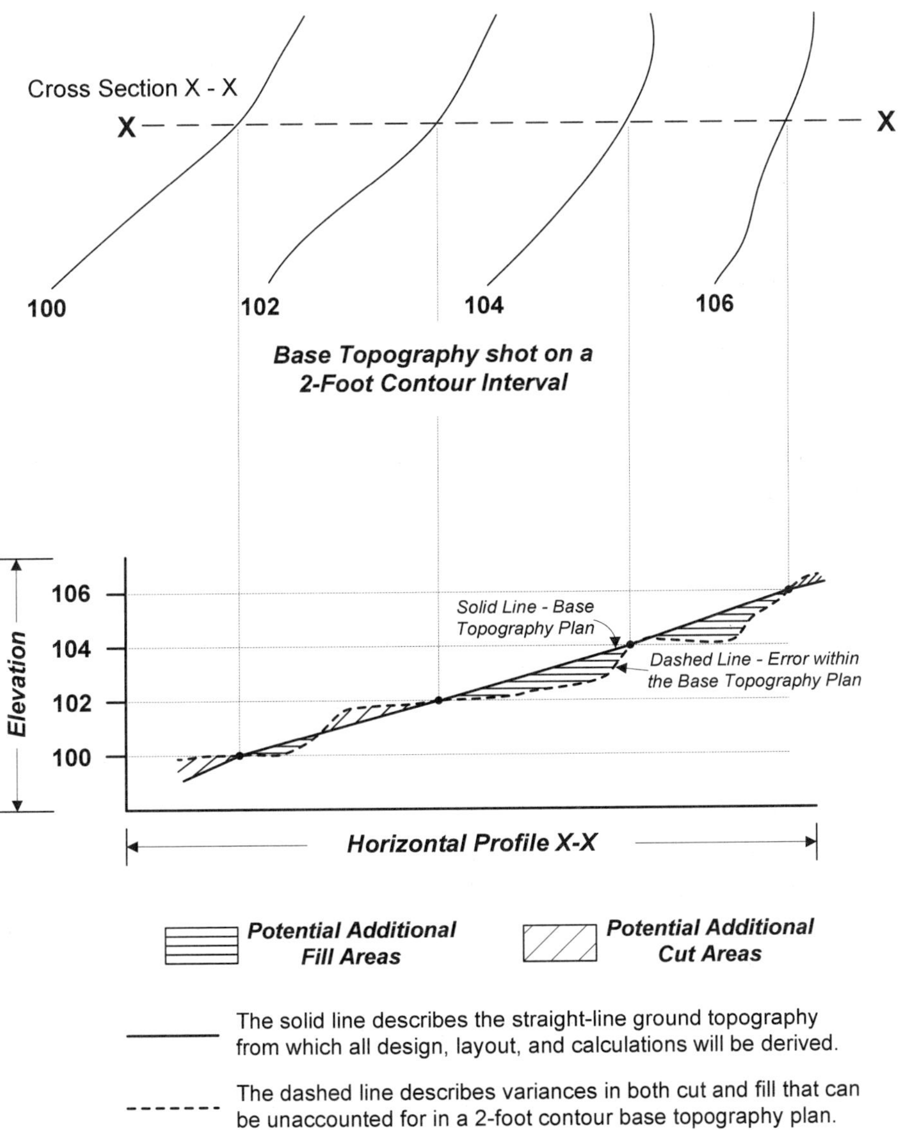

Figure 4-1

8. Monitoring excavation quantities from the outset. If quantities appear to be higher than expected, check the topography plan and it's relation to field staking. Adjust the topography plan if necessary. Then—adjust your budget.

One surveying firm or multiple firms?

Good question. Only use firms that possess ample resources and crews. Using a survey firm already overextended may result in being unable to get layout or staking when needed. Also avoid having multiple survey firms collect data concurrently along common borders. Costs can needlessly double, and data may not match, be in conflict, or possibly overlap. Remedies include using one firm's data over the other or in interpolating numerous sets of information into one contiguous plan. Differences in style, philosophy, and personnel of competing survey firms can exaggerate problems making splicing information cumbersome and potentially wrought with error.

Owners who expect to use multiple surveyors must define the process and procedures with each firm before beginning work. Otherwise, once construction gets underway, schedule disruption, change orders, and disputes clearly related to initial surveying may never get solved simply because an owner may have no idea which firm to hold accountable. So—guess who pays? This underscores the fact that owners are better off paying for a complete surveying job up-front by one firm.

Other problems can linger when using multiple survey firms. If an owner isn't currently using or has rejected one of the firms, it may be tough getting it to produce project records and information when needed. Or, owners may have to pay a premium price to have information pulled from a firm's archives and delivered in a timely manner. Not all firms utilize electronic data transfer so costly communication and information flow problems can arise. A solution to this problem resides in not allowing survey firms to be the sole holder of your information. Owners need to continuously keep in-house and available duplicate electronic and paper copies of all survey work.

For reasons related to insurance, licensing, and legal entanglements, surveying firms may not be able to endorse each other's work. Other concerns about quality, accuracy, and format may prevent some firms from accepting or basing work on another firm's data and control. These examples provide additional reasons why one firm should perform surveying. Owners who change surveying firms will likely suffer from a short-term loss in continuity, higher and extended surveying costs, and lost time. Meanwhile, the owner may end up assuming liability that would normally rest with the original survey firm. The benefits of changing surveying firms must out weigh the costs.

But this isn't the end of it. I've had to intercede between competing firms (and not just survey firms), in order to get an open, non-confrontational dialog going just so that we, as the owner, could get something accomplished. Why chance getting caught in the middle of an ongoing feud between a couple of firms with genuine dislike for each other? Don't and instead, do your homework and hire one quality firm to perform preconstruction surveying and mapping.

Design and surveying

Owners should pursue surveyors who are part of a larger firm offering civil engineering services. If this isn't possible, at least allow your civil engineer to manage and subcontract out third-party surveying. When hitched together, surveyors and engineers frequently communicate and share more information. This openness and availability greatly benefits road, utility, and grade designers and engineers who occasionally require doses of "control reality" from field surveyors.

Likewise, engineers who control surveying can design better roads and grading plans when they have immediate access to up-to-date field information. If field verification is required, in-house surveyors can be quickly mobilized and sent to collect information. The benefits work both ways since surveyors with field questions are more likely to communicate with an engineer or designer that's on their team.

Transferring data between an engineer or design firm and a surveyor is also facilitated when both entities are linked. A civil engineer's work must be calculated into useful terms before work can be laid out and staked in the field. When changes occur, engineers with an up-to-date and correct design can quickly calculate work and relay it to the field for layout. When under one roof, the interaction between engineer and surveyor is greatly economized resulting in less room for error or passing stale information. This efficiency saves owners from paying by the hour each time that an engineer and surveyor speak to each other.

But equal to any reason for coordinating civil engineering and surveying exists minimizing owner liability. When both entities bill from under the same roof and sit at weekly meetings under the same corporate name, problems related to engineering or surveying are easier to trace. For instance, if mistakes and disputes arise, neither party is going to incriminate the other without undeniable cause. The engineer is less apt to hide behind poor surveying and the surveyor cannot easily claim that he received late or wrong calculations or design information. Consequently, engineering and survey firms have an incentive to quickly solve issues. Contractors and other consultants also benefit when lines of communication are cleaner and more direct.

Civil engineering firms that possess their own in-house survey teams are attractive for site-work. As terrain steepens and potential error in design and layout becomes more of a factor, the relationship between civil engineering and surveying becomes more valuable. Afterall, the surveyor is typically the first entity to test an engineer's work via staking. When both firms are equally responsible for mistakes, mistakes decrease. For instance, surveyors are usually the first to discover:

1. Inaccuracies in topography
2. Overclearing and logging encroachments
3. Potential for violating sensitive area boundaries
4. Inconsistencies in vertical and horizontal control
5. That new construction doesn't match existing conditions
6. Deviations, flat or reverse piping, and potential design flaws during utility staking
7. That the civil engineer is pursuing a design (particularly with utilities) that may not work

When mistakes occur, designers and surveyors acting as one cover for each other. Things get fixed directly and quicker than when having to go through an owner, architect, or contractor to process information. Fewer entities in the loop expedites communication, saves mark-up costs, and decreases the risk of something being dropped or forgotten.

Aerials

Also known as *fly-overs*, aerial photography is an efficient way to collect raw data, to verify with some degree of confidence, the accuracy of existing topography, and to take pictures of a site. Digital satellite imagery, another option for gaining large-scale high-resolution ground data, is also available and can be competitively priced when compared to aerial photography. Satellite imagery offers larger areas with fewer seams, consistent color, and less geographic and flight distortion. However, for most projects, aerial shots taken by use of a LIDAR (Light Detection And Ranging) sensor system from an airplane or helicopter can provide highly accurate measurements of the ground and canopy.

Prior to logging, aerials can be used to measure timber volume, locate boundary lines, and delineate ground features such as roads, buildings, fence lines, and some sensitive areas. Taking aerials after logging and clearing but before topsoil stripping accomplishes two objectives. First, it reduces the impact of surface distortion and second, it provides another base map for calculating earthwork and checking field-surveyed topography.

Under certain well-defined conditions, survey information drawn from aerial topography can be used to create base maps suitable for design and engineering. Unfortunately, this option fades as foliage, trees, and slopes increase and available light decreases. Even so, when relying on aerials, it's a good idea to invest in field check surveying in order to verify their accuracy. If a property is large enough to justify the minimum fixed cost of flying, it's generally cheaper to map and collect topographic information using aerial photography than it is to utilize field surveying. Again, depending upon conditions and property size, topographic plans and boundary surveys already produced by field surveying can also be economically back-checked for accuracy using aerials. By now owners should know that they cannot have too much good data upon which to base engineering and design. On some projects, that can mean utilizing both aerial and ground topographic mapping in order to gain accuracy and total coverage.

Since aerial surveys depend upon camera shots taken from an airplane, there are limitations to its use. Aerial surveys can be negatively affected, lose accuracy, or rendered useless by any of the following:

1. Increases in ground slope—Steep slope that breaks away can lead to huge inaccuracies. The steeper the slope, the larger the margin for potential error in aerial topography.
2. Ground cover, logs, and surface boulders—Here, the ground is either not visible or distorted, or surface features are mistakenly interpreted as actual ground. This leads to a multitude of errors including grossly inaccurate earthwork calculations.
3. Tree cover—The same goes for trees and foliage. If the camera cannot "see" through a forest or thick vegetative cover, ground conditions cannot be photographed or mapped to any degree of accuracy. Cutting tall grass, shrubs, and trees, and performing fly-overs when leaves and needles are off deciduous and certain coniferous trees will enhance mapping.
4. Weather—Aerial photography should be acquired under clear "blue-sky" weather. Poor lighting caused by clouds situated above the craft taking shots acts to lower accuracy and effectively reduce map detail. Intermittent clouds lying between the craft and the ground will at best prevent continuous and clear panoramic photography. Complete cloud cover below the craft will result in losing total visibility and being unable to take shots or perform mapping.
5. Uneven and variable terrain—As opposed to ground-shot survey data collection, photogrammetric data is normally acquired in a much denser pattern. More data points taken equates to better imaging. Thus, ground undulations and irregularities including ridges, creeks,

depressions, and mounds of material are more likely to be detected when aerial photography is used. However, the accuracy of data collected on uneven and variable terrain depends on whatever precision the photographic scale will yield as it relates to the desired base map contour interval.

6. Shadows—Shadows play havoc with aerial photography. Site aspect, time of year, and physical features such as trees, cliffs, and deep draws all create or influence shadows. Because they receive direct sunlight, south-facing slopes offer the best opportunity for taking accurate aerial shots. Conversely, north-facing slopes are the worst, and east- and west-facing slopes vary with time of day and changes in latitude. Aerial photography is most effective when shadows are at a minimum, which means taking shots when the sun is at its zenith or on days that are close to summer equinox. On timbered sites of deciduous trees, schedule aerials just after leaves begin falling or in spring before leaves emerge being careful to avoid the long and pronouncing shadows and clouds of winter. Edge effect, caused by vertical walls of standing timber adjacent to cleared areas and ground that is steep enough to throw a shadow, can also adversely affect aerial photography.

Owners wanting to create base maps by use of photogrammetric techniques should remain patient and wait for a clear day. Clear "blue-days" provide the sharpest site images and allow collecting accurate data from whatever altitude photography is being shot. Anything less will result in reduced detail and decreased accuracy. Owners should also know beforehand, the desired mapping accuracy required for planning, design, and engineering. This includes understanding that not all photogrammetric work and mapping is priced the same. Increased detail and greater accuracy are more costly to produce. For most owners, they must decide how-to balance funding with their expectations of what aerial mapping and ground surveying can offer. Before setting budgets and commissioning work, first discuss your project needs with your designers and engineers and then meet with both a certified photogrammetrist and a survey firm. Where photogrammetric mapping is challenged or not possible, pursue traditional ground survey techniques for collecting data and creating project base maps.

The value of investing in profiles

Investing in a few ground-shot profiles is a great way to check the accuracy of an existing topography plan while gaining more information about a property. I am a firm believer in running and plotting profiles at a one-to-one scale (1:1). For beginners, a one-to-one scale is reality. It's an actual picture of what ground really looks like perpendicular to contour and relative to both the horizontal and vertical axis. When using a 1:1 scale for profiles, there is less chance for interpretative error. Planners, builders, and contractors can get a feel for actual slope and terrain. But the true value in running 1:1 slope profiles is that anyone can understand them. Designers, accountants, marketing people, and financiers can all "picture" what's being proposed because the plan isn't exaggerated in any one direction. Real plans for real problem solving. Good stuff.

When there are more flags than trees

Today's projects can attract visitors with pockets full of colored flagging. Some bring rolls of blue, others like orange, many favor glo-pink, and someone lately has been hanging red and white checkerboard marked flagging. This practice leads to confusion and renders important flagging practically useless. Preventing uninvited guests from hanging flagging is tough but you can at least control members of your team by drafting a flagging chart for all consultants to abide by.

Most firms have their own color codes for flagging. Owners wanting all firms to use consistent flagging from the beginning should designate colors by related task. For instance wetlands can be flagged in glo-green, waterway buffers can be marked with blue flagging, and bright orange or pink may work for grading. Candy stripe patterns can be effective when flagging logging skid roads and hiking trails through thick cover. Discuss flagging requirements with prospective consultants and contractors and then include the flagging chart in their contracts. Drawing up and enforcing a flagging scheme early makes a statement concerning regimentation and order. It also lays a foundation for project control.

Squeezing more out of your initial surveying investment

Here are some tips to consider when contracting out surveying:

- The need for field shot topography increases as ground steepens and foliage thickens.
- Collect and store in-house, complete paper and electronic copies of all survey information.
- Surveyors expect project owners to provide adequate vehicular access. Budget accordingly.
- Dirt volumes and retaining wall quantities are only as accurate as the topographic plan measured from.
- Cross sections can be generated directly from accurate topographical plans without the need of reinvesting in additional fieldwork.
- Use the highest intensity (smallest interval), most accurate topography plan needed to support design and construction. The greater the

contour interval, the greater the inherent error.

- Retain one firm to perform boundary surveys, to collect original field topographic data, and to compile all related base maps. Don't enlist multiple surveyor firms to collect original base topographic information for any one piece of property.
- Crisp weather combined with bare trees are optimal conditions for field surveying deciduous sites. Schedule base topography and boundary surveying early enough so that survey crews have adequate slack time should poor weather and ground conditions deter fieldwork.
- Ensure that the surveyor is basing elevation control on established datum (base elevations) and coordinates that are accepted and in-use by local public jurisdictions and other consultants. Avoid carrying equations throughout a project that translate measurements from one format to another or that correct mistakes or differences in base data.
- Contained features such as bridges, larger retaining walls, storm vaults, structures, water reservoirs, larger culverts, pump and lift stations, extraordinarily steep buildable ground, and points matching existing utilities and roads leave little or no room for error. Design them from base topography with a contour interval no wider than one-foot. Sloped property destined for sale to a builder or buyer shouldn't exceed a two-foot topographic contour interval.
- Vary contour interval according to land use. Concentrate close interval contour surveying where elevations and grades are critical. Survey the interiors of steep unbuildable slopes, land giveaways, and open space with a wider and cheaper contour interval. Afterall, if you're not building on them, why process more topography shots than necessary? Invest in surveying only where it counts—in areas slated for development. Also, make sure that the consultant considers jurisdictional requirements. Increasing the contour interval should decrease field surveying and plotting costs while expediting turnaround time.

Chapter 5

METRIC OR NOT

If you are building outside of the United States of America, are dependent upon federal construction funds, or would like to pursue using the metric system for design and construction, this may be an opportune time to move on to the next topic. Otherwise, this chapter explores metric as it applies to private, for-profit site-work and land development.

Let's begin by framing site work and its dependence on materials deliverable in metric units. Short of pipe, retaining wall material, precast structures, and lighting, manufactured materials delivered on a flatbed truck *don't* drive site construction. Granted—these products are important but site work is more about changing terrain and developing raw materials through excavation and grading, recycling waste products, and forming or placing volumes of bulk material delivered in a dry, loose, or semi-liquid state. These types of materials include concrete, asphalt, crushed rock, topsoil, and fill material. Unlike building construction, neatly packaged materials are typically a relatively minor cost component of site work.

This leaves us with predominantly non-material labor and equipment type cost centers including, but not limited to, logging, clearing, grading, utility trenching, and landscaping. Grading requires loose material from either on-site or off-site sources and landscaping requires growing stock, usually measured by count, square area, per size or diameter. Of the remainder, only logging produces a solid measurable product. But there's only a consequence for not cutting to metric when mills require logs delivered in metric length—which few if any do. Site work can also involve the services of smaller specialized subcontractors who work by "feel" or who are happy *only* when working in an imperial unit environment. When they're unhappy, costs escalate.

Hard costs aside, any discussion evaluating metric versus imperial measurements wouldn't be complete without also exploring soft cost services such as engineering, design, and layout. The imperial system by nature is complicated and at times, cumbersome. Mistakes can easily occur when architect meets engineer or viewed another way, when inch/fraction values and decimal values converge in the same set of plans or in construction staking. Typically, architects work in inch/fraction and civil engineers work in decimal scale. As a result, site work is usually designed, staked, and understood in decimal scale.

Problems can also develop where architectural designs and civil designs meet or overlap as in foundation building and perimeter yard work. I've had to routinely check the carpenter's work to ensure that cut-staking information and its relationship to form work has been correctly converted. Site-work crews don't normally encounter inch/fraction values until they reference the architectural plans. This grand event signals the end for most building-perimeter site work and the beginning of structures construction.

From this point forward, evaluating imperial versus metric units will only be in terms of decimal equivalents and not inch/feet. Also note that the ensuing conclusions are based on the assumption that project owners want to produce the highest quality product at the lowest possible cost with the goal of realizing a profit. Since systems of measurement don't in themselves guarantee or infer quality (they are instead, a means to an end), we'll focus on cost, profit, and project management implications.

The advantages of using metric in site work include:

1. Comfort. Some architects and designers prefer to work in metric.
2. It works. Metric can be as accurate as any other system of measurement.
3. Adaptability. Once all parties understand metric, it's a very user-friendly system to employ.
4. Connectivity to metric. Metric easily ties into existing metrically dimensioned infrastructure and piping.
5. Global understanding. Owners can freely use foreign trades people and off shore consultants who understand metric.

Here are the potential disadvantages of and implications for using metric in site construction:

1. Design-review and permit issuances may lag in jurisdictions unfamiliar with metric.
2. Guaranteeing a steady supply of metric materials may result in sole-sourcing work to certain suppliers.
3. Since metric is accepted in urban environs more readily than in rural, the farther out that a project is geographically, the less likely that metric will be the best option.

4. If metric work entails arcane methods, lack of plans, or unknown information, contractors uneasy with metric may feel comfortable enough to proceed only on a time-and-material basis.

5. Competition decreases and concrete infrastructure costs can increase if contractors who install curb and gutters, median barriers, edge-type bridge barriers, and concrete paving don't possess metrically-sized slip form molds or paving kit attachments.

6. Suppliers may not stock adequate quantities of metric materials. This can be an acute problem for owners of projects in rural areas. Short supplies and special orders usually mean higher costs. The same applies to fabricators who need metrically sized fittings or materials.

7. Contractors who understand metric are best equipped for metric projects. Professionals reared on the imperial system and who rely on intuition, a sense of feel, and sight estimating may not be as effective when dealing in metric. Frustrations could mount and costs could increase.

8. Mistakes happen and they happen more often when people work in a system that they don't fully understand. Turnover, different trades coming and going, and fast-track work in a system not universally understood may lead to breakdowns. In site-work, where materials play second fiddle to measurements, breakdowns can lead to cost and schedule impact.

9. Owners don't want their project to be an experiment and don't want to pay for schooling. Still, in an indirect way, they may end up financing the education of those not familiar or comfortable with metric. Unfamiliarity with metric can also narrow the prospective list of contractors available to work. Some contractors aren't interested in and possess no desire for learning to work in metric.

10. Some manufacturers express construction equipment volumes or capacities in imperial units. Converting a closed pool of equipment to metric that doesn't leave the site shouldn't be a problem. Volume busts arise when trying to quantify an open pool of equipment where rentals and trucks with various box sizes and capacities come and go daily from a project. Problems may also occur when converting bulk

material deliveries, such as asphalt, concrete, and loose structural material to metric.

11. Hopefully, everyone's bids will reflect plans issued in metric units but it can't be counted on. If an owner uses a contractor who has submitted his bid in either imperial or, worse yet, a mixed bag of both systems, significant budget problems, cost variances, and schedule impacts can result. Straight-line converting of unit prices to metric isn't an easy solution since it can produce erroneous pricing. The only person who should convert unit pricing to metric is the estimator who authored the original unit price schedule.

12. Converting imperial standards and specifications to metric can be arduous and drawn out. Further complications can arise because few if any materials convert to metric on a one-to-one basis. As a result, owners typically end up overdesigning and purchasing materials larger than necessary to satisfy the minimum imperial equivalent of materials as specified. Or, they push to overhaul and re-define the standard. Either option is more time consuming and expensive than working exclusively with imperially sized materials.

13. Rarely do metric projects take place without having to convert some pre-existing survey or engineering information into metric equivalents. This task may include both having to update an engineer's software and having to convert boilerplate files to metric. If conversions aren't precise, rounding errors proliferate. Other issues include potentially placing decimals erroneously, and referencing the wrong imperial coefficients when converting performance criteria for items such as storm-water calculations, pavement analysis, and structural equations.

14. Metric can impact site-work in a variety of *unexpected* ways. Here are a few examples:

    - Dumpsites may not charge by metric weight or volume
    - Truck weigh scales calibrated in metric may not be available
    - Metric work may require using a testing lab geared to metric
    - Landscape stock and tree caliper are not normally metrically-sized

- Finding appropriate metrically-sized street signage may be problematic
- Street furniture, lighting, and fencing may not be available in metric sizes
- Purchasing off-site dirt and rock deliverable in metric units may not be possible
- If required for use, pre-existing or recycled materials may not be available in metric
- Not all specialty contractors, such as well drillers and pile contractors, are comfortable working in metric
- Cruising and measuring timber in cubic measure versus by the board foot may lead to haul, scale, and mill payment errors
- Metric projects should begin in metric—This means performing all upfront environmental, geotechnical, and engineering studies in metric. Property surveying, aerial photography, and topography mapping should also be conducted in metric terms.

Complications can arise when an owner tries to sell land described in metric terms to a builder, retailer, or commercial buyer expecting imperial units. If metric doesn't limit the number of prospective buyers for lots, it may repel smaller homebuilders and developers who lack a staff versed in dealing with both measuring systems.

Regardless of which system an owner chooses to employ, never simultaneously use metric and imperial units when preparing for site work or when performing site construction. Create and maintain plans, specifications, and field-staking in one system. (For that matter, also avoid mixing inch/fraction measurements with engineering decimal scale. Engineering decimal scale is the standard measuring system used in site construction.) Owners committed to using metric must use it from the beginning or risk the expense of handling errors, omissions, and spending time to convert plans and specifications to metric. Although expensive, engineers can always check single calculations from metric to imperial and vice versa, but it is better to complete construction drawings in metric from the start.

Owners opting to use metric should allow ample budget time and money for periodic meetings as the project unfolds in order to address questions and concerns from contractors, inspectors, and consultants. Ultimately, the decision to go metric has to make logistical and financial sense. You be the judge.

Chapter 6

WETLANDS & OTHER SENSITIVE AREAS

Sensitive areas include, but aren't limited to, wetlands, steep slopes, habitat for endangered species, rare and unique vegetation, former mining activity sites, bodies and ribbons of water, hazardous waste sites, archeological features, cemeteries, and historic landmarks.

First and foremost, owners must map, catalog, and include sensitive areas in the grand scheme of planning before beginning site construction. The task is to protect, modify, or remove them as stipulated per plan and permit. Archeological sites, someone's bones, or previously unknown features, such as mine adits discovered during construction, might trigger special regulations regarding how to proceed while protecting the find. In site work this means schedule impact and higher costs.

In the project budget account for anticipated costs associated with constructing around or through known sensitive areas and detail all work in the plans or specifications. Owners of projects in areas known for surprises should budget a contingency to cover the costs of handling possible disruptions and possible mitigation. Unexpected incidents are more likely to occur on in-fill and redevelopment sites. Contractors—particularly those working under fixed cost and performance specifications—experience elevated levels of stress when affected by unexpected finds—and who can blame them? There isn't much that an owner can do except to cope and try to minimize impact in anyway possible. The most practical way of dealing with out-of-scope or "custom" work is to proceed on a time-and-materials basis.

Although site construction doesn't occur in areas slated for preservation, set-asides are still at risk of being trespassed and encroached upon. Protection includes fencing off sensitive areas and installing adequate erosion control measures to keep turbid and unnatural volumes of overland flow and construction runoff out of save areas. Unfortunately, owners are accountable for sensitive area-deterioration even if they follow the plans and remain innocent of intentional wrongdoing. The solution is monitoring sensitive areas to ensure that no damage or impact occurs during construction.

Tips for developing sensitive areas

When developing in and around sensitive areas, proper planning, cautious management, and use of competent contractors and consultants can go a long way in avoiding trouble. Here are some tips for dealing with sensitive areas.

Do it right the first time. Returning and rebuilding a dysfunctional wetland is more expensive than designing and building it right the first time.

Trade them. Transfer development rights (TDRs) from undisturbed environmentally sensitive land over to property that is suitable for development.

Rely on the right people. Consider assembling a team that can adequately address botanical, wildlife, hydrological, environmental landscaping, and wetland construction issues.

Check them. Check any FEMA or other government broad-based flood or sensitive area mapping for accuracy before investing in concept or design work within or near restricted zones.

Acknowledge them as control points. Don't delay planning on how to protect, interface with lots and parcels, how to cross and breach with roads and utilities, or how to mitigate these sensitive areas.

Monitor them. Unstable soils and slopes should be monitored for movement well in advance of development. The longer the monitoring period, the better. To ensure long-term survival, plan on routine monitoring, maintenance, and inspection of wetlands after they have been established.

Design with them. How can wetlands enhance or add value to a project? Will residents benefit from recreational or nature trails incorporated around wetlands or other sensitive areas? Can newly created wetlands serve as biofiltration areas to treat stormwater prior to it flowing into pre-existing wetlands?

Retain hydric soils. When legally removing a wetland as part of project work, owners should save, stockpile, and protect all excavated hydric soil. Hydric soils are valuable for re-use as substrate material when creating new wetlands, enhancing an existing wetland, or for resale to an off-site party interested in creating a wetland. On-site hydric soils perform best when used to support the growth of new wetland plantings indigenous to your site.

Consider grading out steep slopes. Owners in need of either structural fill or holes for losing topsoil, or who want to increase net buildable area should consider grading out steep slope areas. Of course, this strategy won't work in areas of bedrock or unstable slopes, where lay back distances are a problem, or where cut volumes would be too great, but if it's possible that the benefits may outweigh the costs, grading out steep slopes is worth considering.

Contour them. Besides being visually interesting, varying water depths and multiple degrees of soil saturation resulting from contouring the side slopes of wetlands (or storm ponds) will better ensure the survival of wetland plants located at different elevations. Placing plantings along contour where water is most likely to exist leaves them vulnerable should the water table unexpectedly rise or drop due to increased runoff or changed upstream topography.

Move quickly when designing wetland or habitat mitigation plans. When unprofitable cost centers, such as wetland mitigation, fall on the critical path, it can mean high expenses for project owners. Design sessions, multiple conceptual drawings, and prolonged meetings focused on, what may be subjective issues, all add time and expense to a project. It's best to deliver persuasive and workable designs to government agencies as early as possible and procure permits for work.

Avoid them. Avoid building in and around sensitive areas. Sensitive areas penalize developable acreage, trigger the need for increased upfront spending on studies and site assessments, result in more complex permitting, can strain erosion control criteria, and, depending upon where they exist, restrict construction and vehicle circulation. Sensitive areas rarely complement lot layout, they can limit roadway alternatives, and, when mitigated, they can be very costly. In extreme instances, sensitive areas can render a project unbuildable.

Time buffer delineation. Typically, all sensitive areas must be located, flagged, and surveyed prior to gaining project approval. When it's not a requirement to also mark sensitive area buffers, it may be wiser on some projects for owners to flag buffers after gaining entitlement. The reason? An owner desperate for entitlement may not be in a position to challenge a non-governmental organization (NGO), an individual, or regulator from re-examining buffer widths and appealing that they be made wider than specified prior to project approval. It may be better to wait.

Keep all permit agencies working together. Include all permit agencies in sensitive-area meetings and keep them equally informed. Meeting with all parties will gain an owner witness, improves the odds of getting by with one plan, reduces exposure to multiple interpretations, and helps to guarantee obtaining one directive under which to proceed. Managing permit agencies also buys owners a degree of trust and cooperation. If something goes wrong, owners will appreciate working with allies and friendly government officials who possess first-hand knowledge as to what has transpired.

Find them. Survey property under wet conditions and identify all possibilities for intermittent streams and surface-runoff collection points. Since sensitive and wetland identification can be subjective, consider seeking multiple opinions if flagged boundaries seem excessive or if you believe you're at risk of needlessly losing net developable area to open space. As shown on *Figure 6-1,* the penalty in preserving more buffer or land than required could be great in terms of lost developable area. The objective isn't to compromise genuine wetland or sensitive-area boundaries, but rather to ensure their accurate delineation and surveying. Leave nothing to doubt.

Consider using a mitigation bank as a way to compensate for the filling of existing wetlands. In this scenario, developers pay a portion of the cost of developing or enhancing a larger off-site wetland in exchange for removing a smaller wetland

Potential Loss of Net-developable Area due to Overly Conservative Buffer Delineating and Surveying

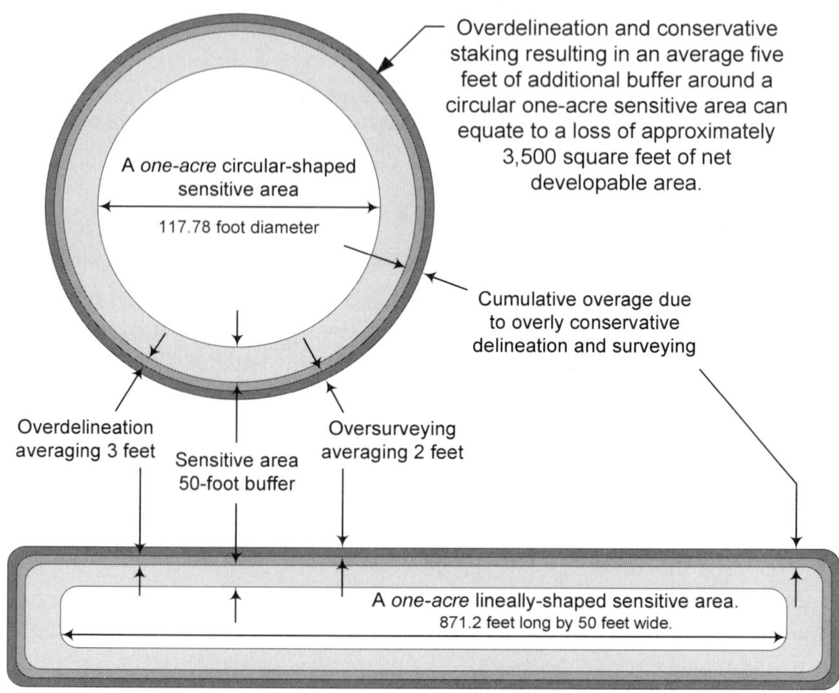

Overdelineation and conservative staking resulting in an average five feet of additional buffer around a linear one-acre sensitive area can equate to a loss of approximately 11,312, or more, square feet of net developable area.

Figure 6-1

on their project. This maneuver may actually reward owners with additional developable area if, in trading their wetland area to a bank, they're able to also save the loss of land that would have otherwise been sacrificed to wetland buffers. Another perk for owners is that they can avert costs and risks associated with guaranteeing establishment of both a wetland *and* wildlife habitat. This can also include costs associated with elaborate irrigation systems and prolonged maintenance and monitoring to confirm both animal and plant survival.

Enhance rather than create. Given a choice between removing and recreating, or enhancing an existing wetland, enhancement is usually cheaper and more effective. Why? Owners who enhance wetlands avoid larger investments related to procuring land, performing site-work, ensuring that the ground holds water, and guaranteeing plant growth in an area that may not naturally work as a wetland. Enhancements capitalize on a wetland's existing and natural location, which in turn increases the likelihood that plantings will survive and attract suitable wildlife. Since wetlands require continual moisture, water becomes another issue. Piping in potable water can be expensive and, once there, owners have to maintain and monitor irrigation systems. Whenever an owner can minimize or eliminate the liability of providing water, it's a good move. Soil type, soil amendments, regime, topography, aspect, and shade also play important roles in determining new wetland survivability. With so many variables to satisfy, it makes sense to enhance wetlands in remote and unnatural locations rather than to try to create new ones.

Find land for wetland mitigation early. When locating areas suitable for establishing or enhancing wetlands and wildlife habitat, search for land that is:

- Not in a rain shadow.
- At grade or can easily match future elevations of adjoining developed areas.
- Large enough for mitigation including land for buffers and setback distances.
- Shaded or exposed to the same amount of sunlight as those sensitive areas being replaced or modified.
- Already adjacent to greenbelts, buffers, or landscaping. These types of areas add depth and a sense of wildness to newly created natural areas.
- The appropriate contour. The wrong piece of land in the wrong place can be difficult to irrigate naturally or dewater sufficiently in order to maintain adequate moisture conditions. It must be ecologically functional.
- Self-contained and separate. Wildlife tunnels, special bridges, or continuous aboveground boxes and structures can be very expensive. Encouraging wildlife behavior can also backfire if, for instance, conditions allow beavers to build dams that flood a project. There's no guarantee that animals will migrate or behave as planned.
- Slowly becoming a wetland. Instead of fighting it, join it! Instead of spending money, take credit by writing off portions of a property that are slowly becoming wetlands. Allow beavers to create backwater or contain subsurface water migration and runoff enough to promote wetlands. Use or sell the wetland credit elsewhere.
- Easy to service with utilities. Install J-boxes and run conduit, or direct-bury power and phone lines to wetlands requiring aeration, controllers, and remote monitoring. For the sake of hydration, the earlier that wetland

mitigation areas are located, the better, since wetland location can dramatically affect storm-sewer and water-system design. Inflow and outflow storm piping and water mains needed for irrigation may have to be located specifically to serve a wetland.

- Already compromised or lost through regulation. It's more efficient to preserve or develop sensitive areas where setbacks or buffers already exist, such as along drainage courses, adjacent to existing wetlands, and along property lines. The idea is to use an existing land set-aside for two purposes. Provided that they're wide enough, existing setbacks along property boundaries can also act as buffers for a new wetland. Similarly, an existing sensitive area front-side buffer can double as the backside buffer for a new wetland. The result yields more interior land for development.

Note: At times, an owner or contractor may be asked or required to cut certain trees or pile logs, stumps, and woody debris for another entity so that the material can be used for aquatic restoration, wetland creation, or habitat enhancement else where. Besides absorbing the additional cost and sometimes awkward logistics of collecting and working around select piles of unsightly woody material, owners need to minimize their risk of further disruption by finding out where piles are to be located and the quantity, condition, species, and size of material required. Too little of the wrong material saved can result in the need for additional out-of-sequence clearing. Too much material and piles of orphaned clearing debris may end up lingering on-site indefinitely. Agree on a schedule for when such material must be removed from the project. Clearly spell out which party will be responsible for re-locating material on-site as well as for loading and hauling it off-site. Piles of logs and stumps can also become a liability and safety concern if they become dry enough to pose a fire hazard, attract weekend firewood cutters, or if they remain for extended periods locations accessible to children. Factor these issues into any agreements.

If you make a mistake, don't do it in a sensitive area. Besides the threat of costly mitigation and possible fines, remember that project opponents seeking to disrupt or halt development may find sensitive areas ripe hunting grounds for overclearing, construction intrusion, erosion control violations, or mistakes in boundary delineation. Here are some guidelines to help control site-work activity around protected sensitive areas:

- Don't store hazardous materials near sensitive areas.
- Maintain predevelopment wildlife habitat conditions.
- Maintain hydrologic connections into and out of wetlands.
- Don't allow construction vehicles beyond protective fencing.
- Keep sensitive areas clean of garbage and stockpiled materials.
- Don't allow vehicles to fuel or be maintained near sensitive areas.

- Prevent construction runoff from cascading onto and down steep slopes.
- Don't work in wetlands without Corp of Engineers and other agency approval.
- Be cognizant of sensitive areas and endangered species existing on adjacent property.
- Install water-monitoring equipment where it is accessible, yet hidden to minimize threats of theft and vandalism.
- Ensure that stockpiles and fills are stable and that they don't shed abnormal amounts of runoff or silt-laden water into adjacent sensitive areas.
- When undercutting or excavating below a wetland or waterway, monitor the cut to ensure that water doesn't drain out of its natural course and into the job site.
- When taking photos and video of the project, include incidental shots of sensitive areas with work activity going on nearby. You never know if and when you'll have to defend yourself for an act impacting a sensitive area caused by another party.
- Time-restricted permits, allowing site work to occur in sensitive areas only during narrow periods within the construction season, must control the construction schedule if such work will affect vital activities such as grading, road building, or utility installation.
- Do not rely on loose flagging hanging from branches and tall brush to protect sensitive area boundaries. Instead, survey-delineate sensitive areas and immediately follow by planting steel stakes (not wooden stakes!) and colorful fencing along the line.
- Do not remove fallen trees out of wetlands. Instead, cut whatever portion of the crown extends into the work zone and seek further direction. Also, promptly seek permit agency direction for removing hazard trees that pose an immediate threat to workers and equipment. Don't risk lives.
- Before setting fencing, it's good practice to walk sensitive-area boundaries with your inspector in order to discuss removing hazard trees. Do this while hazard trees are still easy to fell and remove before ensuing vibration and adjoining overstory removal due to logging and clearing begins. Absolutely pursue the removal of hazard trees before grading and well before beginning structures.

If a contractor discovers something archaeological—stop work! Layers of environmental laws and review processes at the federal, state, and local levels protect archaeological sites. Known archaeological sites take priority early during planning as alternatives unfold to remove or protect them. The fact is that most archaeological sites aren't excavated. They end up being protected with buffers and setbacks, being designated as part of open space, or being overfilled according to predetermined site-burial or engineering procedures. Only when in-place options prove unfeasible do archaeologists resort to excavation and removal.

How should owners handle sensitive areas? Begin by protecting mapped and located sites as required per plan and permit. Hold special sessions with all contractors and consultants explaining the importance of any sensitive areas. If something is unearthed during site-work, immediately stop work and provide notice to your permit agency and proper authorities. By promptly halting work, unexpected archaeological finds are likely to be considered an accident. Prompt and responsible action can also help owners avert possible charges grounded in or based upon gross negligence. (For more information on this subject, see your attorney.) Here are some tips to remember should you unearth any hidden wonders.

1. Call the proper authorities.
2. Take dated photos and video of the site.
3. Stop work immediately and vacate the intrusion.
4. Take necessary precautions to protect and isolate the site.
5. Provide a written report describing what was found and noting what construction activity was occurring at the time. Include the date, time, weather, and copies of pertinent photos and video.
6. Organize a meeting with appropriate archaeological interests, consultants, government officials, contractors, and the owner in order to discuss direction, mitigation, monitoring, and conditions, if necessary, for continuing site construction.
7. Carefully measure the costs and benefits if faced with alternative archaeological site mitigation. Although the expense and loss of schedule to issues of recovery, replacement, and testing may initially appear ominous, the loss of net buildable area due to in-place preservation and setbacks may prove much more costly to owners in the long run.

It is best to develop an archaeological treatment and monitoring plan that addresses any inadvertent discoveries or known conditions prior to beginning work. This could also include preparing a photographic documentation and historical report of significant structures or features known to exist on-site. Include procedures for handling sensitive-area intrusions in all site construction contracts. Of course, unexpected discoveries that impact scope of work and that delay schedule will probably lead owners to an array of change orders and the possibility of disputes.

Chapter 7

FIRE-CONTROL PLANNING

Wildfire, whether caused by natural conditions or human error, has always been a potential threat to land development, structures, and personal safety. The problem is worsening due to reasons rooted in two growing trends that, when they overlap, produce conditions ripe for wildfires.

The first trend involves population growth. Increasing numbers of people are choosing to live in rural areas surrounded by brush, timber, or grassland. The second trend involves land management, or more specifically, the lack of it. Over time, timber and shrubbery die off naturally. Similar species may take their place, or a community of different plants and trees may emerge. The result is that trees weaken or die due to competition from other trees or because the forest itself slowly changes the soil moisture and nutrient content and the conditions within the canopy enough to instigate the establishment of new and different tree, shrub, and grass species. This process is called *plant succession.*

Dead organic material on the forest floor collects or decays depending upon shade and moisture. As a naturally occurring event, fire cycles through dead and dying organic debris cleansing the land and turning this material into ash and base nutrients. Unmanaged, mismanaged, or areas preserved in territory prone to fire are seasonally at risk of igniting and taking out surrounding homes and structures where no defensible space has been created or maintained. To the extent that we:

- abstain from suppressing wildfire;
- ignore insect and fungus invasion;
- discourage activity that opens up a forest;
- outlaw silviculturally-responsible logging;
- allow overly mature forests to thicken, age, and decline in vigor;
- abandon forest roads which serve as safe exit points, wildfire breaks, and access for fighting fire;
- pursue policies that encourage leaving natural areas to accumulate dead vegetation and increasing fuel loads;

wildfire will continue to be an ever-present threat to forest communities, property, and lives.

With these factors in mind, it's vital that owners shopping for ground evaluate land in terms of it's natural resistance, current susceptibility, and history of wildfires. Here are some ways to assess property relative to wildfire.

- Ask local agricultural or forestry departments about recorded wildfire history in the area.
- Look for fire scars on stumps. Old stumps and logs that have stood the test of time and standing mature trees may reveal blackened marks indicative of past wildfire.
- Search for areas on a property that appear natural but exhibit mineral soil, no topsoil, and either no vegetation, or invader-type weed species. In the past, these areas may have experienced ultra-hot wildfires resulting in sterilized ground.
- Aspect, steep topography, and canyons can greatly influence the advancement of wildfire. Concerning aspect, the same attributes that work for you in keeping graded soil dry with maximum sunlight on south-facing slopes also help to promote conditions for wildfire.
- What types of trees exist on the site? Coniferous trees (needle bearing, softwoods) are more susceptible to flammable crownfires than are deciduous trees (leaf bearing, hardwoods). In comparison to deciduous trees, conifers are also less likely to die or incur injury because of wildfire.
- Inventory existing access roads above and below a property. If wildfire threatened the project, would surrounding roads provide adequate access for firefighters, vehicles, and heavy equipment? Is water available? Do firebreaks in the form of roads and open areas exist around the property and especially, downhill of it?
- Determine how close the land is to surrounding areas of significant fuel load, preserved parkland, fields of invasive and dormant plant species, or unaltered and over-mature forests. Dead and dying trees and brush, and accumulated sticks, branches, leaves, and needles covering the ground or in the understory indicate fire potential.

What can project owners do to minimize the impact and threat of wildfire? In short, the answer is plenty. Although choosing land with low potential for fire and specifying fire-resistant building materials with ample setbacks from vegetation is important, also consider the following additional steps when attempting to lower the risk of wildfire:

1. Remove all standing dead trees.
2. Don't orient potentially flammable wooden fences in a manner that could result in having them "wick" fire to a structure.
3. Design road systems to also work as an internal and defendable firebreak. That is, maximize the design of roads and hardscape perpendicular to "fall-line" or slope.
4. Layout a recreational trail system that can double for fire fighting purposes. Pay particular attention to adding trails to steep slopes and through dense brush occupying open space.

5. Plan your project so that fire trucks can access the site from at least two directions. Provide adequate turnarounds and install waterlines and hydrants along these routes. Clearly mark the locations of dependable water.
6. For ornamental landscaping, encourage your landscape designer or homeowner's association to assist in maintaining protective plantings aimed at reducing susceptibility to wildfire. Fire-resistant plantings include slow-growing, drought-tolerant shrubs and groundcovers.
7. Reduce the "ladder effect." Keep all natural vegetation retained on-site trimmed of dead branches and limbs. Prune all limbs and raise tree canopies from the ground up, making them less vulnerable should they be exposed to a fast-moving ground fire.
8. Remove combustibles. Clear debris from the forest floor including dry brush, logs, limbs, woody, and other organic debris, but be careful not to remove forest *duff*. Duff is the dark moist layer of topsoil found in forests. Usually present under organic debris, duff works to retard heat.
9. Thin natural areas so that forest and shrubbery aren't continuous. Thin tree canopies so they're horizontally clear of each other, and maintain vertical distance between shrubs or small trees and the lower limbs of taller trees. The more open space left between tree canopies and shrubs, the better for countering fire.
10. Include as part of the logging and clearing plan, a safe and cleared zone around the community. Reduce the visual impact of these zones by filling them with roads, utility easements, hardscape, swales, ballfields, recreational amenities, or open park space. Build structures with fireproof materials and surround them with fire-resistant plantings.
11. Upon completing portions of logging and clearing, promptly remove slash, chips, and woody debris from your site. Resist burying woody debris as it may ignite belowground. Abstain from shoving excess woody debris into a perimeter forest. With permission to burn slash, do it only in proper weather and only under the supervision of your local fire protection agency.
12. Install on all gates fire-department-approved "lock boxes" or other approved devices to allow emergency personnel access to gate keys. Lock boxes for securing gates are available for single or multiple users. In addition to providing security, use of lock boxes can also "clean up" the appearance of gates by eliminating the need for chains, pins, and numerous locks.
13. With the guidance and resources of your local power utility company, execute a plan that clears power-line right-of-ways of branches and potential windfall (hazard) trees. Fires can result when limbs or organic tree debris fall across power lines. Where possible, keep power lines

underground. Underground lines will not cause wildfires and are safe from the affects of them.

14. When planning, remember that clustered housing is better protected from wildfire than housing spread out over a larger area. Also, consider slope. Depending upon fuel loads and air movement, steeper ground promotes faster moving fires and thus, higher wildfire danger. Homes built on flatter ground are better positioned to resist wildfires than those perched on hillsides.
15. Integrate emergency-access alternatives for pumping non-potable surface water from ponds, streams, and storm-detention facilities. Design auxiliary-power generation into on-site water reservoirs; pump stations, and other supply systems. In high fire-danger country, consider installing emergency water wells with backup-power generation. Wells drilled and cased for geotechnical or other exploration and that have intercepted significant subsurface flows may be useful later as sources for emergency water. Any of these actions also mandates having enough hose, fittings and small trash pumps available in order to adequately distribute water.
16. Remove from buffers and setbacks those native tree species that are not fire-resistant. Their presence normally indicates the end of the growing cycle for that plant community while trumpeting that conditions are ripe for change. A forest in this state can easily succumb to wind, insects, fungus, and fire en route to producing a new, healthy, and young forest. Conifers usually rank high on the successional ladder because they contain oils, resins, close-cropped needles, and thin bark. Examples include certain pines, hemlocks, and cedars. Such high (or late) successional tree species typically germinate, grow, and thrive in the shade and moisture of surrounding trees. In addition to having shallow root systems, their foliage is tends to be flammable and thus easily engulfed by fire.
17. Consult with your local forester, state agricultural or forestry department, fire department, landscape contractor, certified arborist, or professional forestry association for more information. These professionals will be able to assist you in identifying fire-resistant trees, shrubs, and grasses that exist in your geographic area.

The rewards for fire-control planning are numerous. From a marketing angle, people are becoming more aware of the value in protecting homes and communities from wildfire and smoke damage. Environmentally, saving open space and vegetation benefits wildlife, conserves soil, enhances water quality, and ensures forested conditions. But in the end, smart design and strategies aimed at preserving property and life are really what building communities is all about.

Chapter 8

ROUTE ANALYSIS

Off-site access

On most projects, access locations are obvious. The lay of the land or surrounding infrastructure determines fixed points of entry. Other projects may require an engineered or planned solution for ingress and egress. Smaller, urban, and in-fill projects commonly possess fewer options. Here, access is secured wherever it exists, and is available as dictated by utility stubs or by municipal decree. Flexibility in locating access increases on large-scale, rural, or remotely located properties except as constrained by water, steep terrain, and sensitive areas. Regardless of project size or location, access can influence how a project is developed.

Owners are most vulnerable to access issues when consultants mobilize drillers, heavy equipment, and vehicles to perform pre-entitlement exploration and ground studies. Contractors then follow and improve these trails or create new accessways that can also be in the wrong location. In their exuberance to get underway, capitalize on good weather, or, perhaps, buckle under pressure from contractors, owners may ignore analyzing permanent and efficient access alternatives. The solution is to identify permanent road locations *before* consultants or contractors mistakenly fell trees and cut temporary access through sensitive areas, future green space, or setbacks. Unless there's no way of knowing future permanent points of entry, it can be a colossal waste of time and resources to establish and then tear out temporary roads and the likes of security fencing, truck washes, and erosion control, and to regrade building pads or replant vegetation.

As a rule, avoid investing in temporary solutions and throwaways. Instead, maximize returns by "building once." Out-of-location temporary roads are a sunk cost and they can create tension with adjacent landowners eager to move traffic into its permanent alignment. Buyers will want to know when the "real" points of access will be open and they may condition purchase and sale agreements accordingly. Temporary roads crossing buildable area prevent lots from being improved and placed on the market. And, there's never a good time to move a road, detour construction and delivery trucks, or funnel traffic past open trenches and utility installation. When situations in land development warrant the cost of temporary access, interim utilities, and construction roads, do whatever it takes to get a project going without sacrificing preparation. Those blessed with time to prepare are foolish not to which brings me to another favorite old saying of my dad's. That is, "act in haste and repent at leisure." He was a wise man.

Here are some options for owners to consider when planning hasn't progressed far enough to identify permanent road locations:

1. On tough sites, establish access where the least amount of damage might occur.
2. Try establishing temporary roads only across disturbed areas or over existing trails.
3. Instigate upfront engineering and geotechnical work in order to identify workable access points that *could* coincide later with permanent infrastructure and, from there, establish temporary ingress.
4. Invest in a study that identifies all possible access options surrounding the project and choose least-impact relatively inexpensive routes that align best with future utility connections and on-site permanent infrastructure.
5. Coordinate the needs of consultants and contractors before inappropriate access is established. First accommodate the most restrictive vehicles. These commonly include loaded trucks and trailers, lowboys, fuel trucks, and trucks delivering cranes, modular buildings or trailers.

Of the above, options one, two, and five result in owners having to pay to build temporary access, remove it later, restore the land, and then pay again for permanent access. Options three and four better position owners to place access once in the most probable, future permanent locations.

Although they differ, the above five options still involve penetrating the project boundary somewhere in order to link with existing infrastructure. There are techniques for evaluating access alternatives on projects that possess multiple entry points *and* incomplete on-site road layout and utility design.

Mechanics of analyzing access alternatives

Analyzing perimeter-access alternatives begins with collecting off-site and on-site data for each route considered. Generate off-site data by evaluating surrounding roads, infrastructure and utilities, distances to sources of materials, and neighboring properties. Ask yourself; where is the best place to connect this project to the world? Examine on-site access points by studying terrain, sensitive area locations, sight distances, shape of property, potential road grades, and elements described in an ALTA survey. Try to link each access alternative with a "most-likely" backbone road laid out with respect to control points, such as open space set-asides, sensitive areas, water, crossings, and targeted buildable area. Also remember that it's easier and cheaper to grade roads and to haul loaded trucks downhill. This analysis by itself will wash out some alternatives.

By combining information, it may be apparent that multiple on-site routes can connect with one or more off-site access points. Final conclusions however, cannot be drawn solely in the office. Complete analysis requires field observations. Locate, flag, and field-truth each route observing potential problems with adjoining private property, water, slope, rock outcroppings, and any ground constraints. Inspect the condition of on-site trails or roads including culverts and bridges. What works and what doesn't? Do any routes complement construction that is needed to get the project off the ground?

Establishing temporary access usually occurs before owners have a chance to explore soils, calculate roadway earthwork, prepare for runoff, fix centerline locations, or consider wall or slope requirements. Not knowing land set-aside locations, permit requirements, or what easements or land will be required for access can defeat making informed decisions. The result is that temporary access and on-site routes may have to be located quickly on the ground. Only individuals possessing a deep understanding of pioneer road construction, its costs, and the logistics of how the project is going to be built, should be relied upon to locate temporary access in this manner.

One method that allies itself with subjective decision-making involves using a detailed checklist specifically tailored to each project's conditions. A well-built access-alternative checklist can go a long way to helping sort uncertainties and provide a base for decision-making. By default, enough positives or negatives one way or the other per access alternative is enough to either discard that alternative or keep it alive for further evaluation. Reference the checklist shown on *Figure 8-1.*

The checklist is nothing more than a collection of questions used to equally assess each access alternative. Rate each alternative according to a scale calibrated by the owner. The result is a scorecard that ranks each access alternative relative to the others. Examples of criteria that may be used to build custom checklists are detailed below.

Evaluating existing off-site access route criteria:

- Is the route direct?
- Is the route paved?
- How long is the route?
- Where will runoff flow?
- Is two-way traffic possible?
- Are overhead wires an issue?

- Is year-round access possible?
- Is ingress and egress favorable?
- What is the road's current condition?
- Does the route entail using private roads?
- Do you have legal access and utility rights?
- Will raising or lowering power lines be required?
- Can the route handle large construction equipment?
- Is this route reconstructable and, if so, at what cost?
- Will signs directing construction traffic be an issue?
- Does this route course through private neighborhoods?
- What is the condition and rating of bridges and culverts?
- Are suitable wet and dry utilities available to connect to?
- Will neighborhood or private property require restoration?
- Can the grades accommodate heavy hauling and suppliers?
- Where can construction-material lay-down areas be located?
- If walls, cuts, fills, or utilities are necessary, will easements be required?
- How will projected construction traffic play against existing traffic flow?
- What upgrades might be required to improve and protect neighboring property?
- Would more than usual road maintenance of dust, mud, and snow removal be required?
- Could permit authorities require upgrades to this route as a condition of occupancy or final project acceptance?
- How are you going to handle off-site erosion control? Where will temporary or permanent ponds be located and how will they release water?
- Is damage to existing infrastructure highly probable? Always consider that construction traffic has a way of finding and punishing weakness in subgrade or pavement.

Evaluating new on-site access road criteria:

- Are there sight-distance issues?
- Is any bedrock or boulders present?
- Is this route constructible and at what cost?
- What and where are physical control points?
- Does this route provide strategic elevation gain or loss?
- Can this route also work as part of the future project infrastructure?
- Do the road grades work for ensuring on-site to off-site gravity flow?

- Where can the road go in order to avoid crossing sensitive water courses?
- Are sensitive areas such as wetlands, streams crossings, and habitat impacted?
- Are on-site borrow and waste areas available without creating another cost center later?
- Does this route disrupt or impact future developable area or eliminate potential open space?

Concerns common to both off-site and on-site alternatives:

- Does the route affect any utilities?
- Is the route politically acceptable?
- Is the route isolated from the public?
- Can hauling occur anytime of the day?
- Are traffic control and flaggers required?
- What potential permits may be required?
- Can access be restricted? Are gates legal?
- Will noise-abatement fencing be required?
- Are project signs acceptable along the route?
- Are any wetlands or sensitive areas impacted?
- Will the general public share in using the route?
- Is it acceptable to build a truck wash on the route?
- Does the route pose any extraordinary maintenance issues?
- Is the route well-screened or in full view of private residences?
- What number of private residences and properties does the route affect?
- Although always interesting, does the route negotiate any private junkyards?
- Does the route harbor any currently or soon-to-be listed endangered species?
- Will any public or private parties stand to benefit if the owner chooses to use the route?
- Are there any suitable nearby storm facilities available to collect, detain, and treat runoff?
- Does the route negatively impact rush-hour traffic, school traffic, or nearby intersections?
- Does the route offer dense downhill vegetation and forested conditions suitable for absorbing road runoff?
- Besides construction access, what other potential opportunities may

exist for the owner by using the route?

- Does the route stay dry and well lit, or is it likely to stay wet, to hold snow and ice, and to pose visibility problems?
- Will the route have to be abandoned once the site is developed and then restored to predevelopment conditions?

Other tips when considering off-site access:

- Pick up a field office. As allowed by zoning or conditional-use permit, expendable structures on land purchased for access can serve as field offices. Check to see if the structure possesses ample parking, septic, water, and dry utilities.
- Consider using one-way, narrow road access. One-way roads can make otherwise unusable access perfectly acceptable. Traffic is easier to manipulate, it's safer, construction costs and impact decreases, and there's less area to shed runoff and concentrate turbid water. One-way haul roads provide uncommon flexibility and project control.
- Know the condition and weight-holding capacity of existing bridges and culverts. Bridges and culverts are grossly expensive to upgrade or replace. Procuring permits for reconstruction can take years. Survey and inspect bridges and crossings to ensure that construction traffic will safely pass over them and under any aboveground bracing. Compare your research data with whoever owns and maintains the structure.
- Break-even points. What are the monetary and schedule break-even points for each alternative? For instance, what is the break-even point of purchasing access further upstream to avoid the cost of crossing the same stream downstream? A potential trade-off may include weighing the purchase of access rights from a number of private owners to the construction cost and time needed for an expensive elevated crossing. Another form of break-even analysis is to compare the cost of completely rebuilding a longer existing road against the cost of establishing a more expensive and new access road that is significantly shorter in distance.
- Avoid expensive road-rebuilding projects as compensation for access. Existing pavement is hard to beat if it's thick and supported by solid subgrade. On old pavement that shows signs of decay or where no record of construction drawings can be located, core the pavement to determine exactly what's there. Meanwhile, videotape roads approved for access *before* running trucks or construction traffic over them. You'll want to record pre-construction conditions in case regulators later try to collect damages for something unrelated to your work. For dead ends, the video should capture at least to where the road forks or

where it's being impacted by similar traffic from another site. On through roads, video coverage should capture no less than a quarter mile in each direction from the project access point. Videotape the entire road right-of-way including landscaping, trees, curbs, aprons, sidewalks, and utility lids and valves.

- What works for ingress may not work for egress. Either may be more favorable for construction traffic depending upon sight distance, how existing traffic is crossed, the number of lanes available to trucks, and existing curve radii where the access point joins an existing roadway. For example, egress becomes a problem when public traffic congestion prevents trucks from leaving the site at regular intervals. Trucks can block driveways and impact access when waiting to flow into traffic. These types of delays also add significantly to the cost of construction. Don't forget that older and "looser" trucks also tend to drip lubricants when idling. Oil dripping from waiting trucks may prematurely degrade asphalt leading to road repairs and possible complaints. Lubricants certainly do nothing positive for concrete or storm runoff. Egress also poses it's own set of problems when trucks leaving the site must immediately accelerate downhill or painfully grind uphill in order to merge with traffic. As awkward as they are to maneuver, construction trucks, particularly lowboys and long, loaded dump trucks can make transitioning into moving traffic problematic.

The checklist

Again, refer to *Figure 8-1.* This type of analysis considers existing and potential conditions pertinent to each route. It allows owners to sort access alternatives according to each route's characteristics. If an owner or consultant has a biased or preferred access route in mind, an analysis of this type tests that partiality thereby allowing him to make clearer decisions.

As shown in *Figure 8-1,* color-code each route on a master plan and then grade all against the same set of criteria. Subjective numeric ratings set from 1 (most favorable) through 9 (least favorable) are used to value criteria. The subjective nature of this analysis underscores the need for having qualified and experienced people involved and to collaborate when ranking each alternative.

The analysis is complete after each alternative is scored, totaled, and ranked. The example shown in *Figure 8-1,* portrays the highest-ranking alternative—Willow Brook Pit access—as the most favorable access route with lingering but moderate concerns regarding ingress and cost of abandonment. In contrast, alternative two (Alt. 2), which accesses the project from Victory Hill Road, appears to be the worst choice for access because of major construction, cost, and permitting concerns.

Owners can formulate off-site access and haul agreements with surrounding landowners and jurisdictions once they've chosen a most favorable route. They're then free to concentrate on on-site planning, route analysis, and design issues.

On-site route analysis

On-site route analysis is precursory to designing project infrastructure as it relates to fixed, terminal, and/ or intermediate project control points. Control points by definition regulate design. Engineering mistakes resulting from failing to meet or honor fixed control are inexcusable.

Owners shouldn't proceed with planning and design of on-site access until they have corralled their priorities and expressed them to team members. For example, a list of objectives that an owner may wish to meet by way of on-site access design may include:

- Maximizing views
- Optimizing circulation
- Minimizing road length
- Minimizing construction time
- Accommodating future growth
- Maintaining lowest possible cost
- Accommodating hauling and deliveries
- Providing maximum street side parking
- Hiding infrastructure from off-site view
- Maximizing net buildable area or lot yield
- Accessing lots and parcels slated for initial development
- Generating cut material needed to fill and grade buildable area
- Providing interconnectivity with adjacent properties or infrastructure
- Minimizing separation between roads resulting in uniform and predictable grading
- Strategically positioning the roadway to also serve adjacent off-site land in order to collect future "late-comer fees"

Route analysis involves more than fitting alignments to terrain. It includes planning for the movement of liquids, gases, and wire commonly referred to as utilities. Gravity lines, such as sanitary and storm sewer, in particular, can greatly affect road location and design. Product type also influences road layout and design. And, successful route analysis is more than an office exercise. Field reconnaissance and walking proposed alignments and road locations are critical.

Rating Off-site Construction Access Alternatives

STEP 1 - RATING EACH ALTERNATIVE:

Item	Factors Affecting Access	green Alt.1	white Alt.2	yellow Alt.3	red Alt.4	navy Alt.5	orange Alt.6	teal Alt.7	Criteria for Judging
1	Ease of construction traffic ingress	4	3	1	5	7	6	2	Ease of access considering number of right-hand turns, curve radius, and geometry.
2	Ease of construction traffic egress	1	9	2	5	7	6	3	Ease of leaving project relative to traffic control, cross volumes, and sight distance.
3	Potential environmental impact	2	9	1	4	6	5	3	Impact on stream crossings, wetlands, storm runoff and steep slope cuts.
4	Number of adjacent land owners affected	1	7	9	8	6	5	2	Determined by reviewing property and plat records.
5	Constructability	3	9	1	5	6	4	2	Evaluate grade, soil conditions, earthwork, surfacing, runoff, erosion potential.
6	Potential cost of construction	3	9	1	5	6	4	2	Rough order of magnitude estimate based on experience and ground conditions.
7	Potential cost to maintain	1	6	9	8	5	4	2	Maintenance costs depend on surfacing, weather, and degree of shared public use.
8	Construction & public traffic safety factor	1	8	9	7	5	4	2	Does using this route pose public safety concerns?
9	Probable ease of obtaining permits	3	9	1	4	7	5	2	What permit requirements are associated with constructing this route?
10	Risk of damaging existing infrastructure	1	4	9	7	8	6	2	What's the condition of existing infrastructure and the potential risk of damaging it?
11	Ability to preconstruct future project access	2	7	1	6	5	3	8	Does constructing this route contribute to the future project road system?
12	Probability of all-weather access	1	8	3	6	5	4	2	Is this route conducive to all-season, year-round construction access?
13	Is this alternative ready for use?	2	9	1	7	5	4	3	Is this route ready for use? What would it take to prepare it for construction traffic?
14	Potential cost to abandon later	4	8	1	7	5	3	2	Potential cost to bed and abandon the route back to natural or acceptable condition.
15	Affect on adjacent rush-hour traffic	3	5	7	6	8	8	1	Impact of construction access on peak volumes of traffic, schools, and pedestrians.
16	Probability of sharing with public users	1	5	7	6	4	4	3	What volume of local traffic can be expected to mingle with construction traffic?
17	Ability to handle loaded trucks each way	3	8	1	7	5	4	2	Grades, surfacing, and road conditions as they affect truck speed, traction, etc.
18	** Would use of a flagperson be mandatory?	0	0	0	1	1	1	0	Would additional costs associated with flaggers and traffic control be required?
19	* Potential utility right-of-way permit required?	0	5	0	5	5	5	0	Does this route trigger the need for gas, power, and utility right-of-way permits?
20	***One Way Hauling Options***								
	Best suited for ingress?	✓	✓	✓		✓	✓	✓	Should one-way access be desired, which alternative works for ingress?
	Best suited for egress?	✓		✓	✓			✓	Should one-way access be desired, which alternative works for egress?
	Total Scores per Alternative:	36	128	64	109	106	85	43	
	Rating as Scored:	***1***	***7***	***3***	***6***	***5***	***4***	***2***	{ ** Yes = 1, No = 0 } { * Yes = 5, N0 = 0 }

The Lower the Score, The Better the Alternative: 1 = Highest or Most Favorable Rating, 9 = Lowest or Least Favorable Rating

STEP 2 - RANKING EACH ALTERNATIVE:

Alt.	***Ranking As Tallied Below***
1	***1st Place Ranked, (green line) - Willow Brook Pit access.***
2	***7th Place Ranked, (white line) - Off Victory Hill Road, along the north side of the project.***
3	***3rd Place Ranked, (yellow line) - The Black Bear Road.***
4	***6th Place Ranked, (red line) - Access south off Black Bear Road and easterly along the north side of the project.***
5	***5th Place Ranked, (navy blue line) - Access off State Highway 17 through the Cloud family property, southerly along the power line easement to Black Bear Road.***
6	***4th Place Ranked, (orange line) - Access off State Highway 17 through the Cloud family property and gas line easement to Black Bear Road.***
7	***2nd Place Ranked, (teal line) - Continuation of the Old Black Bear Road.***

Figure 8-1

As stressed throughout this book, nothing good happens when an inferior or inaccurate topography plan is used for anything short of starting a campfire. Only analyze roads against accurate topography and precise off-site control thereby guaranteeing that on-site infrastructure will transition and connect properly with existing off-site roads and utilities. When it's apparent that something under design isn't going to work, connect, or match off-site pipes and roads, ask why—and then correct it. What is the reason for the discrepancy? Is it a case of basing work on inaccurate topography, poor engineering, incorrect base surveying, or are off-site pipe and road coordinates wrong? Before proceeding, immediately identify and rectify defective work and control or you will risk having the problem plague the project indefinitely.

Reconnaissance

While working for a client and collecting data for a next-phase cost estimate, I had the pleasure of walking an "L" line that was in the process of being flagged for logging and clearing. For clarity, an "L" line is the permanent centerline location of a road. The "L" stands for location line where as a "P" line stands for preliminary line. Once a road's "L" line is located and staked, site construction soon follows. While walking this line, I came across water mixed in with leaves and forest clutter. Further inspection revealed that the road intercepted runoff from a spring that was located approximately 30 feet (9 meters) uphill of centerline. Nobody seemed concerned about the water. Even though it was springtime, the presence of water meant that a pipe, ditch or something else would need to be installed later to handle it.

So—why was the spring a mystery? Because the spring was seasonal, perhaps, topography was shot or perhaps environmental studies were completed during the dry season or perhaps it was simply overlooked. Once water was visible, though, it should have captured someone's interest. Was the spring an exit point for a shallow water table, or did it relieve another wet and challenging surface feature further uphill? Did the spring indicate the presence of bedrock or something else impenetrable? All questions worthy of investigation. Once this condition was brought to his attention, the engineer called off the surveyors, and evaluated the situation. As it turned out, the engineer never walked the alignment and no one had explored the upper section of land where the spring originated even though that parcel was on schedule for the next phase of grading. In the end, the road was relocated above the spring and the clearing limits were readjusted.

The point is this—it is easy to underestimate the time and expense necessary to properly plan and locate an on-site road network prior to design and engineering. As terrain steepens and foliage thickens, it becomes increasingly important to stake and locate preliminary road locations before committing to final design. Obtaining sufficient information may entail shooting side slopes perpendicular to centerline at regular intervals. An exercise such as this provides data from which a designer can confirm ground conditions as topographically represented on the plans relative to road location.

For instance, do cuts and fills as field-verified coincide with preliminary calculations based upon initial paper layout? Is the roadway wrapping more quickly into the ridge than expected? Should a planned intersection go in as intended based upon field observations at that location? What about "save" trees? Is the roadway going to miss or conflict with groups of trees intended to remain? Is the roadway heading to a location favorable for a water crossing, or are road elevations going to make it necessary to construct an embankment in order to cover and provide adequate headroom for a culvert? Besides the cost of construction, how much adjacent land will the fill slopes require or, does it make sense to consider retaining walls in order to limit slopes and maximize buildable area? As an owner, it's always better to know this information before finalizing design and committing to budgets.

Some owners expect that they'll receive road reconnaissance services included in the architect's or engineer's scope of work. The problem is—it's often an extra. Field-truth exercises aren't automatically included with proposals for design, engineering, CAD, or plan development services. *Total* engineering budgets set as a fixed percent (let's say 15 or so percent) of projected construction costs also don't often include everything necessary to protect an owner from sloppy or inadequate design work. A less-than-optimum road network may result if design work is based upon a budget that doesn't include reconnaissance and field-truth work. Even when money for reconnaissance is included, engineers faced with a thin budget may end up allocating funds however necessary in order to quickly produce a permitted set of road and utility plans. The ultimate price to owners can be huge.

Besides construction cost overruns, lack of reconnaissance and field-truth investigation can also lead to additional expenses in engineering and redesign, busted earthmoving budgets, building steeper roads than anticipated, tighter tolerances for gravity sewer lines, schedule impacts, and possible loss of buildable area. Require field-reconnaissance of road networks in challenging terrain regardless of cost, weather, brush, or distance from the vehicle.

Road-system design considerations

Projects successfully developed on sloped terrain have one trait in common. That is: roads drive the grading plan—not the other way around. When roads are forced to "chase" or conform to overall site grading, earthwork is likely to increase, utilities may not drain as planned, square area of walls or slopes are prone to increase, lot shape can become irregular, and net buildable area may diminish.

Grading should precede road design in one aspect only. That is to identify a range of minimum and maximum lot grades acceptable *per product type* expected to be built on a project. Although roads may have to diverge in a direction counter to lot grading, knowing what the targeted adjacent lot grades are, sets a baseline for engineers and planners contemplating horizontal and vertical roadway coordinates. Once maximum and minimum grades per product type are established, owners need to stick to them and knit this criteria into project planning as constraints. This information also helps owners to decide how and where to prepare building sites for specific buyers or builders known to prefer certain grades and tolerances for their product. Owners benefit when buyers option or purchase land for a project still in the design phase. Besides the early financial gain, it makes it more likely that the seller will install utility stubs, curb aprons, and dry-utility boxes in locations preferable to a targeted buyer or builder. In hot markets, lots and parcels that are ready for structures are more marketable.

Conversely, when road design precedes lot grading design, there's a chance that both designs can be integrated into one overall mass-grading plan making permitting easier and more efficient. Rough grading roadways *may* then proceed under a mass-grading permit provided roads are treated as gradable area. The key is to grade, or "bench" roads in a manner that blends with mass grading while being careful not to make it appear as though road construction is proceeding without the appropriate permits. Of course, the owner must be confident of road location and design before committing to grading activity. Owners should also level with the permit authorities when attempting this maneuver. Grading roads and lots in one continuous operation eliminates having to return later to separately rough-grade roads or to match lot fronts with the edge of road right-of-way. The road prism gets shaped and fine graded after utility installation. This strategy expedites work, averts costs for redundant staking, and results in much more efficient and cheaper earthwork.

As broken terrain steepens, designers will find it more difficult to synchronize road location with a plan for optimal overall grading. At some point, steepening ground dictates fall lines and overall grading regardless of road location. At the same time, while climbing or losing elevation, a road may reverse direction and "switchback" it in an opposite direction resulting in steeper and oddly-shaped lots. Under these conditions, an owner must decide whether to accept the loss of buildable lot area, or, increase costs and compensate for grading deficiencies by use of heavy cuts, fills, or retaining walls, or, hold site-work costs steady and require builders (buyers) to instead, make up the difference in say, foundations. A buyer forced into making grades work, especially with foundations, will expect lots to be priced accordingly so, no matter which option an owner chooses, it's costly. The point is, matching lot grading to roads is a dependable technique for finding the lowest cost and most usable grading plan consistent with minimal earthwork until ground steepens so much, that it either forces changes in product type or results in over-sized lots or ground being left to open space.

The economics of developing steep ground revolve around road building. Road building on steep terrain requires moving greater volumes of earth. Deep cuts (partial or full-benched) and fills supported by walls or slopes are the result of plowing across slopes rather than weaving and undulating with slope and grade. This results in a more pronounced vertical separation between roads and adjacent grading. Steeper ground commonly demands employing engineered walls and slopes, which can mean increased loss of buildable area to road embankments and cut or fill slopes. Any of these can penalize lot configuration and overall grading. If it's too expensive to build roads, it's usually too expensive to develop lots.

On most sloped property, there exists a break-even point—a degree of slope beyond which development isn't financially or technically feasible. That point changes from project to project, however, it can generally be defined as a function of:

1. Steepness
2. Regulations
3. Increased bedrock
4. Market conditions
5. Your team's ability
6. Builder resourcefulness
7. Ground and soil stability
8. Loss of too much net buildable area
9. Flexibility within your permit agency
10. Increased earthwork, slopes, and walls
11. Sensitive areas, such as steep slopes, wetlands, or streams
12. Cost—Increased construction costs may not produce enough developable area of high-enough value to justify pushing the envelope.

Knowing that a point in slope exists beyond which development isn't practical should challenge owners to identify and push the break-even point as far up or down slope as possible. Creative design *can* transform marginal property into land that's financially feasible to develop. Pushing the envelope may require designing multiple road layouts in order to compare lot count, product layout, and open space. In terms of site construction, design multiple road alternatives, compare earthwork quantities, and estimate costs of construction for each layout. Question everything. Minor adjustments to total road length or grade can significantly alter construction costs. Reviewing multiple road alignments can also trigger new ideas and varying perspectives.

When roads build out steeper than planned

In challenging terrain, it's not uncommon for roads to build out steeper than designed. This problem becomes more prevalent with increased slope. Here are explanations as to why roads can build out steeper than designed:

1. The staking is wrong.
2. The topography plan isn't accurate or the contour interval is too wide for precise design and engineering.
3. Design based upon off-site elevation control may not match actual on-site topo elevations causing a newly constructed road to gain or lose more elevation than expected over the same fixed horizontal distance.
4. Design is pushing the envelope. Topography may have left the engineer with no choice but to design road grade to the maximum allowed slope leaving no room for error. And—as it would happen—errors in staking or construction forced the road steeper and out of compliance.
5. Cumulative error. Earlier and minute engineering errors carried with slope are growing worse with elevation and grade. The surveyors aren't back-checking and reporting actual as-built final road grades to the office, so the road designer is proceeding with incorrect base information.
6. The contractor isn't following staking or he may be manipulating material volumes for other purposes thus affecting grade. This is especially true if flaws exist in subgrade where poor compaction, rotten soil, or spreading of trench utility waste have created "mini grades" within the road causing dips and steep pitches. Contractors may also inappropriately shave materials to make road fills or to complete utility trench backfill.
7. In and of themselves, intersections can greatly influence road grades. They can be big, complex, and can merge many curves and tangents that don't always match line to line per plan. For example, in order to gain additional road run out, queing, or sight distance, vertical curves in and out of an intersection may be forced to absorb more or less

elevation gain than designed, resulting in compressed or flattened road grades. Large intersections also require significant grading. As topography increases, intersections can drive and alter property grades in all quadrants of the intersection.

8. Contractor interpretation. This item relates to number six above with a slight difference; in the absence of horizontal staking, contractors improvise. It's also possible that the surveyors are allowing operators to make vertical curve transitions at intersections "work" instead of properly staking them, or that the operators are just taking it upon themselves to make things fit by eye, completely ignoring staking. Similarly, the contractor may be grading tight vertical curves by feel or as allowed within the limits of his equipment. Items related to road grading that are prone to operator interpretation can include:
   - Channeling or redirecting storm-drainage flow, as needed, to catch basins
   - Forming or manipulating road crowns at the confluence of intersecting vertical curves
   - Adjusting driveways and aprons for improved sight distance, to better access structures, or to skirt utilities
   - Grading super elevation that is void of staking on horizontal curves and where grades transition at intersections
   - Changing horizontal radii on acute intersections to better assist trucking or to match ill-placed valves, catch basins, or manholes

Road grade and design

Construction standards ensure that final road grades don't exceed a certain given slope. I've never witnessed a road having to be torn out because it was built steeper than designed or in violation of approved plans, but that's not to say that it can't happen. Many variables can affect road building, so it makes sense to allow yourself and your designer room for error. Play it safe. If you're building to a road standard that limits grade to 15%, don't design it to 15%. Rather, design it to something that's less than the limit but attainable and in sync with current levels of construction accuracy. Examples of grade-sensitive items include routes designed for the disabled, fire truck and emergency vehicle access, and ball field requirements. In short, prevent engineers and planners from designing to minimums or maximums.

Base contour interval also influences error. If a road or alley is designed at a maximum allowable grade of 15% from a five-foot (152 cm) contour interval possessing an average 2.5-foot (76 cm) error, there's a chance that the road will build out at greater than 17%. The same 15% grade designed from a two-foot (60 cm) contour interval possessing an average one-foot (30 cm) error may build out at more than 16% slope. Remember that topographical maps *at best* are only to be trusted to plus-or-minus one-half of the contour interval.

Although tighter base topography yields less potential error in road design, errors nonetheless can be unforgiving. Most permit agencies and fire departments will not accept onerous degrees of error in road grade, but an owner's woes aren't just limited to government scrutiny. Road intersections, lot grading, targeted product type, walls, boundaries—just about everything—can be profoundly affected by steeper than expected roads that result in heavy cost and schedule impact.

Owners who have to design to maximum road grades in order to make development feasible are probably designing the wrong product on the wrong piece of property, and may later regret pursuing the project. If a project has to be designed to extremes in order to work, plan wisely and invest in enough base control to pull it off.

Control points

Control points are fixed physical features that influence road location and project layout. Identify them early. Unlike roads, utilities are one of the site construction items least affected by control points depending of course, on the character of the control point. Examples of naturally occurring control points may include:

- Steep slopes
- Hills and valleys
- Property boundaries
- Cliffs and rock walls
- Perimeter access or terminus points
- Rock outcroppings and large boulders
- Low points that guarantee gravity flow
- Waterways, wetlands, and sensitive areas
- Pre-existing pipelines and utility easements
- Elevations related to limits of water pressure

The greater the number of control points on a project, the earlier road design should begin. Control points lurking in the rear of a project can affect grades and horizontal alignment at the beginning, so it's important to factor them into layout and design early. Even then, changes in product mix or grading may exert pressure on owners to change road layout in total contradiction to what control points will allow. In the end, a road system may end up contorting in order to satisfy both control point criteria and fresh changes in planning or marketing.

Forcing road design into and around control points has its consequences including, but not limited to, higher costs, extended engineering, steeper and longer roads, more expensive grading, problems with gravity utilities, and grade separation issues. The arrival of buyers and builders typically places an entirely new set of demands on road location and grading. Don't lose sight of the fact that control points influence road location and roads dictate grading. Keep these three elements in their place relative to each other.

Perform thorough geotechnical exploration prior to road layout and design. If belowground control points, such as water, sandstone, bedrock, or boulders, exist, then invest in additional centerline potholing on 100-foot (30 meters) or so centers to verify subsurface conditions. Adjust contract item descriptions and volumes accordingly before grading commences. As we'll discuss in upcoming chapters, complete geotechnical work is high on the list for ensuring accurate cost projections, controlling site-work, maintaining budget, and meeting schedule.

Locating roads

Figure 8-2 displays a route-selection logic diagram. While this diagram should be of interest to road designers and architects, it also provides owners with a clear explanation of the reasoning and hierarchy involved in locating an on-site road system.

Of particular interest should be the tests performed in selecting a viable route. Up to this point, we've discussed topography, access, and control points as they relate to land development and roads. The results of geotechnical testing and subsurface mapping are as important as any other criteria. Geotechnical work provides clues to questions such as:

- Is the ground stable?
- What is the topsoil depth?
- What is a suitable paving section?
- Does the road miss pockets of unsuitable material?
- Is cut material suitable for making structural road fills?
- Will road cuts and utilities encounter subsurface water?
- Is the road being built on a foundation of structural material?
- Are the road and deep utilities in danger of intercepting bedrock?
- What are the consequences of exposing the subsurface rock or clay to air, sun, and moisture?

Evaluate roads also from a development and regulatory perspective. Consider the following criteria to help evaluate whether or not road location meets development objectives. Does the road:

- Provide fire truck access?
- Facilitate gravity pipe flow?
- Require significant retaining walls?
- Exhibit a geometrically balanced layout?
- Promote a usable and friendly grading plan?
- Cross sensitive areas at their narrowest location?
- Facilitate favorable locations for wet and dry utilities?
- Work with the land in promoting minimal earthwork?
- Complement land set up for gravity-flow septic systems?
- Appear economical in terms of minimal length and grading?
- Follow natural benches or does it unnecessarily traverse steep terrain?
- Work to maximize net usable area with economically sized lots or parcels?
- Intersect unusable land at odd angles thus creating irregular and inefficiently shaped lots?
- On multi-year projects, lend itself to being built piecemeal in segments with phased development?
- Facilitate channeling storm runoff into and out of storm ponds, vaults, dispersion trenches, and/or infiltration galleries?
- Leave no room for error around sensitive areas that may later, when field staked, result in encroachments, buffer violations, and greater setback limitations? If so, move it.
- Gain too much elevation too fast thereby making dense development overly expensive, technically challenging, and laden with problems for builders and buyers?
- Coincide or follow future water zone elevations thereby serving the most number of units with available water? This also works to eliminate the need for throwaway fire-zone water pressure facilities and pump stations.

Here are some additional tips to follow when locating roads:

- Respect existing utilities and don't take lightly the cost of dealing with them.
- As terrain steepens, project-wide street connectivity becomes harder to attain.
- Ensure that exposed wet-utility pipe exposed under bridges is either of large enough diameter to retard freezing or protected from freezing.
- Lay out projects so that roads can easily connect through parks and

golf courses should future water availability or market pressure make developing recreational green space a viable option.

- When installing utility conduit in structural elements, such as under bridges, add extra conduit for future capacity. This is especially true if your dry-utility requirements aren't yet final. Plastic conduit is cheap—reconstruction is expensive.
- As slope increases, it becomes more critical to lay out the primary backbone road network in advance of locating secondary roads. This action quickly alerts owners to limitations in grading and, thus, the allowable range of slope, which can affect product type.
- Consider reducing the width of roadway by hanging a paved bike or pedestrian lane along the cut or fill slope. Accomplish this by use of an engineered slope or retaining wall system that is backfilled and paved over. Since bike and pedestrian loads are typically light, permit agencies may approve a design with minimal compaction specifications.
- When designing roads on steep terrain with limited horizontal run, ask yourself if you'd be better off solving access and grading problems through structure design instead of by grading your way out of trouble. It may make economic sense to reduce earthwork volumes in exchange for one-way winding roads and more elaborate foundations.
- Removing and replacing existing culverts in itself, doesn't automatically improve or guarantee fish passage. It does however increase cost. Keep fish passage and the maintenance of natural stream hydraulics in mind when contemplating culverts. "Squash pipe," box culverts, and steel or precast pipe arches are popular options when fish and creek parameters influence stream-crossing design.
- Try to locate new road crossings where streams are the steepest. Steeper stream gradients require smaller culverts to pass the same volume of water and they typically indicate the presence of a "V"-shaped gully or defined ravine. Smaller diameter pipes and confined stream banks that require less fill may mean cheaper road crossings as opposed to flatter and more broad flood plains.
- If only designing and constructing a portion of a permanent road network, consider not stopping work short of the next control point out, such as removing a rock outcropping or making a water crossing. Negotiating the next control point allows the seasonal work that follows to begin without delay, and assures that the current road under construction is on grade and in location to accommodate that control point.
- As discussed earlier, road grades on steeper ground may undulate into a series of less steep mini-pitches and flat runs. This can lead to poor

sight distance, uneven flows of storm runoff, difficulty in locating aprons, and an unattractive streetscape. On sloped ground, carefully control how vertical curves transition into tangents. Moderate grades during construction by grading out steeper areas and filling flat runs.

- When soils vary and road location is flexible, aim roads at veins of gravel and course structural material. Excavating through free-draining structural material may afford owners many benefits including draining runoff, obtaining quality construction material, extending the road construction season, quicker utility line work, and the option to overexcavate granular material and lose less-desirable structural fine-grained material hauled in from other on-site sources.
- On lengthy projects where product type and density remain moving targets, don't overbuild. While still installing permanent and properly sized underground utilities, pursue minimum surface road construction as allowed for meeting immediate traffic demands and density. When the need arises to widen roads, add sidewalks or landscape, ramp-up construction, and complete work. Otherwise, time and expenses get invested into underutilized infrastructure. Fixing infrastructure also increases risk if product, density, or builder demands later result in having to make changes.
- Rather than allowing temporary construction road culverts to be installed by chance or as seen fit by the contractor, plan their locations. Locate temporary culverts where they can freely channel water to sediment ponds, meet erosion-control objectives, and collect and pass runoff from pre-graded naturally contoured native ground and finished lots. When placing temporary pipes, try to locate them both horizontally and vertically so that they can also act as future sleeves for utility and irrigation lines. The ends of temporary culverts can be easily damaged or buried, so have them "as-built" and ensure that this information gets incorporated into both the engineering and dry-utility plans.

Project owners should understand and recognize the mechanics of road system design. The decision tree laid out in *Figure 8-2* portrays criteria that an owner, architect, or engineer should reference when evaluating road system alternatives.

Cost

One item missing in *Figure 8-2* is road cost. Road cost comprises a completely different set of variables more affected by design standards, substituting materials, maximizing use of on-site material, and minimizing overall length and square area of hardscape.

Typically in roadwork, the wildcard is excavation and grading—not wet utilities. Installing no more pipe than necessary, utilizing the least expensive pipe material needed to do the job, and installing pipe as shallow as possible, can save owners money. In general, though, most opportunities for lowering project construction costs are found in *manipulating grading.*

Costs for road excavation can vary widely depending upon many factors. For instance, on uniform slopes, reducing the volume of roadway earthwork is usually more easily attained by manipulating and undulating road grades rather than by shifting road alignment in or out of the slope. Less earthwork means cheaper construction, which partially explains why full-bench road cuts and through fills are more expensive to construct than split cut-and-fill sections. Split cut-and-fill sections occupying the same road stationing cut and place material in the same location requiring less need for importing or exporting material. Shorter cut slopes and thinner embankments on split road sections also skirt the need for retaining systems. As opposed to full-bench-cut excavations, split sections are a combination of cuts and fills that result in shallower cuts and that reduce the likelihood of encountering bedrock or subsurface water. Shorter lay-back slopes are easier to landscape, maintain, and protect from erosion. The folks in marketing like shorter slopes, too. They say they look better.

Broken slopes and variable terrain exhibit ridges and draws which can be used as places to move a road into, if more material is needed, or out of, if road excavation is generating too much cut. Moving a road into the hillside in order to pick up cut material works well on projects short of fill as long as the move doesn't penalize existing lot area or, as mentioned above, result in exorbitant back slope or retaining wall costs. Other avenues available for reducing road costs include lessening the actual length of road and using geotextiles and grids in lieu of overexcavation and import when constructing in with poor soils.

Net utilization

Net utilization defines, as a percentage, the quantity of revenue-generating land available after subtracting all non-revenue generating land, such as infrastructure, parks, and facilities, from total developable acreage. Mathematically, remaining revenue-generating acres divided into total developable acres equals net utilization.

It's no surprise that improving net utilization has everything to do with efficiently designed roads and related infrastructure. Project owners free to plan roads and layout may not possess the same control when siting land giveaways and non-revenue–generating parcels that are controlled by a non-profit organization or a public entity. Examples of such parcels include pump stations, parks, and schools. Once the size, shape, use, and location of a non-revenue–producing parcel becomes known, that land becomes a fixed constant in the

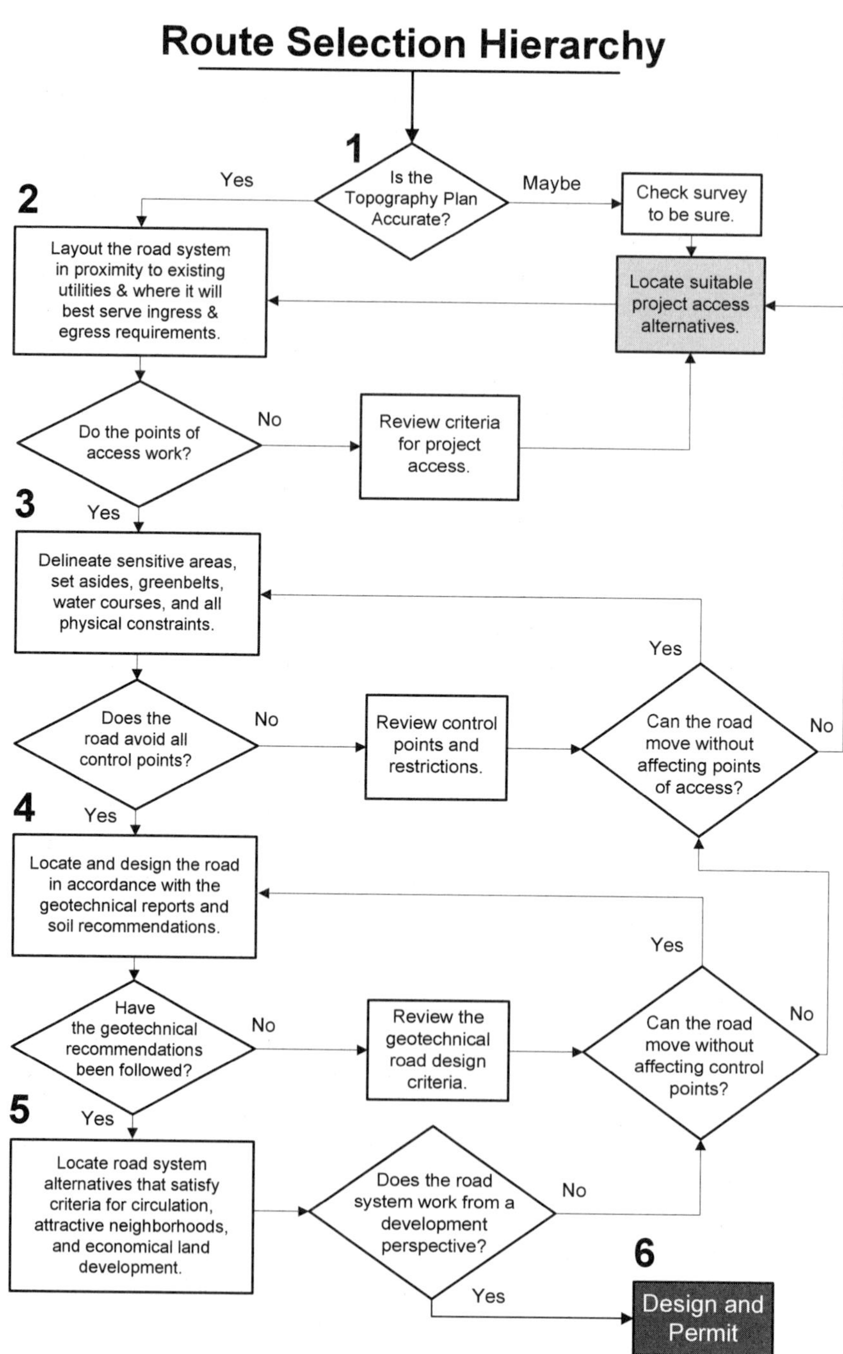
Route Selection Hierarchy
1
Is the Topography Plan Accurate?
Yes
Maybe
Check survey to be sure.
2
Layout the road system in proximity to existing utilities & where it will best serve ingress & egress requirements.
Locate suitable project access alternatives.
Do the points of access work?
No
Review criteria for project access.
3
Yes
Delineate sensitive areas, set asides, greenbelts, water courses, and all physical constraints.
Yes
Does the road avoid all control points?
No
Review control points and restrictions.
Can the road move without affecting points of access?
No
4
Yes
Locate and design the road in accordance with the geotechnical reports and soil recommendations.
Yes
Have the geotechnical recommendations been followed?
No
Review the geotechnical road design criteria.
Can the road move without affecting control points?
No
5
Yes
Locate road system alternatives that satisfy criteria for circulation, attractive neighborhoods, and economical land development.
Does the road system work from a development perspective?
No
6
Yes
Design and Permit

Figure 8-2

net utilization ratio. Owners who understand the size and dimensional requirements linked to land giveaways and entitlements typically place these parcels where they least impact permitting, development, and profitability.

As a project matures, changes in product type and for-profit land use can also alter net utilization. The goal of course, is to increase net utilization and enhance the odds for profitability. Declining net utilization may suggest a problem with project layout and control, increased demands by permit or municipal agencies, or inordinate amounts of land being lost to redesign and plat amendments. Net utilization can act as a barometer to judge a project's direction and health. While it's important to maintain or maximize net utilization, there exists a point beyond which project changes become meaningless and can actually threaten profitability. The break-even point portrayed in *Figure 3-2* describes opportunities for change in relation to savings. The break-even point for any proposed change can be determined by calculating the cost of that change and comparing it to the value of benefits derived. The result indicates whether it's worth making the change. Trying to increase net utilization is no different from any other change. As a project moves forward, making changes may not be worth the cost. After a point, it begins to cost increasingly more to make changes.

The foundation for net utilization is set early when owners formulate agreements for entitlement and development. Once the upfront damage is done, increasing net utilization later can be difficult and expensive. When setting up a project, always consider net utilization.

Septic systems and roads

In land development, there exist two sure-fire ways to kill a project. The first is to limit sources for potable water and the second is to limit sewerage disposal. For this reason, on-site septic systems are valuable commodities. While they can make unbuildable projects feasible, they can also provide unique challenges to planning, design, and construction. Septic fields are control points and their location can alter road alignment and how land is developed.

Beyond issues involving slope, native conditions, and the presence of water, soils are paramount to determining the viability of septic systems. Soils that are too thin, fine-grained, or wet can limit the ability to install septic systems. On poorly drained ground, lots may have to be reconfigured to protect and maximize areas containing soils acceptable for septic fields. Reconfigurations can come at the expense of access, views, proximity to curb, lot layout, and spacing of structures. When septic systems are the only solution, they assume priority. Whereas uniform and free-draining soils favorable to septic systems make locating roads easier, variable soils can restrict the location of septic fields, which eventually impacts lot grading and road location. Exhaustive geotechnical testing and surveying of primary and secondary septic fields can result in higher than expected development costs while confirming for owners that they may have reduced options for maximizing net developable area.

The delicate interplay between house and septic-field elevation can drive roadway elevations. How? When *gravity* septic field elevations closely match the structure's slab elevations, little room for error exists in running pipe from the structure to the septic field. These conditions may require raising the structure to achieve gravity flow. So it can be said that septic elevation controls piping which affects footing elevations and the location of structures being served. Structural elevations and building location in turn dictate driveway location, grade, and alignment, which directly affect driveways, aprons, and road elevations. In this little vignette, evolving control points based upon septic elevation can find their way "upstream" to road centerline. Road location and, more importantly, elevation, can be totally dependent on septic elevation.

Locating roads horizontally in septic areas can be tricky in other ways. On one hand, it's important to coordinate road and lots grades. This works to reduce overall grading costs while minimizing retaining walls and buildable area lost to long slopes. On the other hand, roads must be located to avoid cutting and exposing soil strata that could intercept septic flow and to avoid placing fills that can impair infiltration or trap, consolidate, and plug in situ soils.

Septic systems in combination with poor soils and broken terrain can work to control the location and elevation of roadways. It's important that owners fully understand the underlying geometry and vertical relationship between each lot's septic fields, location of structure, and lot-line layout before endorsing final road location and design.

Chapter 9

SLOPE ANALYSIS & MANIPULATION

Those who master the art of development on steep terrain understand how topography affects roads and how roads affect grading. If success in land development goes through site-work—as is usually the case—success in site-work usually hinges upon grading, which frequently involves manipulating slopes.

First, I want to emphasize the impact of slope. Look at a map of the world and find Nepal. Now, look at the United States. If the vertical landscape of Nepal were laid out flat, it's been estimated that it would comprise a square area *larger* than that of the United States. Hard to believe, isn't it? But it's true. As slope increases, its surface area also increases.

Here's another related exercise. Pick up a 45-degree right triangle and orient it so that the 90-degree right angle is on the lower right and one of the two equal sides is parallel to the wall. Look at the other equal side that's now parallel to the floor. Land is measured along that horizontal edge—not along the longer sloped hypotenuse edge. In real numbers, a 45-degree slope (100 percent) that has a horizontal short-leg run of 100 feet (30 meters) will have a longer sloped side, or hypotenuse, of 140 feet (42 meters). This equates to a 41 percent increase in length that, depending upon the shape of the land, could equate to a similar increase in square area.

Not recognizing increased area due to slope doesn't mean that it goes away. More ground can mean more of everything including additional timber, dirt, pipe, and surface area to protect from erosion. As slope increases, the potential for error in design and estimating quantities also increases.

Horizontal error due to inaccurate topography

Field-staking performed from office calculations based strictly on a flawed topography plan will pass errors on—like a virus. This can lead to overclearing, overexcavations, higher walls, or less buildable area than expected. Design and engineering isn't tested until staking and, moreover, site construction begins. Sometimes the surveyors can see that staking is wrong or doesn't make sense. If errors get past the survey crew, contractors will quickly notice them when connecting on-site utilities to off-site stubs or when trying to match on-site road grades with existing infrastructure. Afterall, on-site-work transpiring within closed borders is self-contained and can only be measured against itself. The act of tying on-site development to the outside world reveals problems that would have otherwise continued unabated. Unless a contractor begins work from existing infrastructure, connecting *to* existing or matching grade will test design. Owners accessing a project from temporary roads have to be careful not to rely on temporary connections to verify design. It's the future permanent connections that count.

When greater than expected clearing or earthwork occurs, the blame or threat of negligence usually migrates to the surveyors or contractor when, instead, the real culprit may be inaccurate base-topographical information. If the surveyors staked work correctly and the contractor performed work as staked, then examine the topographic plan for problems, such as having too wide a contour interval, being based upon an insufficient number of data points, or possessing wrong base-datum information.

To explore how vertical error can yield error in horizontal measurements, refer to *Figure 9-1.* This graph shows how a one-foot (30 cm) vertical error on a 10% slope can result in a 10-foot (three meters) horizontal error. Looking ahead, *Figure 9-2,* shows how construction staking based on a two-foot contour plan with an embedded (and typical) average vertical error of one-foot can impact clearing, grading, and retaining walls. The relationship is constant. The greater the vertical error, the greater the horizontal error. How things can change!

One solution for combating staking errors that originate from data extrapolated directly from an inaccurate topography plan is to have surveyors "slope stake" cuts and fills based upon actual ground conditions. Slope staking provides precise on-ground control *provided* that the areas staked are geometrically correct and accurate relative to the project boundary and limits of work. But, slope staking isn't a silver bullet. It cannot correct a clearing or grading plan that's fundamentally wrong, misshaped, or errantly sized as a result of erroneous base topographic information. Owners should also take note that not all surveyors know how to slope stake!

Somewhere, error within a topography plan must be reconciled with reality. As alluded to earlier, projects that begin in the middle of the site are more likely to carry error than projects that begin on the border. When off-site grades don't match on-site design, the effects are immediate and corrections follow. On-site-work that is not tied to pre-existing regional control can proceed indefinitely until a problem develops but, by then, it may be too late. Prime candidates for this problem include any job occurring in areas free and clear of relative control, such as the logging and clearing boundaries, locating and constructing remote storm ponds, and installing solitary water crossings.

So, how does an owner make up the difference and get control back on track? One way to make up vertical error is to resolve the difference with walls,

Horizontal Consequences of Vertical Error

Vertical scale is exaggerated

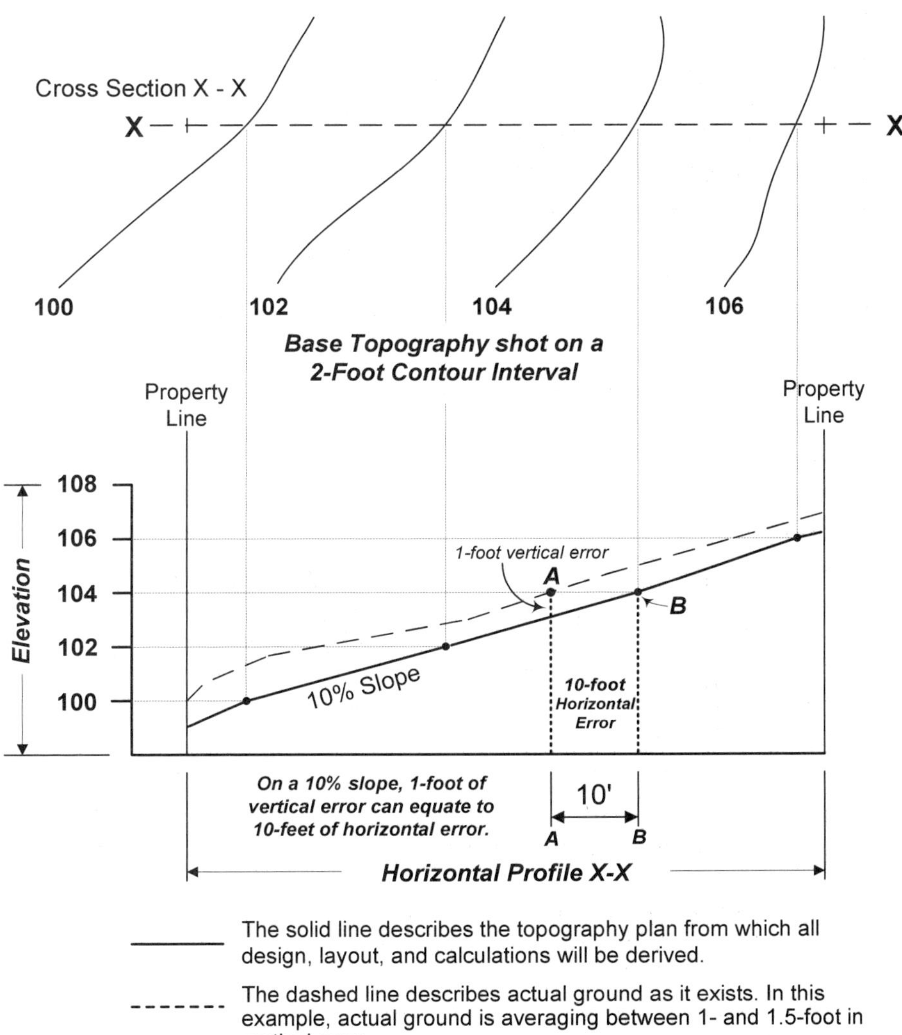

In this case, the contractor or surveyor working from office-derived cut sheets expects to find elevation 104 at Point B, but instead finds out that it actually exists at Point A.

Figure 9-1

Potential Consequences Due to Average Error in a 2-Foot Contour Plan

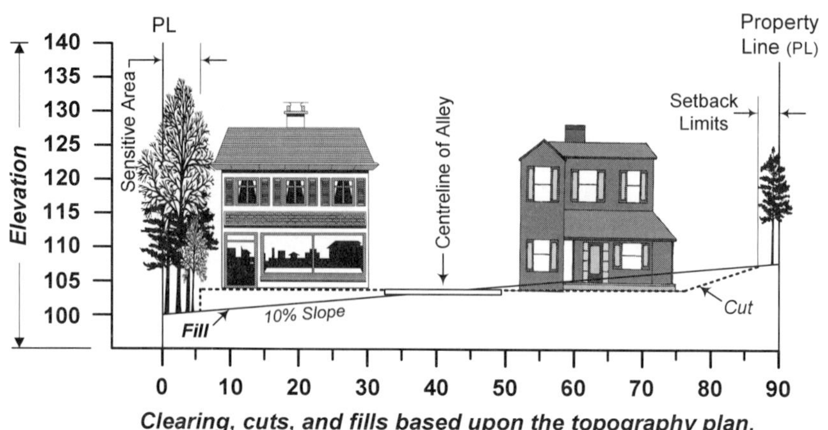

Clearing, cuts, and fills based upon the topography plan.

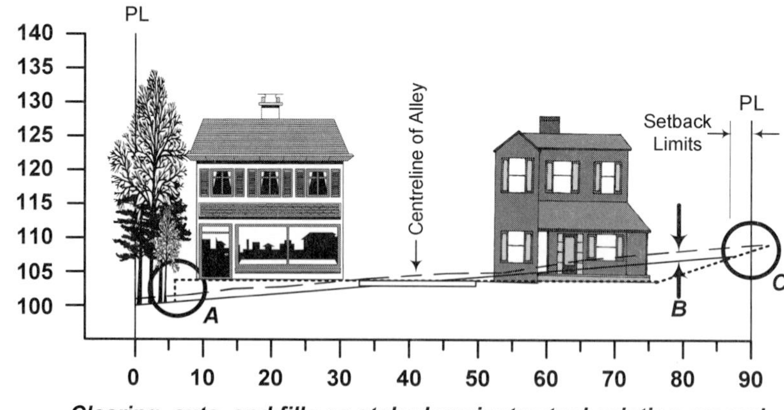

Clearing, cuts, and fills as staked against actual existing ground.

———— The topography plan averaging between 1-foot and 1.5-foot in vertical error.

- - - - - - The limits of grading as calculated from the topography plan.

—— —— Actual existing ground.

Point A - The retaining wall may diminish in height depending upon the depth of overburden required to be stripped.

Point B - Earthwork quantities for cut will increase and fill quantities may decrease depending upon depth of topsoil.

Point C - Unless caught early, overclearing is a real danger. In this example, clearing has violated the setback limit and indicates working beyond the property line.

Figure 9-2

fills, and the like, but, mind you, it's not pretty and can be expensive—but it's a solution. In extreme cases, product type may have to change or owners may have to apply for variances in order to build on steeper-than-expected roads or ground. It all comes back to developing and working with an accurate topography plan. It's that important.

Horizontal error can also create problems beyond busts in clearing and grading. It can negatively impact net-developable area thereby affecting sales and revenue. Less land may equate to loss of net-buildable area resulting in forfeiting prime buildable ground to something else. Costly grading, topsoil import, irrigation, and revegetation may be required if previously cleared ground is to be reclaimed as green space. Other liabilities can include infringing across property boundaries or encroaching into sensitive areas such as wetlands, buffers, and setbacks. Trying to make up lost horizontal area, especially in a traditional neighborhood development and urban environment isn't easy. Dense urban-type projects and small-lot developments usually cannot absorb any loss of area without financial penalty.

Fluctuations in horizontal distance and overall square area resulting from error in the topography plan can potentially impact a project and may result in:

1. Political and permit issues
2. Costs for additional staking
3. Out-of-control grading quantities
4. Mitigating land unintentionally cleared
5. Possible problems with utility pipe grades
6. Additional costs attributed to redesign and engineering
7. Having to push the limits of road design and maximum allowed slope
8. Increased risk of violating boundaries, setbacks, buffer, and land set-asides
9. A narrowed list of interested or capable builders to deal with constricted lots
10. An added burden of longer slopes, higher and longer retaining walls, and deeper cuts
11. Heightened risk of proceeding with smaller and fewer lots or parcels than anticipated
12. Heightened risk and liability in land transactions. Sellers cannot offer documents that represent a set of conditions different from those actually existing in the field. Likewise, this situation can be particularly acute for buyers given incorrect ground information as part of a marketing package when shopping for raw, natural, and uncleared property.

Fast-track grading design specific to product type

Earlier, I mentioned the value in running unexaggerated, real-scale ground profiles for measurement and control purposes. Those same profiles can be put to use in evaluating a given product type and slope as a function of finalizing the grading plan. The prerequisite for utilizing profiles is to tie them into reference points such as roads, property-line setbacks, and sensitive-area boundaries. Afterall, running profiles for grading design purposes without relative control is meaningless. Roads are particularly important because they provide utilities and access to structures. Knowing how grading and product sit relative to roads is critical. With roads in place, product type and profiles can design your grading plan for you.

Begin the process by choosing a scaled section of product (a structure) including foundations that you'd like to consider building on the site. Place the product section against the existing real-scale profile that represents ungraded ground perpendicular to road centerline and through the lot or parcel. If grading is required, lay a profile of the product against existing slope and draw a series of cross-sectional lines through the structure to represent possible grading alternatives.

Compare the product to as many cross sections as the ground will yield. Shift ground profiles through the product at different places. Increase the cut to make a flatter pad. Try sloping the pad. What works and what doesn't? Are retaining walls required? If so, how does wall height and length affect lot cost and sales price? Do walls add sufficient value or should land be sloped? If so, can slopes be cut or filled with minimal loss of land? Or, is a combination of walls and slopes the solution for lowering overall cost and increasing lot value? What about driveway access and overland drainage? Perhaps compound slopes work better than one long continuous slope for this location. How do different slopes interact with road frontage and sight distance? Hold the slab and footings constant to each slope alternative and decide what works best based upon grading, access, gravity utilities, net-usable lot area, and aesthetics.

After comparing product to a number of cross sections against the same profile, a pattern will emerge where certain cross sections work more consistently than others. The favorable cross sections set the control criteria for grading plan design in that area. All that's required is to connect the dots. Run this exercise for each product type considered for a site. Where owners want a flexible grading plan, run this exercise to determine common cross sections that work for a multitude of products and design accordingly.

This approach is an intuitive and economical alternative to spending time and money, and to procuring permits for a grading plan that doesn't complement any product type and, more so, may have to be totally redesigned later.

Multiple product types mandate going through this process numerous times; however, it also gets owners to the optimal result quickly.

It's important to base profiles on topography and boundary surveys that are complete and accurate enough to be useful. Otherwise, this approach is futile. At minimum, profiles should reflect accurate vertical and horizontal locations of all proposed roads and infrastructure. Besides product type, factor in entire streetscapes, underground street utilities, land-use, and views when evaluating alternative cross sections. The greater the number of land uses and product types considered, the greater the number of cross sections required and the more challenging will be the grading design.

One-size fits all minimal grading design

Don't allow a lack of planning or layout lure you into not preparing for grading. Regardless of how far along planning is, every property has a *minimally acceptable grade* that will support a wide range of building types. This may not be an issue on flat property, however, grading alternatives diminish with increased slope.

As discussed in the chapter on road design and diagramed in *Figure 9-3*, the key is to set acceptable minimum and maximum lot-grades per product type. Define this range of grades before laying out roads or performing preliminary grading concepts. Single family, multifamily, retail, office commercial, mixed-use, senior housing, or parks, all possess a range of minimally to maximally acceptable grades suitable for development. Confer with builders, government officials, or potential buyers when setting limits for acceptable slopes. Once identified, set this information as design criteria and use it to help guide road design and project layout.

Project-wide, minimally acceptable grading plans accommodate extremes in slope suitable for all product types under consideration for any parcel or division. For example, consider a site that may be developed as single family or multifamily. Grade to the least expensive and minimally acceptable slope for both products when it's unknown which product will later occupy the site. Boutique and lot-by-lot custom grading can proceed later to match the needs of an individual buyer, product type, or land use issue. What *is* important is providing absolute criteria, such as guaranteed gravity pipe flow and acceptable driveway access, to the lots or parcels. This strategy minimizes the risk of having to move utility stubs, to alter driveway and apron locations, or to change lot sizes and configuration later in order to satisfy sales or changes in layout. Owners' benefit by advancing with a *least-common-denominator* grading plan that serves the greatest number of product types. This allows owners to maintain schedule, undertake basic grading, and pursue the lowest cost grading plan with minimal time invested in iterations.

Figure 9-3 portrays the basic mechanics involved in designing a grading plan. Within *Figure 9-3* are numbers highlighting the major junctures for action. For instance, defining minimally acceptable grade per product type and its relationship to the balance of the decision tree is noted as number 3. Notice the critical interplay between grading and roads. *Figure 9-3* shows how a grading plan ties to planned road locations. Owners can commit to road construction once they verify that grading and road layout will work relative to each other. Conversely, it can be a cardinal error to construct roads in advance of knowing whether or not the road location is going to serve or defeat site grading.

Fitting potential product into variable terrain

As terrain steepens and ground varies, don't create grading plans or decide on product based only upon a minimal number of ground profiles. Working through this issue backwards, a builder, planner, or architect may produce a structural cross section or lot configuration that they would consider building on land that hasn't yet been graded. When compared against one or two profiles, it may appear that the product will work, but that doesn't guarantee that it will work when applied to other slopes across the same hillside. Steep and variable terrain can change infinitely across a project resulting in potentially unpredictable and disuniform grading. Having multiple profiles becomes even more critical when planning larger footprints such as clustered units accessed by an alley or a solo point of entry on steep and variable terrain. Larger footprints and unyielding terrain don't mix without great expense.

Variable terrain isn't the only reason to invest in multiple cross sections before finalizing grading plans. The effect of diverging and converging roads above and below grading can negatively affect how slopes grade out. Exorbitant wall costs and loss of net-buildable area are more common on broken and variable terrain. The deadly combination of variable terrain and a complex road network almost guarantees that a property will retain some degree of steepness after grading. There may also be more long and unbuildable corners than planned. Cross sections provide another perspective on how roads and grading interlock.

Figure 9-4 shows a structure laid against a series of possible cross slopes for the same parcel or division. Laying out product against real profiles in this manner is a productive way to evaluate grading alternatives or to explore ways of fitting product into difficult slopes. Observing *Figure 9-4*, one way to fit structures into the hillside is to mass grade and eliminate enough slope to build product. This approach results in the loss of buildable land to cut and fill slopes

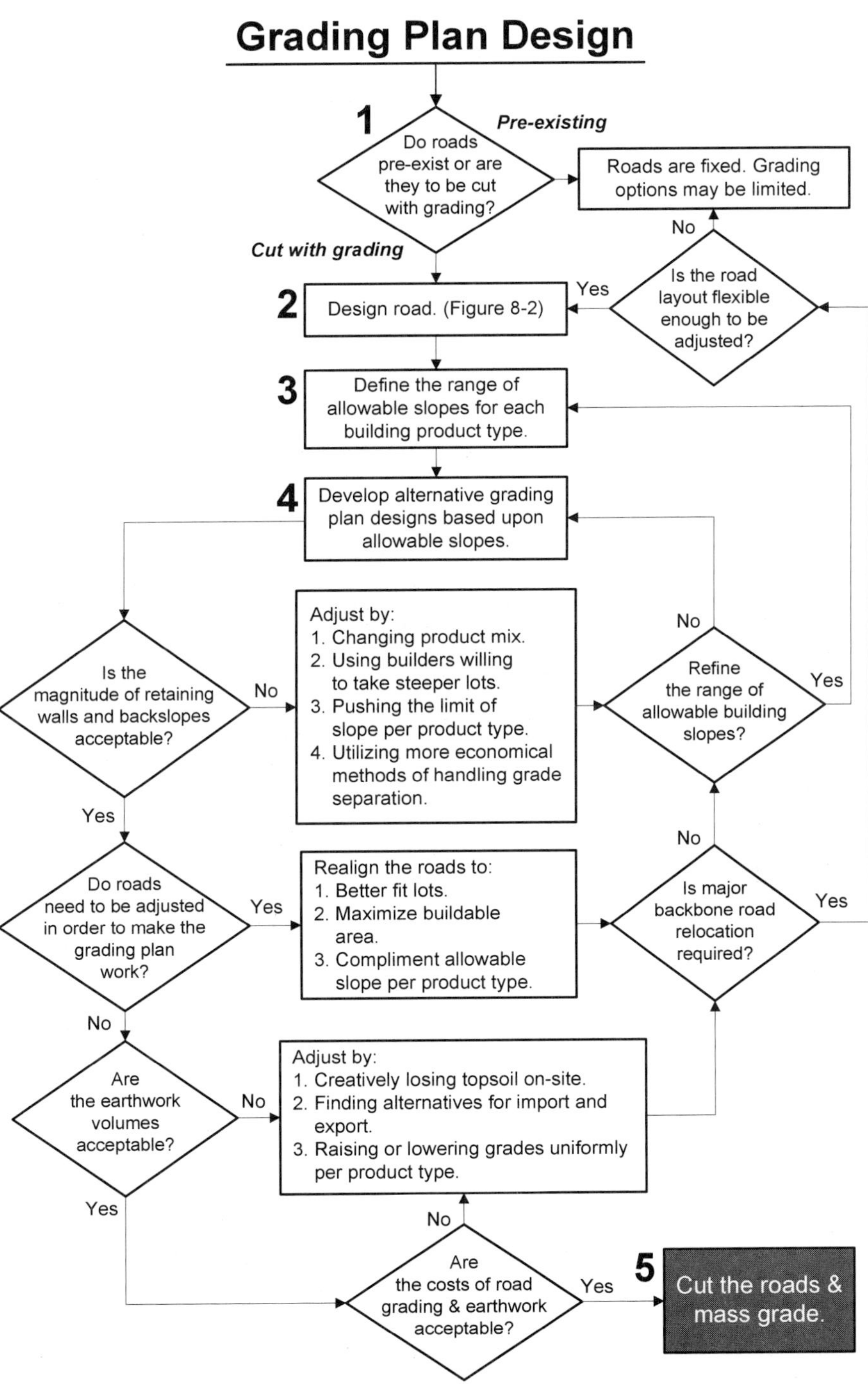

Grading Plan Design
1
Do roads pre-exist or are they to be cut with grading?
Pre-existing
Roads are fixed. Grading options may be limited.
No
Cut with grading
2
Design road. (Figure 8-2)
Yes
Is the road layout flexible enough to be adjusted?
3
Define the range of allowable slopes for each building product type.
4
Develop alternative grading plan designs based upon allowable slopes.
Is the magnitude of retaining walls and backslopes acceptable?
No
Adjust by:
1. Changing product mix.
2. Using builders willing to take steeper lots.
3. Pushing the limit of slope per product type.
4. Utilizing more economical methods of handling grade separation.
No
Refine the range of allowable building slopes?
Yes
Yes
Do roads need to be adjusted in order to make the grading plan work?
Yes
Realign the roads to:
1. Better fit lots.
2. Maximize buildable area.
3. Compliment allowable slope per product type.
No
Is major backbone road relocation required?
Yes
No
Are the earthwork volumes acceptable?
No
Adjust by:
1. Creatively losing topsoil on-site.
2. Finding alternatives for import and export.
3. Raising or lowering grades uniformly per product type.
Yes
No
Are the costs of road grading & earthwork acceptable?
Yes
5
Cut the roads & mass grade.

Figure 9-3

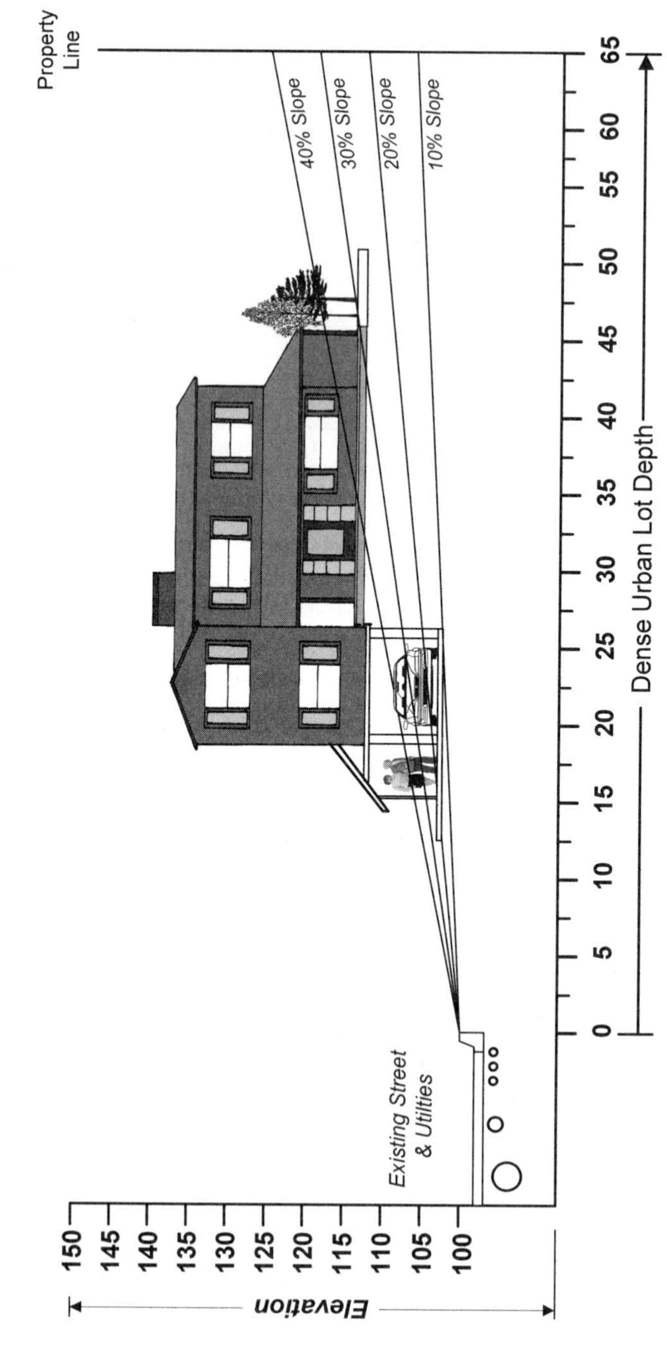
Comparing a Single Product against Multiple Cross Slopes
Property Line
40% Slope
30% Slope
20% Slope
10% Slope
Existing Street & Utilities
Elevation
100
105
110
115
120
125
130
135
140
145
150
0
5
10
15
20
25
30
35
40
45
50
55
60
65
Dense Urban Lot Depth

Figure 9-4

and to move large volumes of dirt. A second solution involves moving the unit in and out of the slope and balancing the foundation and slab with cuts and fills. Shifting the unit in and out of the slope on a lot-by-lot basis according to severity of slope works for owners wishing to minimize earthwork and avoid blanket mass excavation. It can also test setback distances and, visually, misalign the fronts of structures. Either alternative may not work to solve slope retention issues.

Coping with the vertical component of steeper ground is only part of the puzzle. Ground that cross-slopes at an angle from curb line or that inclines contrary to fall line, as portrayed in *Figure 9-4*, creates other dimensional challenges for grading plan designers. For example, how can walls or slopes fit lots in a way that maximizes lot width, remains cost effective, and provides safety for residents without the need of installing costly fencing? How can lots be graded and drained so that runoff doesn't shed down slope into a neighboring parcel? Rear lot lines have to compensate for varying heights in walls and slopes along the sides of the same lot. Unfortunately, the magnitude in elevation changes from one side of a lot to another may not be apparent until well into grading. This suggests a couple of things. First, the budget for walls and grading is going to be off. Next, contractors will end up form fitting and "making it work." Finally, lot fronts may match curb line at a greater pitch than planned possibly resulting in having to change product, garage design, or builders. Answers to compound slope issues lie in analyzing strategies and costs for grading from all perspectives prior to site-work. These types of issues await those who are building dense urban-type small-lot developments on steeper and variable terrain. Profile and cross sectional analysis in all planes can help better understand a property.

Grading plans, earthwork, and opportunity

Earthwork has rendered many a profitable project—unprofitable, and has been the reason why many marginal projects remain—unbuildable. Unexpected stripping, excavation, import, and export costs are usually impossible to recoup. The profit margins in land development just do not allow for it. Although many consider earthwork to be the "Unchained Gorilla" of site construction, it can be tamed and actually turned into a profit center. Opportunity derives from control, and a prerequisite for control is to proceed with a workable and lowest-cost grading plan. Once in hand, resist tinkering with it, proceed with grading, and seek opportunity.

First-draft grading plans are rarely the cheapest and most efficient alternative available. As terrain steepens, it becomes increasingly unlikely that initial grading plans will provide the lowest cost and optimal earthmoving scheme for development. There are just too many interactive factors to consider including alternative road layouts, fluctuating earthwork volumes, maintaining gravity utility flow, comparing costs of walls versus slopes, evaluating net utilization, handling changes in product type, satisfying buyers and builders, and promoting aesthetics. Of these items—at least from a site-construction cost perspective—manipulating earthwork volumes and handling slopes reign supreme in offering owners potential for saving money and accelerating schedule.

Minor alterations to a grading plan can yield dramatic differences in cut-and-fill volumes. Strategic shaving or raising of grades can significantly affect earthmoving volumes and grading costs. There are instances, however where changes to the grading plan will alter elevations and slope but produce only minimal difference in net earthmoving cost. If a number of alternative grading plans work for a project and accomplish the same objective at similar cost, choosing the preferred plan may depend on tactical issues, such as which plan generates more surplus material, or which plan results in ground that is compatible with more product types and allows wider land use?

Choosing the preferred grading plan also entails timing. For example, which grading plan provides an even flow of fill material over a period of years? Or, do any of the plans forestall investing in walls longer than others? It may be wise to capitalize early on off-site dirt sales by overexcavating areas rich in excess and in-demand free-draining structural material.

On projects where dirt stays on-site, develop them in a fashion that strives to balance earthwork seasonally. That is—in one operation—cut, haul, and place material. It can be an expensive mistake to develop cut areas one season, stockpile material, and try to use the same material for next season's fill. Choosing to stockpile and fill later rather than cut-to-fill in one operation results in higher costs, increased construction activity, loss of area to stockpiles, erosion liability, and the risk that the pile will attract moisture and transmogrify into a compacted and unworkable pile of swill. When stockpiling material is the only option, timing affects how and where piles of surplus material should be located for other on-site uses. If dirt isn't immediately needed—especially fine-grained clays, silts, or till—leave it in the ground undisturbed until it can be placed dry and on the heels of excavation.

Another option for owners is to offer the world a dumpsite. Do any grading alternatives lend themselves to overexcavation where excess material can removed, sold, or used for construction elsewhere while, at the same time, creating a depression suitable for fill? Dumpsites offer year-round places to lose material and can generate a stream of revenue from dump fees. When other dumps close or refuse to accept wet slop, an on-site dump is a handy place for owners to lose organic, wet, and non-structural material. Also, evaluate divisions or parcels that sit low and can use fill. These areas provide excellent places to lose slowly over time wet swill and other structural dirt.

Having an on-site pit supports the theory that it's better to be in a net import situation (needing dirt) than in a net-export situation (having a surplus of dirt). Why? Most projects need a place to get rid of excess material. In the right market, developing a hole in a non-structural area that can accept topsoil and other unclassified material can be very lucrative for owners. On-site builders on larger projects need a place to dump waste material from utility work, foundation excavation, and grading. That place could be your next division or parcel requiring fill. While projects short of material can charge dump fees around the clock, and possibly improve land at little or no cost, having too much dirt yields owners fewer options. Their goal is to export material the cheapest way possible and that could very well be to another developer's on-site dump. It takes time to procure permitting, set up accounting, and develop a system for billing, but creating a pit is worth the effort. If generating additional revenue isn't a possibility, then owners must work to achieve balanced earthwork since there's only *one* option with surplus dirt and that's typically export. For more information on pit development, visit the Chapter 16, recycling.

The term "balanced site" is a misnomer fluently used in land development. Sites rarely balance. If and when they do, it's more by mistake or by an edict to prevent export than it is by design. The odds of cutting a project perfectly and ending up with just the right amount of structural fill to meet exact design grades, or even come close, are low. Topsoil, overly optimum structural material, shrink-and-swell factors, grading waste, and trench displacement are some of the many factors that can render volume calculations inaccurate. At best, they're approximations. And—that's before the builders cut foundations, extend lot utilities, and complete fine grading. Balancing a site is more inclined to mean leaving some areas slightly lower than design due to a lack of fill or a desire to curtail cuts to prevent overgenerating material. Balancing earthwork on multiyear projects where site work continues through wet weather or where structural cut-and-fill volumes fluctuate with planning is even more remote. How, for instance, could an owner ever accurately calculate excavation volumes if product type fluctuates with market conditions or, if grades are to remain flexible for future buyers and builders? When asking consultants to balance a site, understand what that means as well as what might have to happen grading-wise in order to produce similar results.

Septic systems and grading

Simply put, septic systems restrain grading. When the economic viability of land hinges on finding suitable soils for drainfields, septic can drive grading, layout, and, as mentioned in the previous chapter, road location. Suitable drainfields, which exist randomly with no respect to lot lines or boundaries, are control points for development. To counter this, owners must set priorities on, first, how lot layout will be determined in accordance with septic locations. Then, other considerations follow. Here are some examples of priorities that can influence how and where lots are configured in "septic country":

- Option 1—*Road location is priority.* As allowed by septic location and elevation, building sites are secondary to optimal road location. In this case, roads take priority over building sites.
- Option 2—*Lot count is priority.* Reconfigure the lots around acceptable septic fields in order to maximize lot count. In this case, septic drives lot layout and roads follow.
- Option 3—*Flexible lot layout and home sites are priority.* Forget about maximizing lot count and instead, pursue higher quality lots. Place lots where they best fit the land and complement structures. In this case, septic has less influence over larger lot layout and road design.
- Option 4—*Views reign supreme.* Forget about lot count and lot layout. Orient and group lots in order to maximize view potential. Reach for septic if necessary. Roads go where they can in order to serve view lots.

Naturally, any of the above criteria may be mixed and matched or amended depending upon a project's particular character. Septic systems also influence the location of haul routes, restrict logging and clearing, and dictate where fill can and cannot be placed, and how embankments channel water across undisturbed areas.

Changing lot configurations

Changing lot configurations can influence grading plans. Depending upon market conditions or as requested by builders or regulators, land use might change between low, medium, or high density or lot use might change between single family and multifamily or between retail and office. Such changes can profoundly affect grading, slopes, walls, utilities, and infrastructure. Areas that are already paved and landscaped are more costly to reconstruct. Reconfiguration typically impacts shallow utilities. Deeper utilities able to serve various configurations and pressured utilities, such as waterlines, tend to be more forgiving.

Vacillating between land uses, drawn-out builder negotiations that freeze site-work, and ongoing lot manipulation is particularly troublesome to dry-utility providers. Dry-utility system designers must know all the facts well in advance of formulating a work plan. When loadings, type of facilities, density, and other on-site details remain in limbo, service providers shelf work and concentrate on other jobs. And, who can blame them? Owners must vehemently work to avoid this. Without dry-utility construction drawings, utility staking cannot happen, dry-utility crews cannot be scheduled, and lots don't get service on schedule. Severe changes in on-site planning may also trigger the need for additional land to place hand holds, switch gear, transformers, and pedestals. Remember that on most projects, dry utilities are a hidden long-lead item and that service providers work at their own speed. Besides, nobody is happy when vaults, pedestals, and transformers double as entry monuments.

More helpful strategies to consider when developing grading plans

1. Given a choice, it's better to be dirt poor than dirt rich.
2. Smart grading plan design pays off in many ways including being easier to permit.
3. Pursue buffer averaging around sensitive areas to improve the shape of lots and parcels and to maximize net-developable area.
4. Don't look for water. Minimizing cut reduces the potential for finding hidden seams of perched or flowing water. The more you dig, the more liable you are to encounter water.
5. When planning park grading, remember that land too steep for multi-family housing will also be too steep for a managed park. Ground that is suitable for most non-revenue producing amenities such as parks is usually ground that could also be developed for a profit.
6. Require your engineer or contractor to run cut-and-fill volume calculations on every grading alternative and to color-code cuts and fills. The deeper the cut of fill, the more intense the shading. Colors make it easier to understand the magnitude of grading and both buyers and builders appreciate the graphics.
7. Use open space. Place parks adjacent to and in front of greenbelts and sensitive-area buffers. This strategy adds depth to parks while avoiding having to plan grading and lots around irregular boundary lines and difficult geometry. It also makes park area appear larger while capitalizing on the visual use of unproductive open space.
8. Geometry. Always consider geometry. Long corners, bulging tree lines, and narrow areas look great, but they're tough on grading, deceptively difficult to wall, and can kill developable area. Builders want nothing to do with odd or poorly laid out horizontal or vertical geometry. Points that are inordinately high or low can cost more to grade and

they inhibit surface drainage.

9. As terrain steepens, design difficulties and expensive road construction and grading makes using foundations a viable option for overcoming obstinate ground and irregular terrain. However, depending on the value of what's being built, the window for investing in foundations as a means of compensating for harsh terrain is typically narrow beyond which, development doesn't pay.
10. Remove the topsoil "wildcard". On projects destined for mass grading that have a consistent depth of native topsoil, consider developing grading plans without topsoil. Just remember to budget for stripping and, if needed, for importing some of the same raw topsoil back into select areas. Topsoil is useless for structural purposes so including it in usable cut-and-fill calculations will later lead to a costly earthwork mistake.
11. Make the grading plan appear *less* intense. Grading plans can be made to appear less formidable when shown or submitted with a wider contour interval. On steeper ground, wider contours may appear less obtrusive than tight and numerous contour lines and, for some, easier to understand. For instance, a five-foot contour interval plan is less intimidating than a one-foot contour plan of the same ground. Plan reviewers may find a simple and uncomplicated plan more palatable to work with than a more complex and steeper looking layout, even though both plans result in exactly the same slope and grading.
12. Take what your project's outer-boundary control will give. Developable land common to control points or fixed buffers, such as wetlands and waterways, usually translates into irregularly shaped parcels, inefficiently sized lots, and tough lot configurations. On some projects, it makes sense to plan layout, design grading, and pursue road design from the outer edge in, instead of from a central road location out. This approach involves confronting ragged and fixed edges, such as sensitive area buffers and property boundaries, first—taking what they'll give. The result is a road system and grading plan that accommodates edge control while leaving the softer, inner parts of your project to absorb and make-up whatever irregularities remain. It's surprising how much buildable area can be picked up by using this technique.
13. Resist unrewarded re-engineering. Not every buyer, builder, or big-box retailer will think that your base-grading plan will work for their design or product, and that's okay because as a speculative land developer, it's not possible to design and build for everybody on the first cut. The result is that a parcel may require some additional engineering. And that, too, is okay, however, forcing redesign in response to a buyer's particular building product without promise of a sale or reimbursement can prove expensive and can cost more in time

and expense than the exercise is worth. Things can get further out of hand when your planner or consultants decide with a buyer to ignite redesign efforts without your blessing or knowledge. Taken to an extreme, re-engineering can result in changing road locations, moving utilities, and all kinds of counter-productive activity.

14. It's easier to surround building envelopes with dedicated, fixed-width native-conservation easements on flat ground than it is on steeper ground. As ground steepens, owners wanting to maintain fixed-area or fixed-dimension conservation easements on the same-sized lots may have to make choices. Here are some options:
    - Change product type.
    - Grade out clusters of building pads and consolidate open space.
    - Fix building pad size and terrace lots with cut-and-fill retaining walls.
    - Fix building-pad size and compensate for slope with more expensive foundations.
    - Abandon or make-flexible, the use of conservation easements above a certain percent slope.
    - Create split-section building-pads. Use cuts, fills, and back slopes to establish building pads consequently reducing areas of disturbed ground.
    - Similar to wetland buffer averaging, vary conservation easement widths as necessary in order to place fixed-sized building pads on the mildest, most sensible terrain per lot. This only works when easement widths are flexible and where buildable terrain exists.

15. Density on steeper terrain mandates using precise grading and wall plans based upon accurate topography. The combination of high density and increased slope can lead to problems for owners solely relying on retaining walls to compensate for differences in lot elevation. Density results in smaller backyards, setback and boundary restrictions, and, with increased slope, a reliance on walls and daylight excavations in order to make things work. Under these conditions, inaccuracies in topography or an unrealistic grading plan may result in:
    - Grasping at solutions to lower wall costs. Last minute actions such as raising grades or steepening cut-and-fill slopes in order to lower walls that separate lots or parcels, can produce drainage issues if runoff passes storm and yard drain systems and ends up in structures or running across lots.
    - Unexpected foundation work. High density nestled into steeper slopes may leave little or no tolerance for inexpensive solutions should problems related to product type or foundations develop. It doesn't take much error in grading to result in lots that are overly

expensive to build on or difficult to access.

- Mishandling wall runoff. A classic cost-saving maneuver reduces or eliminates the washed rock drainage field and piping normally designed for the bottom face of non-cast-in-place walls. In this application, the purpose of washed rock is to collect and convey drainage that is seeping through wall systems. Developers who skip this detail must think that landscaping or topsoil will absorb all excess runoff. Water that is allowed to flow from a wall directly into a backyard or landscaped buffer, will eventually course it's way downhill towards whatever is below, which is usually a structure, driveway, or parking area. The result is a waterlogged site, waves of overland runoff, or a wet crawl space. Owners may have to invest in French drains and other expensive post-construction solutions or possibly face charges of negligence.
- Problems with retaining walls planned at grade breaks and along lot lines. The allowable margin of error on sloped ground where walls separate smaller lots usually doesn't allow walls to shift in order to better match terrain, to reduce wall size, or to maximize lot area. Owners wanting to limit the magnitude of walls find themselves entrenched in constraints around minimal lot size or shape, safety, cost, and aesthetics. Furthermore, a dense lot layout based on an inaccurate grading plan leaves little room for modifying walls later should an owner seek to improve overall development. In fact, modifications can worsen the situation. Ultimately, something isn't going to be built per plan. Lots may end up steeper, walls may end up higher, gravity sewer may have to be pumped, or fencing may have to added. In extreme cases, grading costs can spiral out of control leaving the owner with a finished product that is too steep, unsaleable, or that has unsafe and unattractive retaining walls.

Chapter 10

TRAILS

Projects commonly provide trails for non-motorized use in non-right-of-way areas. Whether intended for pedestrians, horses, or bicycles, designing and installing trails across tame and open country is relatively straightforward. The rules, however, change with increased topography, across sensitive areas, and as foliage thickens. In this book, we'll only cover trail planning and construction as it applies to tough country with dense vegetation.

The first step in conquering trails is to think about them. Yes, I know, it's hard to believe, but trails are more than a passing fancy. In steeper foliated terrain, properly establishing them requires more work than most owners realize, that is, if owners want a usable and quality trail system where trail elevations and locations work with pre-existing parking and infrastructure.

In the realm of building communities, trails don't produce revenue. They aren't a priority and usually come into existence towards the end of a project after budgets become inelastic. Faced with shrinking revenue and growing debt, an owner may look to trails as a place to cut costs only to discover that trail standards and specifications typically aren't pliant. Looking at it from the perspective of a permit authority or homeowner association, why should they forego quality trails to compensate for a lack of owner planning or to improve an owner's bottom line?

So—for owners—herein lies the first issue. Trails shouldn't be an afterthought. They need to work from both an engineering and cost standpoint. Waiting too long to address trails may result in squandering opportunities to build the best trail system at the lowest possible price. Site construction as it relates to trails entails focusing on the following areas:

1. Trail terminus points
2. Construction methods, materials, and surfacing
3. Trail location, earthwork, and retaining systems

Trail terminus points are almost always slave to existing infrastructure. Logistical and permitting issues arise when a trail doesn't end as planned, or when it fails to terminate at existing roads or parking lots. An example would include a trail of fixed length planned in order to negotiate a water crossing, to meet wheelchair requirements, and also to match fixed coordinates of a parking lot or road. Proceeding without proper planning may result in a trail that is unable to meet all three conditions. That is, it may not cross the creek as designed, stay within maximum grade, or terminate at predetermined trailhead locations. Sure—it's possible to randomly meet all three criteria, but I wouldn't accept the odds for myself or on behalf of a client.

Locating trails in tough terrain requires more than a passing glance or the talents of an inexperienced planner. In fact, one could make the case for locating a section of project infrastructure (streets and sidewalks) solely to serve trail locations. For this reason alone, investigate trail locations early during planning to determine how they link relative to planned infrastructure. Designers need to know trail locations in order to design trailhead connections for proper drainage, traffic speed, parking, and sight distance. They also must ensure that trails merge correctly and at the right angle with infrastructure. This includes providing ample and safe trail run-out area, room for message boards and kiosks, and trailhead access controls such as fences, gates, and bollards.

Trails expected to traverse broken and steep terrain deserve the same level of care and engineering as roads. But engineering alone is too rigid for locating graceful and weaving trails. One obvious solution is to rely on local wildlife. What routes are critters using? Using existing game paths is one of the smartest ways to locate a trail. Animals know how to travel with the least amount of effort. They also know the quickest routes for viewing, avoiding predators, and negotiating uneven and stubborn terrain. Abundant game trails are a blessing offering multiple options for use. If it's possible to take readings, invest in having someone walk and map game paths with a handheld GPS receiver. Plot this information onto base topography *before* laying out trail alternatives. Then, field-truth trail locations, test connectivity, and evaluate grades in and out of the project.

Next, as always, take what the land will give but don't take too much. Here's a list of suggestions that will help when laying out trails:

- Avoid damp areas.
- Always consider features that will attract or repel trail use.
- Don't allow trails to lose or gain elevation without purpose.
- Lay out trails after the leaves have fallen when visibility is greatest.
- Wider trails are easier to manage and are less likely later to be overgrown with vegetation.
- Consider pruning. Adjacent and remaining trees will require trimming and possible disposal of limbs.
- Don't cross roads on an angle. Trails that cross perpendicularly to roads are safer, easier to see, and more user-friendly.
- Don't concentrate elevation gains or loses. Instead, design trails that are

steady and uniform possessing minimal "grade shock."

- Avoid areas of concentrated, loose rock. These areas may indicate a potential safety hazard or a probable ongoing maintenance issue.
- On already steep sections of trail, avoid straightening alignments since such action will shorten the trail contributing to yet steeper grades.
- Closed loops with single access avoid "backtracking", better distribute users, and allow more trail run to be established in areas of limited space.
- Switchbacks that are too tight or intervisible encourage hikers to take short cuts. While this may be the only option available on steep terrain, switchbacks should be as wide and far apart as possible.
- Areas of blow down may indicate perpetual maintenance headaches. Before locating a trail through a blow-down area, determine the reason for the blow down and evaluate whether or not future blow down can or will affect the trail.

Lastly, visit the site with a preliminary trail plan laid out against existing topography. Does the topography plan used for trail design match on-site topography or are there discrepancies in elevation between open green space (trail area) and developable areas? Check elevations according to road stationing. On paper, do trailheads connect with infrastructure as they should? If they don't match, it may be due to differences in field topography. Frugal owners pay for accurate field topography, tighter contour intervals, and profile checks *only* on land destined for grading and development. Unless trail locations are known beforehand, it doesn't make sense to pour money into developing accurate field topography that may be of no use later. Trails that fail to connect correctly with roads may indicate that the local topography plan is off or that the road hasn't been located exactly per plan. Verify that trails connect as expected with infrastructure before committing time to further trail location and design.

Upon confirming trailhead connection points, field-compare possible trail locations with actual on-site animal game trails. How closely do they match? Whole or portions of game trails that work with reliable trailhead locations become control points from which the remainder of the trail can be located.

Locating trails requires someone to slap on a pair of boots, walk the ground, and explore all possible route alternatives before committing final alignments to paper. The person performing this exercise must be familiar with using a compass and perhaps an inclinometer, must understand how to flag centerline, and must be skilled enough to ensure that he or she returns for dinner that evening. (Hand held GPS is great but heavy tree canopies and problems with accuracy can negate their effectiveness.) Proceed with design and engineering once a preliminary route has been field-verified to work.

Unlike typical project design and layout, owners should understand that trail designers might at times be flying "solo." Unless they walk the route, most municipal plan and permit reviewers won't be able to visualize how the trail sits in context with actual ground conditions and, therefore, will be unable to offer constructive feedback. Trails are one of the few items in land development where plan review and permitting can occur quickly. But—buyer beware. If your permit agency isn't scrutinizing layout and design, your trail planner had better be right.

Choosing materials

Although many factors influence cost-effective trail construction, few are more important than the type of materials specified for surfacing. Here's a listing of potential trail surfacing materials and their relation to construction. Bear in mind that the following descriptions are general in nature. Length, grade, width, weather, conditions, and type of soils can dramatically affect each material's application in trail construction and its cost.

1. Dirt—Existing dirt surfacing is cheap and dirt trails are the cheapest to maintain. Issues involve aesthetics, dust, mud, and problems with sprouting forest vegetation. As expected, native dirt is the easiest trail surface to construct upon.
2. Organics—Wood chips and nutshells are light and the most economical trail-surface material to transport through restricted green space. They rake out easily and that's good because until these materials pack down and start to decompose, they're prone to displacement and being kicked everywhere. This type of trail material is forgiving, goes down easy, and cheaply covers shortcomings in subgrade. Unlike dirt, organic material added as trail surfacing may require shallow sideboards to hold it in place and shape the trail. Issues involve the tendency for material to dislodge and be wasted, to resist compaction, to shape irregularly, and to have to be reapplied every couple of years. Certain products may not be available everywhere.
3. Gravel—Rock and gravel possess advantages similar to organics with a few differences. Namely, it's more expensive to purchase, it's heavier to transport, but it lasts longer than organics. Of these differences, rock and gravel weight can affect trail construction the most due to the need to haul material into place. Vehicular access can elevate trails standards to that of roads if subgrade support and additional width is

required for hauling rock. Upgrades to an otherwise dirt or organic type trail may entail additional clearing, rebuilding subgrade, exporting material, using fabrics, implementing or increasing the compaction standard, revegetating shortcuts to the trail, and the possible need to replace culverts, bridges, and pipes to support trucking. Most permit agencies don't envision building a highway in order to construct a trail. Owners shouldn't either.

4. Asphalt and concrete—Hard surfacing almost always requires stricter trail standards and subgrade specifications. Although some permit agencies may not require structural subgrade soils and compaction to support hardscape, investments in concrete and asphalt mandate placing material on suitable subgrade and fabric and adding a grading course that can double as a capillary break. In terms of remote trail location and construction, here are some characteristics of asphalt and concrete worth pondering.

Characteristics of asphalt-trails construction.

- Asphalt goes down in one pass and can quickly accept trail traffic.
- Being dark, exposure to the sun allows asphalt to absorb heat and gradually melt snow and ice.
- Asphalt can be placed manually on narrow trails, however, problems may persist in transporting and placing hot mix without using a screed.
- Trails are typically paved using small paving machines. This may equate to higher in-place prices per ton or square yard compared to street paving costs.
- Use of a paving machine means maintaining some level of production and a subgrade suitable for uniform and proper placement. This may require building a trail wide enough to accommodate small equipment capable of grading as well as accommodating trucks delivering hot-mix asphalt.
- Asphalt tends to flex and adjust with minimal subgrade movement. It's more forgiving than concrete but less forgiving than any of the loosely laid products described earlier. Use of flexible pavement material may be advantageous on trails where subgrade excavation, compaction, and grading cannot be performed as thoroughly as in easily accessed road work.

Characteristics of concrete-trail construction.

- Although easily controlled, wet concrete can bleed near water or in excessively wet areas.
- Concrete must be protected from the weather and traffic while it cures.

On trails, this means critter control.

- Unless an expensive slip-paving machine is used, concrete trails will need to be formed by hand and stripped after the concrete hardens. Concrete placed by hand can result in a rougher surface than machine-placed concrete.
- Concrete can "bridge" minor subgrade settling. Although this may at first appear to be a positive attribute, the amount of "bridging" that concrete can perform before failing is limited, depending on the size of the underlying gap, the strength and thickness of the concrete mix, and the amount, if any, of steel reinforcing used.
- Concrete mix must be transported to the trail for placement, however, unlike asphalt, wet concrete is flexible and can be pumped. Depending upon distance and elevation gain, if a hose can reach the point of construction, it may be possible to deliver enough concrete mix to build a trail. Consult with a concrete contractor in your area before committing to concrete trails.

Asphalt and concrete trails create an edge that will have to be matched with fill (topsoil) and then vegetated or seeded. When building projects a distance from processing plants, it can be a challenge to deliver concrete and asphalt in a workable state. Additional time and handling that may be needed to get mix from the delivery truck to the trail can further stress the integrity of concrete and asphalt mix. Weather and temperature also affect placement making scheduling work a priority. A more comprehensive look at the characteristics of asphalt and concrete in construction can be found in Book 1 of the Project Logic Series titled *Road Repair Handbook*.

Owners working through development agreements and entitlement should commit to trail construction standards only after carefully reviewing the options. Impacting minimal area during trail construction means that less time and money will have to be spent later on additional design work, revegetation, and remediation. Constructability issues and the consequences of proceeding with an ill-advised design or of working out of season can cause trail building costs to multiply. When constructing trails through open space, pursue simple trail layouts, amenable standards, and materials that transport and install easily.

Other trail considerations

- Efficiency—Aim trails at already cleared, graded, and maintained corridors such as utility easements.
- Drainage—Provide grade breaks to expel water runoff. On more even ground, rolling grades in combination with sufficient side slopes

provide interesting sight lines and prevent water from concentrating on the trail.

- Trails need variety and suspense—Exploit vistas and open areas as well as hidden jewels like boulders, old stumps, and mossy areas. Instead of removing trees, it may be cheaper and more effective to prune them. Another cost-cutting scheme involves dressing up trail terminus points with rocks. Trailheads are wonderful places to decorate with large and unwanted boulders wrenched from excavation.
- Trail design should complement use—Question design efforts that appear stalled, produce diminished results, or lack focus. For instance, if a trail is to accommodate the elderly, the design should focus on rails, grade, rest areas, and trail length. Interpretative trails displaying signage and centers for education should be located near specimen trees, plants, or water. Horse and equestrian trails have their own set of preferred specifications. Remember that trail purpose influences location and design.
- Control degradation—Consider ways to prevent trail degradation including potential abuse from motorized vehicles. Designers can accomplish this, in part, by minimizing the footprint needed for construction. Through design and construction management, control damage to tree root systems, incidental compaction, and erosion. Create a planting and revegetation plan that promotes interim growth before native species can reestablish pretrail conditions. After initial hand seeding or hydroseeding, native species will eventually establish themselves along the trail edge and survive maintenance-free.
- Anticipate environmental constraints—Strike before costs escalate, environmental restrictions become onerous or prohibitive, or public opposition stalls construction. Although it may not appear to make financial sense, it can make political and tactical sense to construct trails through open space or sensitive areas once permits are in hand. This becomes imperative for approved *portions* of trails that cross drainages or wetlands. Where evidence suggests that "scope creep" is happening to other phases of a project, accelerate trail design and permitting to lock into something predictable and within budget.
- Retaining walls—Wall systems are expensive but owners can take steps to help defray their cost and promote a successful trail. Consider fabric-wrapped retaining systems that can utilize waste materials including topsoil and uncompactable inorganic soils. Retaining systems that employ on-site organics provide both a dumpsite for construction and woody debris as well as a medium for establishing vegetation. Gabion or wire baskets may work well on projects rich in easily

transportable rock. Substitute logs and cribbing that naturally resist decay, such as cedar, for wood types that will quickly rot.

- Compaction—Unless justified, avoid transferring road-compaction specifications into the trail standards. Agreeing to attain compaction in areas that are too narrow or remote to move heavy equipment into, may not only be unwise, it may be next to impossible to honor. Compacting native inorganic soils implies removing topsoil and replacing it with structural material. How will contractors accomplish excavation and compaction? Without knowing trail design, how can anyone agree to backing trucks in, emptying them, and sending them out reloaded with waste material? Lastly, has any geotechnical work along the trail been completed that verifies soil conditions? If not, how will the trail designer, engineer, or permit authority know the condition of in situ soils? Dismiss unattainable and, more importantly, unneeded trail-compaction specifications early.

Don't buy into expensive trail obligations as a basis for obtaining project approval without first evaluating the consequences. Not all trails are buildable as initially proposed and the choice of materials used in trail construction can greatly affect budget and schedule.

Chapter 11

THE GEOTECHNICAL WALTZ

Surprises, particularly in dirt, aren't rare; but when they occur, they can send a project team into disarray complete with finger pointing, bickering, and increased tension. Mix in some poor weather and a dash of schedule impact and soon the owner, consultants, and contractors may end up at enmity with each other. What's the fuss about?

It revolves around the word "*geo*" meaning earth. In the context of site-work, it's linked to your project's soil or—the lack of it, or too much of it, or the wrong type of it, or it's being managed poorly, or it's too wet, or it has been incorrectly specified in the plans, or something else related to earth. Something that is germane to grading, excavation, or utility work is wrong and it's showing up, cloaked in red ink. Mismanaging soils is common in site-work and it leads to a galaxy of problems; which introduces us to the next consultant who specializes in studying and working the earth—the geotechnical consultant.

By choosing the right geotechnical consultant, owners can make their lives much easier. An engaged geotechnical consultant will enhance owner control, provide ample and reliable base information for setting accurate budgets, recognize early warning signals in both the field and in design, and increase the odds of meeting or beating the same budget and schedule. Few owners realize how important this decision is until they get burned, but by then—it's too late.

Acting as a geologist, a geoscientist, a hydrogeologist, or a consulting geotechnical engineer, the geotechnical consultant can perform a wide array of tasks. In terms of everyday site construction, a full-service geotechnical firm can usually supply at least the following services:

1. Value engineering
2. Slope stability analysis
3. Foundations and pile design
4. Laboratory soil analysis and testing
5. Subsurface soil and rock exploration
6. Engineered earth and retaining-wall design
7. Instrumentation installation and monitoring
8. Subsurface-water measurements and control
9. Construction monitoring, documentation, and inspection

10. Recommendations for protecting exposed soil and stockpiles of material
11. Recommendations for construction specifications, plan details, and procedures
12. Historical surveys of prior land uses and conditions as recorded in aerial photographs

More specifically, geotechnical consultants are key players when it comes to foundations, surcharging soils, drilling, piling, shoring, caissons, soil-movement studies, seismic engineering, percolation testing, subsurface-water monitoring, and sensitive-area analysis—just to mention a few. Regardless of the task, geotechnical consultants offer services aimed at demystifying and working the earth.

Risk

But before proceeding, it's essential that we discuss the topic of risk as it relates to earthwork, grading, utilities, and soils exploration. As an owner, how much risk are you willing to accept? Depending upon the magnitude of earthwork and soil variability, owners naturally assume a degree of risk. But how much risk is tolerable and, more importantly, at what cost is risk worth reducing? As an owner, how much more are you willing to spend for predictability? Before answering this question, understand that risk as it relates to dirt and site construction can include at least the following:

- Developing unstable ground
- Encountering soil conditions other than anticipated
- Encountering and handling unexpected subsurface water
- Paying the price for a poorly-managed earthwork contract
- Encountering unexpected quantities of bedrock or boulders
- Creating ground instability on your or your neighbor's property
- Working moisture-sensitive soils under cold, wet, or dry conditions
- Excavating unexpected contaminated or unclassified material, or a previously filled area
- Miscalculating earthwork volume based, in part, upon lack of sufficient exploratory data
- Designing or engineering with limited or poorly characterized geotechnical data and under relatively unknown conditions

While it's possible for a project owner to fall prey to any of the above circumstances, risk when working soils can be controlled. How? From the beginning, choose the right geotechnical consultant, consolidate geotechnical activities under that consultant's control, and invest in a thorough program of soils exploration and mapping. We will take each one at a time.

Choosing the right geotechnical consultant

Everything as stated in Chapter 1 concerning consultants holds true for the geotechnical consultant but with a slight twist. That is—owners should employ a geotech who is understanding enough and flexible enough to pursue design within a corridor that is defined on one side by onerous, unnecessary, and overly conservative recommendations, and on the other side by irresponsible or negligent consulting work. One side of this corridor is prone to produce unnecessary costs, and the other side of the corridor, in spite of possibly saving an owner money, may inflict incalculable damage by placing a project at-risk and making owners liable for shoddy, incomplete, and porous geotechnical work.

But, choosing a geotechnical consultant involves more than considering cost or risk. A geotechnical consultant needs to be sensitive to and smart about potential impacts that his recommendations can have on overall project schedule. Before proceeding, let me be clear. I am not espousing or condoning the use of geotechnical consultants who may cut corners, exercise poor judgment, practice flawed engineering, leave owners dangling, or promote substandard techniques. However, project owners, both public and private, should not, and need not, incur the extra costs of overbuilding when it's not necessary.

Well then, how do owners get the right geotechnical consultant? Begin using the techniques described in Chapter 1, "Choosing Consultants." After interviewing prospective firms, recognize and react to early warning signals that may indicate that a firm is short on resources and creativity, interest, or talent. Is the geotech unsure about handling the scope of work or the project? Certain types of projects may attract or repel geotechnical firms depending on whether or not that firm has been the subject of a lawsuit or targeted with charges of negligence. Similarly, geotechnical firms that have a history of negative experiences or general inexperience may compensate and shed responsibility early by limiting their scope of work. Or, they may simply specify unreasonably high construction standards that cannot possibly fail even if the earth collided with another planet—regardless of cost or convenience to you, the owner.

In addition to the tips offered in the chapter titled "Choosing Consultants," here are signs to look for early on that may suggest that you're not be dealing with a geotechnical firm geared to your project:

- The lack of in-house laboratory facilities
- No seasoned construction field inspectors on the payroll
- Firms that bring an entourage of people to every meeting
- Firms that constantly change their name, ownership, or scope of work
- Firms that issue studies when a report will do and reports when a memo will do
- Firms whose portfolio suggests monotone solutions to earth, slope, or wall problems
- Firms that push metric regardless of your basis for funding or your project objectives
- "Geotechnicrats" who appear lofty, unreachable, or that exhibit a Father-knows-best attitude
- Even with adequate notice, firms that cannot totally commit to being present when required by contractors
- Firms unfamiliar with permitting and inspection requirements, or the local geology in your jurisdiction
- The first sighting that you're about to inherit gold-plated specifications written higher and beyond those necessary to complete work or as required by your permit agency
- For private projects, firms that possess an expensive or complicated public-works construction frame-of-mind, and for public-works projects, firms that lack experience on public-works jobs or that don't have a framework in place for detailed and continuous documentation and recordkeeping—regardless of the owners' needs
- If the property has been purchased, review property records, environmental studies, and past geotechnical reports. Do prior geotechnical documents appear frightening, unrealistically dark, or suggest grossly expensive solutions to problems that may exist on the property? Since permit agencies commonly impose specific construction methods, design specifications, and other project conditions based, in large part, on these types of preliminary reports and studies, do you want to use a geotechnical firm that may be defining its future scope of work through its initial review and studies? Before agreeing to unnecessary or overly conservative regulations, standards, and mitigation, it's wise to procure a few more geotechnical opinions and options. The right geotechnical firm will emerge after owners scrutinize multiple opinions.

Another pitfall to avoid is allowing low-bid to color your choice in geotechnical firms. Part of this warning originates in the fact that geotechnical work can be somewhat subjective in nature. There is more art to geotechnical work than most people realize. For instance, it's possible that various geotechs within the same firm may interpret a soil boring differently and even reach different conclusions concerning that boring's relation to subsurface conditions. Or, they may differ on how to best stabilize an unstable slope or even how much topsoil stripping is adequate prior to accepting structural fill. There's nothing odd about this. Examples including land with high soil variability, fluctuating moisture content among similar soils, or how and where to take "grab samples," reinforce the fact that geotechnical work is subjective in nature. It can be an inexact and dynamic science. But, given the play inherent in geotechnical scope, what *is* critical is how the geotechnical consultant balances good science and the owner's needs. Therein lies the challenge. It's an owner's responsibility to evaluate a geotech's demeanor early on and decide whether that consultant will save money or spend more on the project than is required.

Differences in geotechnical bids can also pale in comparison to what a capable geotech can save an owner in the scheme of site construction. Understanding estimated costs, including a firm's fee schedule, means of operation, its estimate of personnel and hours to do the job, are important but it's more important to understand a firm's overall plan for value engineering and whether it offers smart site-work strategies. Based on sound practices, geotech's that exercise the art of mixing, swapping, creatively working, and losing dirt on-site can create far more profit for owners than the amount of money that owners will save in pursuing low-bid. Conversely, firms that miss important soil or groundwater conditions because they assumed an inadequate, "low bid" scope of work may later be able to shed responsibility for any cost overruns, schedule impact, or re-design resulting from their lack of better subsurface information.

From a management perspective, owners who engage in hands-on site-construction management will find that their choice of geotech ultimately weighs heavy on whether or not they meet their development goals. Owners who rely on a contractor to manage their site construction should choose their site contractor and geotechnical consultant with equal care and scrutiny. However, it's usually better to have the geotechnical consultant report directly to the owner regardless of who the owner chooses to manage site construction.

On large projects, for the sake of competition and to guarantee having enough resources to handle exploration, design, and construction monitoring, consider using more than one geotechnical firm. But, from the very beginning, separate them on-site. That is, allocate a division or project section to each consultant allowing each to control that piece of ground from initial exploration through construction monitoring and punch list. Don't mix recommendations or reports from different geotechnical firms for the same ground.

Consolidating geotechnical activities

As mentioned earlier, geotechnical work can be more art than science. When answers don't lie objectively on a chart, in a pat formula, or as defined exactly per text, subjective solutions or interpretations remain, and they can be error-prone. This is why it's difficult for one firm to accept the prior and potentially subjective work of another firm. Subjective analysis by nature is only as good as the person performing the analysis. If, for example, an individual responsible for a subjective or provocative analysis leaves a geotechnical firm, the source of interpretation also leaves. Owners can view this in a couple of ways.

First, using the same geotechnical firm to control work from project inception through completion protects owners because firms are bound to honor the work of their employees whether they remain with the company or not. However, quality, trust, and continuity may suffer. Not to mention, the possibility of acquiring a replacement geotechnical agent that may disagree with the initial engineer's work or who may take the project geotechnically in a new direction. If the first geotechnical engineer's recommendations were erroneous and his style of conducting business was miserable, this can be a good thing for owners. If not, then an owner may suffer. Finding a suitable replacement may also absorb time that can result in setting the original firm back with no hope of catching up. So, it may be worth considering changing firms in order to keep original key geotechnical professionals engaged on a project. This can be risky for owners unless the original geotechnical consultant can guarantee that billings, performance, and work will continue as is, without disruption. Owners must also consider timing and type of work. The deeper into design, engineering, and construction that an owner is, the more critical it becomes to maintaining the same consulting firms on site. In similar fashion, if the original firm specializes in a certain type of geotechnical work, owners would probably be wise to retain the first firm regardless of which individuals they lose.

Switching geotechnical firms without maintaining continuity with one or more individuals can create other issues for owners. Geotechnical people are often reluctant to accept a prior geotechnical firm's work, reports, and conclusions. When a firm assumes the role and responsibility of another firm, costs can escalate due to needs for re-education, to overlapping and redundant exploration, to redistribution of plans, documents, and information, and to the loss of owner time in coordinating the changeover. New consultants want assurance that they are not assuming responsibility for something flawed or for another firm's mistakes. Owners pay for this "assurance" policy. Even then, geotechs assuming another firm's work remain skittish until they've invested enough time into the project to assume "ownership" of work. Changing geotechnical firms that perform in-house laboratory testing means that laboratory testing will be discontinuous. For instance, it may be difficult if not impossible to acquire soil samples that match those which were analyzed and tested by the first laboratory and used to develop original design parameters.

The element of time also relates to risk, responsibility, and reward. As excavations dwindle and the ground surrenders its mysteries, geotechnical risk diminishes and predictability increases. Earlier work such as interpreting soil borings and test pits, studying ground conditions, and recommending strategies for exploration, gives way to more mundane but lucrative tasks including construction monitoring and testing. It's okay that some geotechnical firms consider construction monitoring and job closeout as easy money with less exposure to liability. From an owner's perspective, blessing a geotechnical consultant with construction monitoring is a form of compensation for doing a great job on upfront portions of a project. Ease of construction monitoring can reflect outstanding upfront geotechnical work, sane and workable recommendations for compaction, and specifications for competitively priced materials.

A discussion on continuity and risk wouldn't be complete without mentioning disputes. When two geotechnical firms share responsibility for the same work, things can get murky. Cracks or discrepancies in documentation, interpretation, or recommendations are apt to become full-blown crevasses should the combined work of two or more geotechnical firms end up in dispute. For example, consider a case where consultant number two assumes construction monitoring based upon consultant number one's interpretation of soil logs and its subsequent recommendations for native compaction. If compaction cannot be attained without importing structural material, who's to blame? The firm that wrote the specifications, the firm taking compaction tests, the contractor, or the weather? Taken a step further, if after meeting with the second geotech, the contractor, and the owner, the permit agency agrees to modify the compaction specifications to better suit existing soil conditions and a section of roadway later fails due to what's deemed insufficient compaction, who's at fault? Potentially five candidates are responsible. If the contractor performed and weather wasn't a factor, the blame will shift to the two geotechnical firms. But since the owner changed firms, neither firm may be an easy target for accepting responsibility, so guess who pays? Usually the owner.

Due to the need for continuity, ongoing geotechnical work is one of the few disciplines in land development that's conducive to "sole sourcing." Owners of projects with variable subsurface conditions *want* the same geotechnical firm that performed upfront exploration and soils interpretation to follow through with construction monitoring. Remember that some geotechnical firms will only assume another consultant's work after receiving, in writing, an owner's waiver of responsibility or by having an owner assume at least partial accountability for the first consultant's work. Whether inferred or not, owners may bear responsibility for problems that develop after changing consultants or when blame for something cannot be clearly placed on either consultant. To this end, it's imperative for owners to decipher where one consultant's work ends and another consultant's begins. This presents two challenges: first, in building a project correctly to specification, and second, in getting the second consultant to accept liability for the first consultant's work.

Another reason to use the same geotechnical consultant is that owners stand to end up with a complete, uniform, and unbroken chain of documentation. Continuity protects owners. If an owner decides to replace a geotechnical consultant, the earlier that they're replaced, the better.

Subcontracting geotechnical work

Until now, I've placed much of the burden for executing cost-effective and accurate geotechnical consulting on the geotech, however, other parties can also cause a geotech to flourish or fail. Owners of large projects who rely on third-party consultants to control the geotechnical consultant, or who expect them to proceed only after they receive enough base geotechnical information, may be helping set the table for future geotechnical problems. Third-party consultants and contractors know that good dirt usually equates to structural support, but they rarely understand the relationship of test pit or boring sample size and subsurface variability. This is because most of them don't know the level of geotechnical work required to provide design and support construction and they shouldn't—that's the geotech's job. Still, on many projects, budget often governs the extent of geotechnical exploration more than what's required for adequately describing a site. When based on a percentage of hard construction costs, geotechnical budgets are arbitrary and may be irrelevant to the degree of geotechnical work actually required. So, third-party consultants or contractors given the task of subcontracting out geotechnical work are usually concerned only with cost and collecting parameters for design.

Unless an owner subcontracts geotechnical work through a contractor or third-party consultant who verifies that the geotech's test pit, exploratory techniques, and sample size are adequate, the owner may carry on completely unaware that his basis for budgeting geotechnical exploration is erroneous. Owners proceeding with the thought that other consultants are covering them geotechnically may be shocked to discover that their consultants are doing little more than waiting for the geotech's soil parameters so that they can begin design and engineering.

So, a circular pattern is set into motion where the owner thinks that he's receiving comprehensive dirt sampling from a geotech who is under contract to a third-party consultant who isn't cognizant of the fact that the degree of exploratory work underway is insufficient, or who is stuck with the number because he bid the work without getting the geotech's input, or who knows better but, for whatever reason, has decided not to petition the owner for additional funds. The geotech samples the site by collecting as much data and information as budget will allow and the project unfolds. This example illustrates a primary reason why owners should contract directly with geotechs rather than allow third-party consultants to filter information and direct geotechnical work. Cost savings aside, project owners and geotechs need to understand each other and to discuss scope.

More reasons to procure good geotechnical information

Buyers of raw land can put themselves in a perilous position if they put complete faith in the seller's geotechnical findings. Buyers who rely too heavily on a seller's geotechnical findings may receive dated, insufficient, or inadequate information. And, there's no guarantee that exploration will have occurred in a location useful to the buyer. Plus, a buyer may not know *why* the seller has a geotechnical report. Find out if the seller has collected data for reasons other than sales. Since geotechnical studies are performed for different reasons, exploration may have been performed to collect information for a completely different reason than land development and, therefore, not pertinent or relevant to the buyer's needs. However, when sellers do perform geotechnical exploration for a prospective buyer, they should survey, locate, and structurally backfill all holes. This prevents potholing in the same location later and ensures that a buyer will be purchasing ground ready for construction. Otherwise the buyer may inherit settling, compaction, and other soil-related issues.

Of course, proceeding with bum geotechnical information can also set the stage for lots of negative interaction with a site contractor who had to bid in order to get the work, who views the owner as a one-time client, or who has a seasoned nose for disputes and change orders. Some contractors knock on your door continually for change orders and many like to knock with a hoe-pack. Prepared owners base site construction contracts on sufficient and accurate geotechnical information.

When projects lack sufficient funds for a thorough and complete soil exploration and characterization, the burden falls on the geotechnical consultant to determine, specifications and recommendations from a minimal amount of raw information. Armed with geological maps, past land records, and field observations, a geotech may have to devise a plan for exploration that, hopefully, represents a project's subsurface conditions. But, there are no promises—owners get what they pay for.

The mechanics of exploration

Geotechnical engineers perform exploration to discern a project's soil, rock, and groundwater characteristics. Digging test pits with an excavator or backhoe, drilling and extracting core samples, or other minimally invasive techniques (geophysics) are methods commonly used by the geotechnical consultant.

Geotechs rarely perform exploration with an open checkbook. Instead, they gauge their frequency of sampling and subsequent testing for physical properties commensurate with a project's suspected subsurface-soil variability balanced against cost. Technically, exploratory strategies may hinge on a calculated minimal threshold of accuracy and level of confidence established to produce statistically coherent and meaningful data. Or, a geotech may conduct exploration as dictated by regulatory agencies or as practiced locally. Some geotechnical professionals prefer to lay out exploration by experience and feel. Regardless of the method employed, the number of exploratory tests completed should increase with subsurface variability. A lack of samples can cause owners to error in reasoning and to underestimate risk.

Factors that influence sample size and techniques for data collection can include the following:

- Budget
- Statistics
- Expertise
- Site access
- Anticipated soil loads
- Knowledge of what works
- Special requirements for soil performance
- A prior history of development on that site
- Suspected or observed variations in soil type
- Expected depths of trenching and excavation and magnitude of fills

The age-old question still persists. How many soil samples are enough? On highly variable ground or where builders must miss bedrock, expansive soils, or groundwater, individual lots or parcels may require two or more exploratory holes. If an owner is developing say, nine lots per acre, this could mean taking a minimum of 18 test samples per acre. Conversely, subsurface conditions that are uniform, homogeneous, and geomorphically consistent will usually require less exploration. Good geotechnical consultants earn their salt by guiding owners and other consultants to the "sweet spot" defined by getting enough information without oversampling. *Figure 11-1* illustrates this concept.

Typical pricing includes charging individually for each test pit dug, or changing by the hole or lineal foot for borings. Cost per pit or per boring should decrease with an increase in the number of tests performed. Unit pricing is based on costs for hiring an operator with heavy equipment, performing field-

Relationship of Soils Exploration to Levels of Subsurface Variability

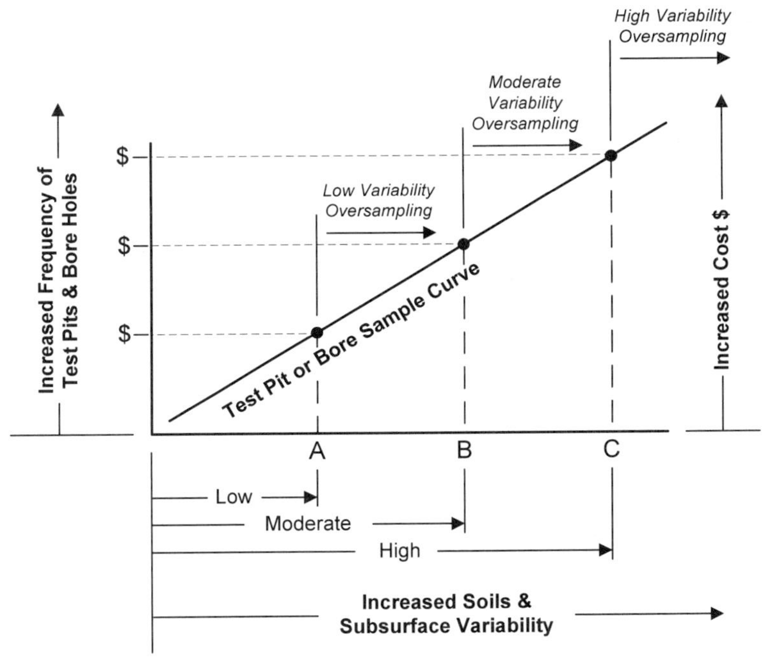

Figure 11-1

work, handling the excavated material, collecting and analyzing samples in the laboratory, and producing a report complete with mapping. It's predictable and it shifts performance risk to the geotechnical consultant. Exploration can also proceed on a lump-sum or time-and-material basis with or without a not-to-exceed clause. The owner handles costs for vehicular access, surveying, and plotting exploratory holes separately.

Unless managed correctly, proceeding with a not-to-exceed contract for exploration can penalize owners. Here's how. Areas of unexpected high soil variability, previously unknown bedrock, saturated soils, or abundant moisture may warrant additional exploration. Regardless, geotechnical consultants and hired drill operators working towards the end of a not-to-exceed contract may elect to pull off the site in order to perform scheduled work elsewhere. They may not want to risk downtime while waiting for an owner's directive to proceed beyond the contract limits. And, geotechs, contractors, and drillers may not always be available anyway since the profitable ones schedule work continuously.

Clearly, not-to-exceed contracts can be counter-productive and may not serve owners as expected. Instead, owners need to monitor exploration and "stay on top of things" in case additional holes or pits are required. The geotechnical engineer often won't know this until the last minute. Besides, why would an owner want to pay for multiple mobilizations, want to break continuity in the report, want to keep designer's waiting, or want to allow productive time to pass unnecessarily? There's a project to build!

When setting budgets for exploration, allow for more holes than thought to be required, but contract only for the number of holes, borings, or work as agreed upon with the geotech. This gives owners latitude to easily expand the geotech's contract later as needed. With geotechnical work, stay flexible and leave all doors unlocked. Here are some ways for owners to save money and facilitate soils exploration.

1. Do the work once by sampling as much area as budget allows.
2. Backfill holes or test pits daily. Leave nothing open overnight.
3. Dig pits or borings deep enough to make the effort worthwhile.
4. Avoid areas of vegetation and trees that may later be preserved.
5. Schedule additional lead-time for drilling rigs as opposed to common construction equipment.
6. Owners should perform and control all surveying, making sure that it's accurate and relevant to existing project control.
7. To avoid encroachments, identify and clearly mark all sensitive areas before establishing access and performing exploration work.
8. Consolidate work. If an excavator is going to be used to dig test pits, what other work can that piece of equipment accomplish while mobilized on-site?
9. Provide adequate access. If possible, find out beforehand the types of equipment that are going to be used and improve access accordingly. Trucks and rubber-tired mounted drilling equipment require better access than does track-mounted equipment.
10. Hire your own heavy equipment and operator. This applies when establishing access and digging test pits. When it comes to boring soil logs, geotechnical engineers know what they want and where to get the right driller. Geotechnical engineers should direct and be responsible for drilling.
11. Consolidate needs. For instance, when it's required to install inclinometers (to monitor ground movement) and drill water-monitoring

wells on the same site, consider drilling wells where inclinometers need to be placed. This way, one drilled hole will serve two purposes. The same logic applies when installing piezometer tubes (to monitor water levels). Decide on inclinometer and piezometer locations *before* beginning drilling for other purposes.

12. It can be a colossal mistake to first identify test holes or borings on a plan, have them survey-located in the field, establish access regardless of terrain, vegetation, or surface conditions, and expect exploration to proceed without problems. Instead, field-locate test pits and borings where a truck-mounted auger, or piece of heavy equipment can gain easy access, set up, and operate. Good choices typically include well-drained flat areas large enough to stockpile excavated material. Survey and record test pits and borings *after* work is complete. This strategy saves costs by establishing access once, getting the equipment in faster, reducing surveying by at least half, and minimizing the possibility of surveyors having to return later to locate alternative test locations because previous points were inaccessible. Plot test-pit and boring locations on the base grading plan. Knowing where borings lie helps to protect them during logging, clearing, and grading. Test pits that fall in structural areas will have to be structurally backfilled. If structural areas aren't known at the time of exploration, structurally backfill everything to be sure.

Working the dirt

I cannot overemphasize the importance of and responsibility for making subjective field decisions that specify and regulate limits of stripping, excavation, and compaction. In short, the leash on geotechnical consultants is long and the opportunity for them to save or to cost project owners considerable time and money during site-work can be overwhelming.

For instance, consider that vegetative roots exist in some form of organic medium or topsoil. Normal construction practice mandates the removal of all organics when grading for structural support. The question is, where would your geotech draw the line concerning sparse and scattered roots, the largest of which is smaller in diameter than a pencil and extends down beyond organic soil into hard, native inorganic material? Would the geotech require the contractor to overexcavate, remove all signs of roots, and then backfill? Where would the geotech draw the line?

Consider that the cost of additional excavation that is required to chase any and all dental floss-sized root systems may not be limited only to the expense of cut and export. Overexcavation may require backfilling with select import material, crushed rock, or even concrete mix. If minute root systems exist in an area destined to be a gravel footpath or landscaped planter, would the geotech require overexcavation or settle for a less stringent and, perhaps, more practical approach regarding limit of excavation?

The same logic applies when assessing a geotech's call on compaction. Final product or land use drives standards for excavation and compaction. For example, to avoid later settlement and meet acceptable limits for construction, compaction standards for large retaining walls, roads, and foundations are normally more stringent than those required for lawn areas, landscaped buffers, or pedestrian sidewalks. If a pedestrian-only sidewalk detail specifies 95 percent compaction, but due to weather, moisture content, or variations in soil, the contractor, after beating the soil silly with a vibrating plate, can only attain a consistent 94 percent compaction, should the contractor be made to tear it all out, export the 94 percent material as waste, and import material that will meet 95 percent? This problem is prevalent in site-work and, for owners; it boils down to risk, expense, and time. In this situation, how would your geotech react? How *should* he act?

Geotechnical consultants control a long list of potential borderline issues in land development. Because owners don't know when and where such issues will arise, it's crucial that they partner with the right geotechnical consultant early and work together to ensure that the site-work specifications and construction documents are commensurate with what's being built.

Compaction

Hear ye, hear ye! Ninety-five-percent compaction isn't always 95% compaction. What in blazes am I talking about? I'm referring to *Standard and Modified Proctor*. If you don't know what this term means, don't worry—you're not alone. It's amazing how many architects, engineers, contractors, and project owners have absolutely no idea what differentiates these two compaction standards or what the difference constitutes in terms of construction cost, schedule, and product performance.

Soils are typically compacted using mechanical energy. Other options involve using dead weight or water. Meeting compaction may likely involve controlling the soil's moisture content or material composition. Compaction of loose, excavated, or disturbed soils during site construction offers the following benefits:

- Increased soil strength
- Improved engineering properties
- Decreased potential for soil movement or settling
- Increased ability to shed runoff or hold surface moisture
- Predictability of total and differential settlement once loaded

Critical to a soil's compactibility is its moisture content. Moisture acts as a lubricant between soil particles. Too much moisture may displace and limit compactibility of a soil's particles. On the other hand, a lack of moisture may inhibit soil particles from consolidating. As opposed to course-grained soils such as sand and gravel, fine-grained soils, including silts and clays, hold and retain moisture. For this reason, fine-grained soils tend to be tougher to compact and they possess increased potential to swell and heave with increased moisture and cold temperatures. This is as far as we'll discuss soils engineering. Your geotech will happy to discuss more on this subject before he pays for lunch.

The following methods are commonly used to determine in-place soil density:

- Core cutting and drilling—is a technique for extracting a relatively undisturbed soil sample with a known volume and calculating density by dividing the weight of the extracted soil sample by the volume of the cylindrical core.
- Volume replacement testing—This test involves digging a hole and filling it with a measured volume of sand or fluid. The material excavated is weighed and density is determined by dividing the weight of the removed soil by the hole's volume.
- Nuclear and non-nuclear moisture-density gauge testing—The chances are that your geotech will use one of these quick and accurate tools to measure your soil's compaction. Nuclear machines operate on the principle that dense material absorbs more radiation than does loose material. The rate that radiation is absorbed determines a soil's relative density.
- On-board monitors—Monitors installed on mobile compaction equipment work by measuring the degree of bounce (distance and frequency) of the vibrating compaction drum. The result is a reading of the soil's stiffness, but not density or moisture content. On-board monitors allow operators to target weak soils and provide more uniform compaction coverage that can be later verified with point-specific, handheld moisture-density gauge testings. Contractors equipped with monitors can maintain a record of their work. (Incidentally, if your contractor is utilizing this tool, request and file daily copies of their readings.)

Getting back to Standard and Modified Proctor. Owners hearing these terms casually mentioned in meetings or seeing them on the plans may be perplexed as to what standard and modified mean. To find out, we have to begin by understanding what the Proctor Test is. The Proctor Test is a laboratory test used to define a soil's maximum density and associated optimum moisture content. In short, it relates to a soil's compactibility. First, samples of a soil are moistened, volume-measured, and compacted at different moisture contents in order to form a curve. Next, this data is plotted onto a graph depicting dry unit weight versus percent water content. The result is a moisture-density curve for that soil. The peak point of the curve corresponds to the optimum moisture content and related dry unit weight of that soil. So, the Proctor Test results set the compaction standards for soils expected during site construction based upon laboratory analysis. Said another way, soil compaction is determined by reading that soil's density expressed as a percent of the soil's maximum density according to a Proctor Test.

The Modified Proctor test was developed during World War II for use in determining airfield subgrade compaction. It is more stringent than the Standard Proctor compaction test using over four times the compactive energy that the standard test utilizes in determining a soil's parameters. What does this mean to owners? It means that using the Modified Proctor in lieu of the Standard Proctor for the same given percent compaction may equate to increased compactive effort, stricter allowable moisture levels, and the need for soils of higher quality in order to build your project as specified.

For example, specifying 95 percent Modified Proctor in lieu of 95 percent Standard Proctor can easily result in contractors having to attain a three-percent to five-percent increase in required compaction that they wouldn't have had to incur had they been able to utilize the Standard Proctor under the same conditions. Given equal percent compaction numbers, it's true that Modified Proctor yields a higher compactive standard, a harder and denser product, and reduced risk *but* owners have to question if its use is justified. Is it needed as specified or is it overdesign? For owners, it boils down to cost, schedule, and expectations. Specifying Modified Proctor when the Standard Proctor will suffice can affect a project as follows:

1. Overbuilding, although more expensive, creates a better product, results in increased predictability, and produces less potential for ground settlement.
2. Once underway, if an owner decides to challenge the compaction specifications, additional consultant and permitting costs will ensue. It's cheaper to specify rational compaction specifications when developing plans and specifications, and always before permitting.
3. Higher compaction specifications may result in restricting work to optimal weather conditions resulting in a smaller window for construction. Consequently, use of Modified Proctor over Standard Proctor can limit being able to attain compaction especially when working fine-grained soils.
4. Construction costs may increase due to the need to devote more machine and crew time to achieving what may be overly conservative or unnecessary levels of compaction. Extended construction activity and time spent

trying to compact overly moist or dry soil can defeat the purposes of unit pricing resulting in time-and-materials work or change orders.

5. Even under optimal conditions, not all native material may be compactable as specified. Marginal material that would otherwise meet Standard Proctor may have to be overexcavated, exported, and replaced with higher quality imported material in order to satisfy Modified Proctor. Other remedies for meeting specification may include mixing in granular material, "farming" and tilling the soil, or adding chemicals and cement or fly ash, all of which result in escalated costs.

Changing from Modified to Standard Proctor or vice versa simultaneous to site construction doesn't happen quickly. Neither does finding higher quality on-site material that will be more conducive to placement under Modified Proctor specifications. First, alternative locations for borrow materials must be located and evaluated for quality and quantity in light of planning and future grading requirements. This means coinciding unplanned, crisis-type excavations with a project's overall grading objectives. Next, grab samples of potentially usable material must undergo laboratory proctor analysis while the contractor waits for compaction results. Changing Proctors or material can take days and results in higher costs. Contractor downtime, charges of changed conditions, schedule impact, problems with permit agencies, and heightened risk of pushing work into poor weather are all likely. In some cases, it may be easier to keep the Proctor and instead change the percent compaction level. There is, however, a better way to prevent these problems and that is to enter projects with reasonable and workable compaction specifications.

Compaction of parcels and lots

Owners developing lots or parcels for sale to multiple builders may be required to record and provide compaction tests on each piece of land for sale. This requirement is particularly acute for those in the industry experiencing burgeoning insurance and bonding requirements. If soils are homogeneous, cuts are predominant, or compaction testing and geotechnical inspections performed during grading are adequate and representative of site conditions, then multiple tests on each lot can be a royal waste of money. Regardless, it's a good idea to budget sufficient surveying and redundant geotechnical work on lots and parcels since buyers, builders, and their financiers may expect it. A side benefit of surveying is that it verifies that each piece of property for sale *has* been tested as advertised.

Any ground for sale is a candidate for lot testing including building sites and areas planned for sidewalks and driveways. However, since many buyers want to preserve test-pit locations, it's best to place them near but out of the way of foundations, probable utility trenches, hardscape, and landscaping. This way, if future geotechnical issues develop, forensic studies may proceed at a location untouched by the buyer but reputedly prepared by the seller per specification.

As mentioned above, cut sites are generally less prone to failure than are prepared fill sites. Depending upon the soil type and degree of disturbance, sellers can make the case that parcels and lots cut to grade should require fewer tests than filled sites. Conversely, redundant compaction tests and recordkeeping on parcels and lots with deep fills and multiple lifts of material may cause sellers to incur significantly extra geotechnical and survey expenses. Fills, afterall, require more attention than cuts. Sellers who know their buyers may want to negotiate having to record fewer tests and subsequent survey work before completing grading. This works to shorten construction schedules and to save money, but owners contemplating this maneuver should only proceed when competent and reliable consultants and contractors are doing the work. Less recordkeeping doesn't mean less quality.

Lot testing protects sellers by recording a site's conditions before a buyer's contractors arrive. A structures contractor can arrive on-site and methodically work to ruin an otherwise properly prepared lot by using the wrong equipment for excavation, performing site and foundations work in wet weather, or by stockpiling unsuitable material over otherwise competent grade. In extreme cases, a buyer or structures contractor may degrade a lot with the intent to prove seller negligence as a scheme to recoup part of the sales price or gain leverage when negotiating future purchases. Seller beware.

Consider forty 5,000-square-foot lots with the buyer requiring each lot to have two recorded compaction and geotechnical test points. If the average fill per lot is six feet (2 meters) and compaction tests are required every foot (30cm), this work could necessitate as many as 480 test points and the cost of surveying. The question concerning adequacy of tests is always important and relevant as portrayed in *Figure 11-1.* Does twelve tests at two points per lot seem excessive? What about cost? If a *buyer* is willing to absorb the costs, a seller can provide whatever level of geotechnical surety the buyer demands, but this, of course, should be open to negotiation.

Putting the 480 test points off to the side and focusing on cost, how can sellers be protected from sinking into lot work that may not be warranted or money that may not be reimbursable through sale of the lot? Part of the answer lies in saying "no" or by carefully negotiating with the buyer. Another solution lies in changing the rules of the game by challenging buyers on the number of tests pits that they require. A better solution is to rely on quality support and advice from a competent, creative, and owner-oriented geotechnical consultant. Having the seller and buyer's geotechnical consultants concur on same requirements *before* committing language to a yet-to-be negotiated purchase and sale agreement is a smart strategy.

Returning to compaction standards, when a seller prepares land for a buyer, it's important that all parties understand and agree on the compaction standard. Sellers wanting to avoid performing unnecessary or excessive compaction shouldn't agree to compaction specifications that may be unattainable or grossly expensive to meet. This means never allowing grading and compaction to proceed under a generic and vague label of, say, "90 percent". Define percent compaction as Standard or Modified Proctor. Owners working against a known proctor will have less to contend with should problems later arise. A seller who interprets 90 percent compaction as Standard Proctor and proceeds accordingly only to be found guilty of a compaction-related issue later may be accused of purposely not providing Modified Proctor, as "they should have known better". Don't be that seller—clarify proctor.

Here are a few tips pertinent to compaction standards:

1. If a geotechnical firm's cost proposal doesn't identify the number of compaction tests that the owner can expect to receive per lot, acre, or some form of unit, request it.
2. Don't let buyers, builders, or anyone add the language "or better" to your agreed upon compaction specification. If you can reach higher compaction, that's great, but you don't want to be in a position where a buyer frivolously requests it, or interprets such language to imply that they're entitled to higher compaction if they so desire it. Fix percent-compaction specifications with a number and in terms of Standard or Modified Proctor.
3. Don't get hung up on the issue of optimum moisture content. Yes, optimum moisture content is important and owners should understand its unique relationship to soils and compaction but it's not the goal. Optimum moisture content is a means to an end and the desired result is gaining compaction as specified. Geotechs and contractors at times delve into the issue of moisture content when moisture content may not be the issue at all. Instead, the mechanics of their performance (working the dirt too hard, using the wrong equipment, placing lifts too thick, etc.) are inhibiting them from gaining compaction.
4. When contractors fail to meet compaction, owners need to research whether the geotech or testing lab is using the correct standard and that the Proctor test is accurate. This may call for running additional Proctors in order to verify that the original test results are true and accurate, or for calling in a second geotechnical or engineering firm to

verify that the correct standards are being used. A third issue involves moisture-density gauge equipment. If the contractor is experiencing ongoing difficulty meeting compaction, it may mean that the geotech's machine is out of calibration, is being used incorrectly, or is damaged. Consider these troubleshooting tips should your contractor continuously fail to attain required compaction.

The geotech and construction monitoring

What defines a geotechnical consultant's construction-monitoring field capability? Here's a list of items that owners should consider addressing when evaluating or questioning a geotechnical consultant for construction-monitoring services. Find out how the geotech will approach the following?

- Stockpile maintenance
- Preparing for poor weather
- Building a contractor's trust and confidence
- Using a soils probe to judge consolidation and moisture
- Working with the contractor's style of mass excavation and utility work
- Ensuring that stripping advances no more than necessary so that grading can begin
- Identifying opportunities for value engineering, cost savings, and improved efficiency
- Rolling grades during construction to reduce a soil's surface-moisture holding capacity
- Understanding the project schedule and learning how activities depend upon each other
- Dealing with asphalt that arrives too cold and segregated, or concrete that arrives too hot
- Monitoring contractors who may haul partial loads of material and charge for struck capacity
- Altering procedures if excessive compactive effort results in worsening and weakening the soil
- Collecting sufficient compaction data between lifts of fill while performing other on-site observations
- Reading construction and judging trends during excavation in order to make progressive adjustments in the owner's best interest
- Utility trench backfill—How will the geotech avert subgrade settlement and ultimately pavement failure that can result from inadequate or poorly placed trench backfill?
- Working multiple sites within a project and not lose control or miss opportunities to measure and document overexcavation or to record related extra time-and-materials type work
- Operating from a proactive position and not a reactionary one—This

is particularly important as geotechs can incriminate themselves over a series of geotechnical reports by acting more like news reporters than consultants. Owners want documentation but they also need solutions to developing or worsening problems.

Geotech reports

Few papers generated in the field are more important than the reports orchestrated by the geotechnical field inspector. The geotechnical report quantifies work completed daily and notes if work has been completed in compliance with the plans and specifications. It provides a blueprint to backtrack geotechnical issues should something later need diagnosis or research. Though not known for exciting reading, the report should provide an all-inclusive daily snapshot of geotechnical work completed under the direction and authority of that geotechnical inspector. In addition, request that all outstanding issues, incomplete work, and work performed and left out of specification be summarized in a weekly "Grass-catcher" list. A list of this type is monitored weekly and used to ensure that field crews regularly rectify all loose ends, correct poor quality work, and remove and replace any materials or fill considered unacceptable by the geotechnical engineer.

Here's a minimal list of items that an owner should expect to receive when the mailman delivers the weekly geotechnical reports. Each daily report should:

- Be detailed
- Be consecutively numbered
- Include boilerplate data such as the job number and the date
- Note the contractor's equipment, its model numbers, and machine hours worked
- Record contractor's name as well as conversations held, orders received, and directions given
- Flow from report to report as one cohesive story—Daily reports should dovetail with the prior day's work, should fully describe the current days work, and should mention any preparations laid for the following or future day's work.
- Include the test results for that day—The inspector should also interpret the results and state whether or not they comply with the contract documents. If not, the report should mention recommendations or conditions for retesting.
- Be consistent in style, layout, language, and format regardless of who the inspector is on any given day—The reports have to make sense and remain relevant in case an owner later needs to construct a series of events in order to better understand prior circumstances.
- Approximate the quantity of contract and extra work inspected that day, such as estimating the volume of bank cubic yards of material

excavated, the truck count, the truck cubic yards hauled in, the lineal feet and depth of pipe installed, or the square yards of asphalt placed.

- Include maps and sketches indicating where contract work has been observed and accurately note test locations—Dimensions of trenches, excavations, and piles of material should be clear and easily understood. Any assumptions and equations used as backup for calculating volumes or earthwork quantities must be included.
- Chronicle errors performed by the contractor—Items such as using the wrong equipment for the job, safety incidents, lack of management or personnel, or operating void of staking and control can all be included. When construction changes occur, the inspector must be there to report, measure, and document them accordingly.
- Note the weather at the start of the day, midday, and at day's end—If an on-site weather station is installed, temperature, precipitation data, and approximate wind speed should be recorded. Any information regarding weather patterns throughout the day should also be included. When it comes to disputes regarding earthwork and poor weather, owners cannot protect themselves with enough backup data.
- Document unanticipated changes in soil type, subsurface moisture, and rock—If the contractor proceeds with change-of-conditions work, the geotechnical inspector and contractor together should measure or estimate the change of quantities for that day and the inspector should document the quantities in the report. The contractor should also sign the report in tandem with the inspector thereby confirming the quantities as noted. Include three-dimensional sketches of changes.

Some geotechnical consultants may balk if asked to perform detailed inspection work. They may feel that being too thorough may jeopardize their relationship with a contractor. In reality, they may lack sufficient qualified personnel and equipment to perform high levels of inspection and documentation. It's been my experience that most geotechnical firms can and will provide this level of scrutiny if asked and at no appreciable difference in cost. If some readers find their geotechnical consultant asking for additional compensation in order to perform in detail, spend the money and get the service—it's worth it. Not only will owners receive better and more complete documentation, they also receive a thick layer of protection. Besides, if the geotechnical inspector doesn't furnish an owner with this information, whose going to provide it? It's not like having such data is a luxury. On large projects, it's almost a certainty that questionable charges will evolve concerning site-work and mass excavation. For this, owners must prepare.

Here's another tip. Owners should be on the lookout for subtle changes in their geotechnical reports. If the preproject reports called for silty sand and your reports are now calling out sandy silt, question the change. Rest assured that hard bid contractors will notice any minor changes in conditions. Geotechnical changes that come too late in the game can lead to disputes and sometimes work as a basis for scapegoating extra compensation against an owner. It's good to catch and rectify these changes early because if a dispute later arises, you, the owner may be responsible. Demand that geotechnical reports be detailed, clear, and consistent. Anything else falls short of the mark.

Differing site conditions

Differing site conditions can impact cost and schedule far in excess of what most owners expect or are prepared to cover. Construction downtime and delayed work while waiting for new or revised engineering and design drawings is costly. When setting up a project geotechnically, owners can follow one of three paths. They can

1. risk proceeding with inadequate geotech work and blindly hope for the best;
2. spend the money required for a thorough and complete geotechnical exploration and include a contingency for unforeseeable problems; or,
3. skimp on exploration and instead, cover the project with a contingency for costs associated with potential change-of-conditions work.

Number two is the safest policy. Not only is this approach innately smart, but also owners remain in control. They get to design once and possess the highest chance of avoiding cost and schedule impact. Once informed, owners can confidently produce a set of plans and specifications correct for the site. Changes in excavation can be restricted in scope resulting in less impact. By controlling grading owners control problems related to layout and foundations, produce product as advertised, and facilitate marketing and sales. As many readers already know, avoiding entanglements between ongoing site construction and sales and marketing is in itself, a brush of brilliance. Surety breeds control, which in turn allows owners to negotiate land sales from a position of strength.

Who's responsible?

When subsurface conditions change, where does an owner's responsibility end and the contractor's responsibility begin? With the stroke of a pen, some owners believe that they can press inordinate amounts of risk associated for unknown soils and water onto a contractor. In doing so, they may inadvertently set themselves up for years of litigation and haggling, not to mention unintendedly impacting their overall cost and schedule. On the other hand, some contractors actively seek out projects where owners are exposed and may be vulnerable to lucrative change orders.

In site construction, long and drawn out issues related to change of conditions can initially appear aimless and without direction. Unlike change orders in structures, evidence of condition changes often no longer exists having been excavated out, filled over, or changed in ways that are useless for reference or study. Consequently, either party may build or defend his case based on photographs and video, shoeboxes full of unreadable truck tickets, or anecdotal evidence. Many times, arguments hinge on technicalities and clever strategies aimed at snagging the opposition for not filing the proper report, not following procedures, or not taking direction. Still, successfully pinning responsibility on somebody for unknown soil or subsurface conditions can remain elusive.

Ultimately, questions involving change of conditions may fall on the shoulders of whoever was responsible for documenting the soils conditions at the time of bidding. It is here that an owner's defense can crack and costs can escalate. Disputes that evolve into a frenzy where both parties fight until they're punch-drunk or dizzy from deciphering endless streams of paper, incomplete reports, or unrelated counter charges typically end up as a compromised settlement, thus, mercifully ending the issue. More often than not, that owner assumes all or a large percentage of the loss. Why? Besides waging an inconclusive or unconvincing argument, most owners fail to realize that change-of-conditions issues are often the product of insufficient of upfront geotechnical exploration. And, it's a shame because the cost of complete exploration and geotechnical work usually would have been *much less* than the aggregate cost of impacting schedule, legal fees, hiring consultants as advisors, paying damages, overhead, and other expenses resulting from the dispute.

How do owners get into such jams? Many of them proceed into site-work without adequate support and expertise, while others concentrate too much on the architectural details or final results. Meanwhile, everybody may be assuming that the next person is "handling" preparations for site construction. Or, they may be relying on consultants in other disciplines to advise them on site construction and geotechnical issues and, thus, enter projects halfcocked. Some owners when advised that words alone, packaged in a contract, are sufficient to absolve them from unknown conditions, place full responsibility for soils squarely onto the contractor's shoulders. Besides feeding a contractor who may thrive on conflict and change orders, pushing geotechnical and site-verification duties onto a contractor can be practically and logically problematic.

The problem here is that most contractors aren't, and don't pretend to be, geotechnical engineers. Most subcontract out geotechnical work and they somehow must recoup this cost plus markup from the owner. So, the first negative for owners is that they incur additional mark-up costs that they'd normally avert had they contracted the geotechnical work out themselves. Taking it a step further, contractors allowed to use the same geotechnical firm as the owner do maintain uniformity. However, this can lead to complications involving conflict of interest. Finally, if a contractor uses a geotechnical firm, other than the owner's firm, to "fill in the blanks", it can create real problems if the contractor's geotech interprets initial subsurface conditions counter to the owner's geotech. Which geotech's conclusions will be considered correct if issues erupt concerning change of conditions? Here again, the owner is at a distinct disadvantage. It's confusion and controversy.

Ultimately, when owners allow contractors to manage or add to the existing body of preconstruction geotechnical documentation, the contractor can question anything. Owners can forfeit tremendous leverage as the contractors and their consultants position themselves to query anything dirt related such as pavement thickness, compaction, wall parameters, even the owner's site-construction schedule. I shudder whenever I hear of an owner suggesting that contractors add to the geotechnical record after bidding. Owners determined to have contractors perform geotechnical work may be better off pursuing a design-build approach and getting the contractor on board early under different terms.

Asking a contractor to verify or generate subsurface geotechnical information out of sequence with design and engineering can also be foolish. The dynamics of the entire design-and-bid sequence go out the door. What if the contractor discovers a pit of contaminated soil or the remnants of an old dumpsite? Although good in some ways, such discoveries can effectively bring into question everything that design or engineered have been based upon for that site. Finding out that a site isn't what it was thought to be can lead to tremendous cost and schedule impact and, in severe situations, can terminate a project. Simply, a contractor is the wrong party, too late in the process to discover such conditions.

Advertising during bidding or negotiations that all excavation should be considered "unclassified material" isn't a solution either for an owner undertaking site construction without adequate geotechnical information. Only owners who *know* that unclassified soils will be uncovered during excavation should consider advertising as such and budget accordingly.

Continuing on the subject of cost, owners have to consider how contractors will recover their expenses when subcontracting out geotechnical work. Afterall, it doesn't come free. Will a line item be included in the bid to cover geotechnical expenses? If not, what creative devices does the contractor possess in order to gain reimbursement? Are charges buried in unit pricing or lump-sum items? Any contractor who rushes out and spends money on prebid geotechnical work probably sees an opportunity for extra work that he can later capitalize upon. Don't fall prey to unnecessary charges by passing responsibility for unknowns onto the contractor. The best policy is to be above board, do your homework, and spend whatever is necessary to know, completely, a project's subsurface conditions. Dictate known and correct conditions to your contractors.

Geotechnical value engineering

Consultants working under negotiated terms are more likely to offer value engineering than are consultants working under hard-bid. No, it's not a rule, just the nature of the beast. Even if they buried funds for value engineering in their base bid, a consultant brought into the fold via hard bidding has nothing to gain if asked to perform value engineering unless they share in the savings or unless such work has been defined specifically as a line item in their contract scope of work. Hard-bid consultants are more likely to charge extra for value engineering.

Negotiated work, however, still doesn't guarantee open and freewheeling value engineering. With some consulting firms working under negotiated terms, if you don't ask, you won't get. Firms, like many owners, budget just enough to perform whatever scope of work it tales to satisfy a vision, plan, and expected permit agency standards. This may leave little or no money for value engineering. Geotechnical firms know that it takes time and money to analyze and to design for alternative and lowest-possible site-work costs. They also know that to be effective, value engineering must begin from the outset.

Examples of geotechnical items open to value engineering may include, but aren't limited to, changing pavement standards and road sections, evaluating walls and engineered slopes to reduce grading, engineering otherwise unusable slopes for construction, tailoring compaction specifications in line with expected use, and exploring ways to creatively lose or use on-site native material.

The best route for owners is to budget for and deck themselves with creative geotechnical personnel who are capable of squeezing high returns out of site construction while maintaining an acceptable level of risk. And remember, more subsurface information is better. Who then, shall you ordain as your geotechnical consultant?

Chapter 12

DEVELOPING PLANS AND SPECIFICATIONS

The following three bodies of paper constitute a large portion of what is commonly referred to as the site-work Project Documents. While this list of subcategories isn't exhaustive, it does represent a cross section of topics common to many construction and land development projects.

PLAN DRAWINGS AND DETAILS	DOCUMENTATION AND REPORTS	GUIDELINES AND SPECIFICATIONS
Erosion control	Geotechnical test pit results	Reference standards
Logging and clearing	Slope stability analysis	Acceptable materials
Grading and walls	Wall designs	General procedures
Road and storm	Traffic analysis	Sequencing and scheduling
Sanitary and water	Hydraulic studies	Submittals and testing
Landscaping	Environmental Assessments	Conduct and regulations

The nature of land development often leads contractors to having to interpret information when working from the construction documents. Owners can minimize contractor interpretation by offering high-quality, well-prepared, and complete construction documents, plans, and specifications. Remember that site-work enjoys a high degree of success when owners possess:

1. inclusive and accurate documents;
2. solid and comprehensive geotechnical work;
3. a detailed, complete, and presumptive unit price schedule; And,
4. good weather and trusting relationships.

Projects that proceed with incomplete, misleading, or poorly worded plans and specifications are less likely to meet budget and schedule—even if these projects enjoy superior design, qualified contracting, and competent management. The stakes rise on hard-bid projects where inept documents act like chum and attract nothing but trouble for owners. This chapter focuses on the first item in the above list—inclusive and accurate documents. More specifically, it's a study for project owners in the hunt for quality, cost-effective, and well-prepared site-construction plans and specifications.

Choosing firms to develop plans and specifications

Creating site-work plans and specifications is a function of design and engineering. Whoever does the design work develops the plans and specifications based on his design. There is, however a different twist with site design in that surveying and layout come into play. Surveying and layout are more associated with civil engineering than any other discipline, which for many land developers narrows the field considerably. It's no wonder that civil engineers, either working independently or within multidisciplinary consulting firms, perform the lion's share of site-work design including developing plans and specifications.

Hiring *any* civil engineer to develop plans and specifications can be a supreme mistake. One alternative is to scrutinize architectural and engineering firms according to their strengths and history. Judge each shop based on the services offered, which may include, but not be limited to, design, engineering, surveying, management, contracting, and developing plans and documents. Can a firm accustomed to producing construction documents in the private sector remain interested and competitive when asked to compose documents suitable for public works contracts? Likewise, a low-bid consultant having trouble performing on a public works project that's over budget and behind schedule may not attract private owners searching for services. Some owners continually use the same consultants without periodically measuring their performance. Is this wise? A time comes when value trumps comfort and familiarity. Others hire only as directed by bosses. For those owners, it may not be in their best interest to choose who designs and develops documents based solely on political appointment.

Viewed from another perspective, at what point does a firm possess too much horsepower or too much overhead and higher pricing to become unattractive to an owner in the private sector? It's a moving target but the target doesn't move too fast. Making the correct decision here is important in selecting the right consultant to complete design, compose documents, and, possibly, coordinate third-party drawings and shop details.

Owners should also evaluate candidates in light of the type of design work or projects in which they specialize or excel. This includes evaluating the format, clarity, and quality of a firm's plans and specifications. Do the plans look complete? What about quality and thoroughness? Are there ample profiles shown for storm ponds, excavations, and utilities?

Finally, there's the question of value, availability, and resources. The optimal site-work designer will be comfortable with your project, possess ample resources including in-house surveying, be able to produce timely work, remain engaged in value engineering, and—yes—produce quality and complete plans and specifications. On larger projects, owners may better off hiring more than one design or engineering firm to increase the resource base and maintain competition.

The Kando Factor

Regardless of the consultant chosen, there are limitations when developing plans and specifications. A friend who owns a civil engineering and survey firm tells clients that plans and specifications can be developed three ways: correctly with minimal chance of error; quickly with minimal turn around time; or cheaply at rock-bottom price. The catch is, they cannot have all three. They cannot simultaneously have their documents produced cheap and fast with assurance that they'll be complete and comprehensive. While it's *technically* possible to have all three, designers or engineers aren't likely to absorb costs for overtime and weekend work or concentrate higher-than-normal resources to one job at the expense of other company business. Such extra cost wouldn't be commensurate with what most land developers can, or are willing to pay.

The KANDO Factor

Potential Consequences of Each Combination

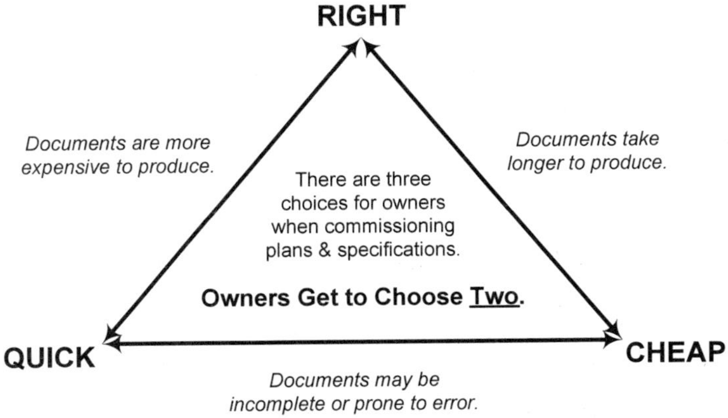

The KANDO Factor is based on the premise that there are three ways for a planner, architect, or engineer to competitively create plans and specifications. That is - produce them *cheap, quickly, or right*. Clients get to choose two.

Figure 12-1

Instead, the engineer guarantees to meet two conditions as chosen by the owner. Two conditions out of three ensure that something somewhere is likely to be compromised in the process. Refer to *Figure 12-1,* the Kando Factor, to graphically view the choices and consequences of each combination. In reality, many owners don't have a choice and end up accepting conditions commensurate with their needs or budget. The following list details the combinations possible when choosing two conditions out of three and compares the process and results of each. Below each possible combination, there is a list of characteristics describing why some owners may be bound to certain conditions as a result of their actions or circumstances.

QUICK & CHEAP	QUICK & RIGHT	CHEAP & RIGHT
Likely reasons:	*Likely reasons:*	*Likely reasons:*
Unreal expectations	Cost is secondary	Cost is important
Schedule is important	Quality is important	Owner is prepared
Likely low-end product	Schedule is important	Quality is important
Could be a crisis situation	Could be a crisis situation	Schedule under control
Quality control is secondary	Likely higher-end product	No premium for quality
Minimal municipal standards	Higher municipal standards	Product could be a factor
Process and results:	*Process and results:*	*Process and results:*
Minimal detail	Complete documents	Highest value
Minimal owner control	Check base information	Minimal pressure
Minimal owner interface	Suitable for design-build	Time to value engineer
Accept all base information	Restricted owner interface	May have to push design
Decreased checking of work	Maximum consultant pressure	Maximum owner control
Risk of incomplete documents	Cost may exceed expectations	Confirm base information

Depending upon their priorities or needs, any of the three possible combinations outlined above may work for owners. When cost is no objective, owners can have it all. Given a choice though, most owners will aim for the best value combining lowest cost with highest quality work. All that's required is enough time for adequate planning and complete scheduling.

Design specifications—the Achilles heel

Like lightning, most problems in site-work course their way through the plans and specifications before testing the construction contract or it's language. Simple-to-solve objective-type issues usually go no further than the specifications and plans. More subjective and messy issues usually pass the specifications and move up the ladder towards the realm of contracts. This is because contract problems usually relate to mismanagement or non-compliance and not clean-cut errors involving design, engineering, or estimating. As an example, dirt issues in site-work tend to be the toughest to track through the construction documents. Dirt problems are usually complicated, compounded with other issues, linked to numerous documents, and are often entwined with the contractor's work style, weather, staking, or subsurface conditions, or they're rooted in base topography and control errors—none of which stand out or are easy to decipher. They can, however, be controlled, and one way to do that is through the specifications. The specifications are the first place where site construction issues are tested. The specifications should work as a filter—a fuse—a source of clarity—and not a breeding ground for interpretations, inaccurate verbiage, or open-ended language. As the first line of defense, write specifications in a fashion that prevents problems or, at least, makes site-work issues easier to comprehend and trace.

Prosperous contractors, consultants, and owners remain successful in part, because they read, understand, and remember their contracts. If one of them is to stumble, the odds are that he'll stumble *first* in the specifications, *second* in the plans, and *last,* by being in violation of his contract. But most owners spend a disproportionate amount of resources and time writing and dwelling on contracts. In all reality, when a problem passes through the documents and drawings and lands on the contract, it's too late. The war has already begun.

Why do specifications tend to be problematic? For one, specifications consist of directives, commandments, and, sometimes, canned information that's cooked into a document and seeded throughout the plans. Specifications usually apply to a project but, occasionally, they don't apply. They may be holdovers from prior similar projects or represent the best, albeit poor, information that the consultant can muster. When specifications don't already exist and there's no protocol to draw from, someone has to create them. Writers may find it especially difficult to create specifications for design-build projects involving unique and challenging site construction where there's no clue as to how work will progress until the job is completed. Writing specifications is an art and not everyone bestowed with the task of composing them is gifted or experienced enough to create them in a coherent and accurate manner. Here are some reasons why problems with specifications can occur:

1. The plans and construction drawings typically take precedence over the specifications. This creates a dilemma for anyone trying to build a case but unable to unearth enough supporting evidence in the plans alone. In response, he may gravitate to, interpret, or contort the specifications in a quest for answers, counter-punches and ultimately, his own safe harbor.
2. Specifications hastily thrown together may contradict each other. Although this happens in the best design, engineering, and architectural firms, it's more likely a result of sloppy work, taking short cuts, lack of quality control, or having multiple people create the specifications. Contractors will gladly perform editorial work on specifications well after they sign the contract. They like to mark in red ink.
3. Some specifications may be irrelevant or missing. When a problem builds momentum, opponents may scramble to the specifications for information supporting their case even though the specifications may not be pertinent to the problem. They may also discover that the specification that they're searching for never existed. Confusion over specifications is an occurrence common for those buried in multiple projects. Necessary specifications can also be missing, left out, or forgotten resulting in holes in the documents
4. Specifications linked to canned details can be a source of problems. In general, canned details are excess baggage and, unfortunately, owners pay to have them included in the contract documents. Most canned details were developed somewhere else for another project completed years ago and now sealed in the annals of time. Landscape and wetland drawings tend to be heavy on canned details that can appear as if they were imported from a bird watcher's sketchbook. Remove unrelated detailing that doesn't belong in the documents and further evaluate canned details that appear as if they *might* apply. Inquire as to whether they are simply the wrong detail, a poor choice for illustrating work, or an imposter for something else more important that's missing from plans. In an extreme case, an owner in dispute can be *harmed* if a contractor or consultant uses a mildly applicable canned detail to his own advantage.
5. Blind ignorance and incompetence aside, some design, engineering, and architectural firms can intentionally use the specifications as a weapon, an antidote to meet their own objectives or to cover their shortcomings. Deliberately crafting ambiguous specifications happens more often than most owners realize. If a consulting firm has the final say in determining the outcome of a contractor's charges pertaining to items such as extra costs, time extensions, or cries of negligence, that

firm may have already prepared for these problems by writing vague and open-ended specifications. With time as an ally, a consultant can deny contractor claims into perpetuity knowing that a dollar-value threshold exists beyond which it's not worth it for contractors to continue pursuit. Break ground only with complete and clear specifications.

Design specifications are normally written well enough to support site construction. However, it doesn't take too many clinkers in the mix to raise the ire of contractors, to invalidate bids, or to set a project back in time and expense. Demand specifications that are pertinent, correct, and immune, or at least logically resistant, to a wide spectrum of interpretation.

Design and performance specifications

There are a few ways to define specifications. Design specifications are made up of detailed and fixed criteria that define the general manner in which site-work is to be performed and completed as submitted and approved per plan. I say "general manner" because there are many ways to adequately perform various phases of site-work, such as logging, excavation, grading, and utilities—to name just a few. The key words with design specifications are *as approved per plan.*

Performance-based specifications differ from design specifications in that they define minimal yet measurable standards that a contractor must meet while working without complete and final drawings. In some cases, an engineer or other licensed individual is required to oversee work and ensure that the contractor works in a safe manner, to code, and follows good construction practices. Performance-based specifications transfer increased responsibility to the owner or contractor for selecting materials, design, engineering, surveying and layout, and methodology of construction. As it applies to site-work, performance-based specifications share many of the same attributes that design-build offers. Mainly—work proceeds without final design.

Some jurisdictions prefer this way of doing business because it shifts greater risk and responsibility to permit holders, owners, and contractors. If a project owner has competent managerial support and able contractors, working with performance-based specifications can be a wonderful strategy to use in advancing the schedule, taking advantage of good weather, and allowing a contractor an opportunity to save money in spite of not having approved drawings. Because it's a progressive way to approve work and reduces burdens at city hall, the demand will increase for performance-based site-work.

Before ever going to ink

Know where you're exposed. Before finalizing design and engineering, know all municipal demands including items such as dry-utility requirements, land give-aways, easements, and forfeiture of possible right-of-way land. This underscores the need to finalize agreements with existing neighbors and to bring to fruition any other vague deals lurking in rumor. Also recognize the difference between existing municipal standards and what has been agreed to for entitlement, and last-minute demands placed by a jurisdiction or private entity. Owners *must know* all of their obligations on all fronts before setting budgets, committing to a schedule, or finalizing project layout. Know where you're exposed and design to accommodate it.

Developing construction drawings

Development of site-construction plans typically follows a routine as depicted in *Figure 12-2.* Beginning with owner obligations, this flowchart shows the dependence of negotiable items such as utility and street design on other less-negotiable items such as net buildable area, fixed control points, and road location. Layout and design shouldn't be refined until all restrictions and constraints have been reconciled and are under control.

Examples of upfront and inflexible restrictions may include agreements to complete off-site improvements before beginning on-site work, land constraints that drive road location, or inflexible grading designs aimed at supporting a limited number of product types or conservation easements. Another way to look at *Figure 12-2* is to observe that the most restrictive elements must be satisfied prior to proceeding with more flexible and dependent design work. Similarly, planners and engineers need to know everything that could possibly affect their decision-making before investing time in design. As one phase of work supports and drives the next phase of work, the process simply builds on itself culminating in land that is developed and ready for structures.

By deviating from this hierarchy and proceeding out of sequence before breaking ground, owners are exposed to schedule impact and increased soft costs. Once under construction, sporadic design and planning increases the risk of added expense and illogical construction. Site-work plans unfolding out of sequence can frustrate contractors and surveyors resulting in significant claims and cost overruns. Even with negotiated work, the danger in pursuing design work and plans out of sequence doesn't expire until a project is complete.

General Logic for Sequencing Site Plans

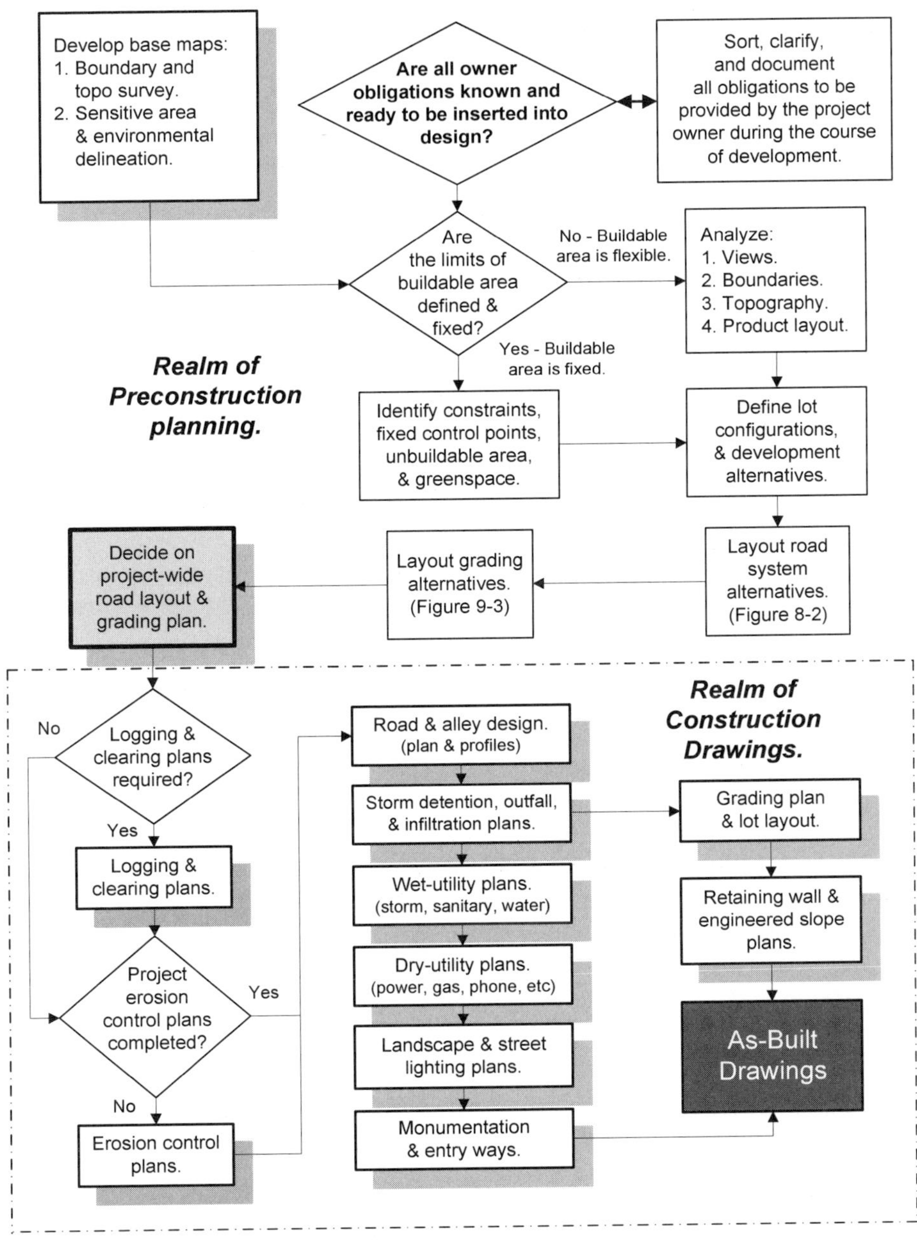

Figure 12-2

Inch/feet and decimal scale

Architects, landscape architects, prefabrication shops, and landscapers tend to design and communicate in inch/feet scale. Civil engineers and surveyors usually work in decimal scale. Structural engineers seem to go either way depending upon what they are designing. For these reasons, it makes sense to carefully choose who will be your master designer and who will create and coordinate project construction documents.

When one consultant designs and engineers site-work in decimal scale and another consultant designs portions of site work or foundations in inch/feet scale, it's important that they coordinate and properly convert information that matches or overlaps such as pipe elevations and final grades. Contractors and operators may find it difficult to resolve inch/feet and decimal scales shown two ways in the plans and, possibly, staked a third way in the field. This leads to construction error.

As a rule, abstain from using inch/feet and decimal scale on the same plan sheets and avoid using different scales throughout the same plan set. It's a mistake to assume that everyone, including designers and contractors, will convert dimensions consistently and correctly. Confusing and mixed-scale plans can also impact surveyors, material suppliers, and inspectors. The same logic applies to metric versus imperial units. Reference Chapter 5 on metric for further information.

Before developing a site, develop your own "in-house" specifications

Some owners prefer to develop supplementary standards that are separate but in harmony with governmental standards and specifications. Consultants, contractors, and builders can all assist in developing in-house specifications. Once drawn up, integrate in-house specifications early on into every facet of design and layout. It's critical to get this information into the hands of your planners, architect, landscape architect, and engineers before design and layout work begins. Here's a list of possible in-house standards that owners may consider developing:

1. Minimum and maximum acceptable range of final cross slopes per product type.
2. List which native trees, shrubs, and species of flora are to be preserved and protected.
3. Preferences for maximum retaining wall heights and length, type of street poles and lighting heads desired, and what species of trees and shrubs are acceptable for landscaping and ground cover.

4. Acceptable conditions and locations for utility-vault farms well in advance of final layout and engineering. List the types of utility vaults acceptable for on-site installation. Know what service providers will accept and demand before committing to final design and layout.
5. The types of retaining walls and rockeries that are acceptable. For instance, owners may specify walls in view to be modular block, decorative block, panel systems, rockeries, cast-in-place, or cribbed with plantings. On hidden walls, owners may specify cheaper wall alternatives or piles and lagging, wrapped fabric, jumbo block, and wire baskets. Also, note acceptable sizes of individual components such as rocks or cast modular units. Specify maximum height, length, and square footage of walls as well as the range of colors, textures, and fencing material acceptable for on-site use.
6. Value engineer and challenge pre-existing standards and specifications that may result in overconstruction and unnecessary expenses. Examples include evaluating cheaper combinations of hardscape and alternative base material more in-line with on-site soils and loading, changing pipe material, and investigating whether or not the compaction specifications should challenged. The goal is to reduce costs without compromising quality and performance, so any proposed changes should be subject to a cost and benefit analysis. If a change is worthwhile and accepted by the permit agency, integrate it into the project in-house specifications.

Deficiencies in site-work documents

Initial plans may be labeled a certain percent complete or stamped "Preliminary". Regardless of how complete plans and specifications are, it's never too early to review and scrutinize them for mistakes and shortcomings. Construction documents can reveal weaknesses and shortcomings at any stage of development. Here are some easy-to-spot warning signals that may suggest trouble ahead:

- Mislabeling
- Nothing is dated
- Sheets are missing from sequence
- Whole design sections are missing
- The design engineer is not the draftsman
- Imperial and metric measurements coincide
- Details cited throughout the plans are missing
- The engineer of record changes from sheet to sheet
- Words are misspelled or they're missing from sentences

- Your contractor immediately spots mistakes on page one
- Repetitious and redundant detailing and specifications abound
- Scales vary from plan sheet to plan sheet and from detail to detail
- Canned, outdated, and recycled details unrelated to the project are included
- Consultant B has designed work on consultant A's "Preliminary" base plans
- The architect is relying on the civil engineer or the civil engineer is relying on the architect to complete design where systems overlap or match
- Details and specifications refer to materials and workmanship to follow in accordance with another jurisdiction's standards and specifications, which aren't included or defined anywhere
- "Not to Scale" detailing proliferates—While a place exists for not drafting to scale, lack of scale improves the odds that a design may not work as intended. In itself, drafting to scale tests the design and provides a degree of quality control.
- The contractor is required to build or install something *unique* without plans or specifications—Instead, the contractor is responsible for relying on a distant specialty manufacturer for consultation, design, installation, and inspection services in order to complete contract work.

Missing information

Check plans and specifications to ensure that they cite and reflect all reports and studies linked to permitting and entitlement or that support engineering. While supporting information such as the geotechnical report, a wall design, or pavement analysis may exist as separate documentation, there exists no guarantee that this information will be cited or linked to the plans and specifications given to contractors. Still, contractors are expected to connect the dots and build in accordance with all documentation cited in their contract. While most contractors can handle this task, owners may need a consultant or manager to interpret and understand how all the construction documents fit and relate to each other before bidding and, certainly, prior to breaking ground.

Consultants, construction managers, or in-house staff should also be capable of constructively perusing a set of construction drawing for missing information, gaping holes, and absent detailing. Here's a head start on some items notorious for being left out of the site-work drawings:

- Dimensioning
- Utility conflicts
- Specified signage
- Tree protection plans
- Yard demolition plan
- Material storage areas

- Curb transition detailing
- Truck-and-haul route layouts
- Catch basin and manhole tolerances
- Sizing of crushed and imported rock
- Pavement and sidewalk alternative details
- Pavement-striping and street-marking plans
- Allowable tolerances for paving and hardscape
- Contractor storage, parking, and job-shack layout
- Detailed and accurate limits of water-pressure zones
- Compaction defined as Standard or Modified Proctor
- Actual and accurate limits of retaining-wall transitions
- Set asides designated as sources for native transplant stock
- Minimum allowable grade tolerances for gravity flow piping
- A base plan showing all benchmarks, control, and survey data
- Location and details for all capillary breaks and fabric placements
- Methods and materials for abandoning or plugging existing utilities
- Clearly laid out utility plans, trench sections, and connection details
- Identification of water sources and locations for disposing non-potable water
- Differentiating where raw, screened, and processed topsoil is to be used on-site
- Areas available for stockpiling dirt and topsoil with provisions for erosion control
- Electrical sources and connection details for lighting, irrigation meters, and amenities
- Provisions for handling gaps in utilities required for fire flow, construction, and occupancy
- Details showing how small-diameter storm, roof, footing, French, and wall drains tie into the storm system

Although this is a partial list, owners should get an idea of what type of information to look for when reviewing the site-work drawings. So far, we've covered the need to scope site construction drawings for both missing, and deficient information. A third category that helps to define the essence of poor and ill-prepared construction documents involves mistakes.

Inaccurate and insufficient data in site-work documents

As a consultant, I regularly perform *cold* reviews of construction documents days prior to bidding. I don't suggest the timing, it just happens that way. The word *cold* denotes the fact that I've had no prior influence on the project. The problem with doing a cold review late in the process is that owners are usually too far entrenched to welcome or approve sweeping changes regardless of potential savings. To an extent, this is understandable. I usually find numerous easy-to-fix problems, deficiencies, and opportunities for improving the plans and reducing the owner's risk to changes. But last minute reviews mean that it may already be too late to save the owner significant money and schedule time. Late changes and suggestions for redesign may also upset other team members and possibility coax the designer or engineer's nose out of joint.

My experience in performing this service confirms that an uncanny number of the same mistakes and inaccuracies occur and re-occur in site-construction documents. Like a paper-bound virus, these items auger their way into the plans and specifications regardless of the product or the planner, architect, or engineer. Here are some examples of shortcomings in regard to construction.

- Holes in the documents. Incomplete drawings lead to inaccurate and incomplete bids.
- Manufacturing firewood on-site. For reasons of liability and common sense, forbid cutting, splitting, or selling firewood on-site. Only allow the logger or contractor of record to cut, load, and haul off logs and firewood. Some jurisdictions even regulate where logs and wood by-products can be hauled to and how they're to be processed or sold.
- Leaving trench backfill a mystery. Don't leave it up to the contractor to decide whether to backfill with clean structural fill, crushed rock, or lean concrete mix. Costs for each product vary widely and the wrong backfill can impact future maintenance work. Clearly specify station-by-station if necessary, materials and methods for placing trench backfill.
- Unclearly defining how things are to be preserved. The term "Protect During Construction" is a catchall phrase commonly dropped into the plans and specifications. To many that pack a lunch box, this expression is meaningless. It's neither safe nor prudent to leave it to a young and inexperienced dozer operator to interpret such commands and react as he sees fit. Clearly specify what is to be protected, how protection is to be applied, and the conditions that are to be preserved.
- Defacto contractor responsibility. Unless you trust, partner, or negotiate an open-ended agreement with your site-work contractor, leave him little or no opportunities for subjective thinking. Pay for detailed documents and be thankful when you can roll out a set of complete and well-thought-out plans and specifications. Using contractors to "fill-in-the-blanks" can be an expensive and counter-productive way to compensate for an architect's or engineer's design and drafting inefficiencies.
- Setting grading tolerances too high. Owners can make the mistake of

setting grade tolerances too high. Plus or minus one-foot is too much tolerance to allow on most site grading jobs. It potentially allows the contractor to leave an average of almost two-feet of variation across the site. Sellers will either have to perform significant regrading or commission additional import or export in order to establish correct elevations. Buyers are likely to request a hefty discount on sites requiring secondary grading.

- No statement that the storm system is clean. Clean the existing utility structures and note it in the documents. Successful jobs leave no stone unturned including closely examining and recording the conditions of *all* existing storm lines and basins prior to and after construction. This means that owners should clean and inspect these items before the contractor gets on the job. In completing and documenting this work, smart owners leave no question as to who is responsible later should storm or sewer systems be damaged or filled in during construction.
- Be careful about defining contractor performance. When it comes to defining how a contractor performs, it's best to let the contractor decide the means for meeting a performance standard. If an owner specifies a method for performing work that ultimately fails, whether or not the contractor properly executed or was even negligent may not be enough to deter placing liability for non-performance on the owner. Contractors can argue that they just followed the owner's orders. That they were unaware that the owner's recommendation would prove inadequate, unsuitable under present conditions, or even negligent. Remember—when specifying methods of operation and procedure, one must not only be right; one must also specify the same methods consistently throughout the documents.
- The same logic applies to recommending use of certain types of equipment. Although some governmental agencies have no choice but to specify the use of certain equipment in order to receive public funding, it's inherently unwise for an owner to suggest what type, make, or year of heavy equipment a contractor should use. Yes, specifying tools and procedures for contractors does display a sense of bravado, but it also promotes risk and unwittingly can make an owner liable if things don't work out for the contractor. And, there's no question that it results in a more expensive project. Let contractors choose their own gear—they get paid to know better.

Here are some examples of shortcomings in regard to design and engineering.

- Specifying materials such as a particular crushed rock that is unavailable locally.

- Unenforceable and unrealistic conditions. Imposing conditions on a contractor or project that are unenforceable is not only a mistake, it's counter-productive, may backfire, and is prone to misapplication.
- Lack of or insufficient erosion-control facilities. At any time, there may be a need for temporary sedimentation ponds—especially in late fall. Include a plan and specification should temporary ponds be required.
- Lack of textiles under crushed rock. Fabric is cheap. It stabilizes and supports erosion-prone soils. Under the right conditions, an investment in fabric under gravel and crushed rock that keeps the rock clean and prevents it from sinking into the ground is money well spent.
- Don't be lulled into believing that sawcutting and replacing existing alligatored pavement as a condition of approval is always a good deal for owners. Accepting blanket responsibility for existing road repair of alligatored pavement can easily lead to an entire repaving job and sometimes wholesale subgrade replacement and utility work.
- Natural topographic and proposed grading elevation lines cannot dangle—they must connect. Lines depicting final grades should tie into existing and ungraded topography. All elevations must be readable, accurate, and remain well-defined with no intermittent merging, crossing over, or ambiguity in topography lines. No gaps, no interpolation—no disputes.
- Micro-area grading design must be predicated on accurate topography. Know that your topography is right by checking field grades, topography, and existing control. Although this problem can occur anywhere on-site, it frequently applies to smaller and isolated sites, such as bridge and culvert crossings, pump stations, storm ponds and vaults, and reservoirs.
- Utility conflicts. Keep wet and dry trunk utility lines on the street-side of curb line. Avoid installing them on a chord, in and out of long road curves and under planters where they may later conflict with future irrigation lines, tree spading, planter work, or street pole installation. This also prevents installing bare lines beyond curbline without placing them in sleeves.
- Unnecessary silt fencing. Pay close attention to silt fence locations. Design silt fencing low enough to where it can catch water and control erosion. Installing silt fence too high performs no function except to collect lunch bags and flying debris. When designers blindly throw continuous fencing everywhere regardless of topography in order to demonstrate that a site won't overflow, an overflow still occurs in the form of money spilling out of the owner's checkbook.

- Unnecessary force mains. Exhaust all options before committing to force mains. Force mains and sewer vacuum systems may be necessary but they can also be the result of lazy design, poor utility layout, uncreative grading, or woeful planning. Not only are they costly, these systems often require access, a dedicated station facility, use of buildable land, a continual source of power, telemetry, back-up systems, monitoring and maintenance, and more piping in the road right-of-way.
- "Storm drain on city plans not found." Owners should verify the status of any and all suspected existing underground utilities, septic tanks, cesspools, wells, and the like. Insist that the civil engineer or architect positively identifies and inventories existing utilities in lieu of leaving them for a contractor to deal with later. This may include potholing or lifting lids and grates to measure existing pipe inverts, sizes and alignment, and inspecting system condition. Include costs for potholing in the budget. For owners, it's patently cheaper and more cost effective to research and locate existing pipes before construction begins. Besides, knowing the status of existing utilities results in better design, ensures gravity flow, and secures being able to connect to the existing system without difficulty. Conversely, designing gravity systems with wrong invert data on flat ground can be disastrous for owners. If conflicts occur or connections don't match, by the time a contractor has laid multiple lengths of pipe, it may be too late to change the pipe slope and the structures without significant cost and schedule impact. Knowing more upfront also helps protect owners later should disputes and claims follow construction.
- Bottom-of-the-barrel service. Some architects, engineers, and designers will gladly accept a pass on quality, timeliness, and thoroughness. When granted, an owner is liable to inherit substandard plans. Owners must work to avoid receiving:
    1. Junior engineering
    2. Incomplete and inaccurate work
    3. Insufficient research of field, as-built, and historic records
    4. Potential claims that information concerning existing utilities was "owner provided" or even worse, was "received verbally from the owner"
    5. Ineptness—You have to ask yourself this: if a designer cannot positively identify man-made objects, how can you expect the same designer to handle natural subsurface anomalies such as dirt, rock, and water?
    6. Increased owner liability—It's wrong and dangerous to expect a contractor to absorb the cost, responsibility, and risk of finding and

accounting for existing underground utilities. This action also puts owners at risk. It can produce high incidents of utility conflict and downtime, high charges from utilities for service interruptions and repairs, and contractor claims against the owner. Moreover, since owners have to pay for this work one way or the other, it's always more expensive when it is performed by production-minded contract pipe utility crews rather than by engineers or as a prebid potholing exercise.

Landscape plans are also prone to being incomplete and are sometimes disengaged from architectural and engineering design. Because landscaping usually occurs later in a project, other site designers may feel that it's the landscape architect's responsibility to remain informed. Or, let the owner worry about it. Some site designers working with a tight budget or having problems with an owner may feel reluctant to spend any time or money on what they perceive as "beyond the call of duty" coordination needed to keep a project moving efficiently. That's provided they've even thought about future planning and coordination. Another problem involves change orders. Unless someone is actively controlling the project schedule and alerting all parties to changes, a landscape architect may unknowingly pursue a design for conditions that no longer exist. Unfortunately, shortcomings in landscape design and specifications occur. Here are a few examples to look out for:

- Metering. Separate irrigation metering from potable water metering if your jurisdiction bases sanitary sewer charges on potable water consumption.
- Proliferation of unrelated information. Eliminate specifications referencing grass, mulch, jute, and erosion control products not pertinent or unacceptable to your project.
- Poor sequencing. Design street landscaping in response to hardscape—not the other way around. Figure out locations for utilities, street lighting, traffic poles, dry utility vaults, tree wells, and monumentation before finalizing landscape design.
- No provision for power or conduit. Irrigation controllers, specialty lighting, and meters all require power. Make sure that ample conduit, sleeves, power (and possibly phone) is available and accessible before final grading and pavement installation.
- Specifying stock that won't survive. Don't design for, sell concepts based-upon, or specify growing stock, such as trees, shrubs, and grasses, that won't survive, aren't readily available, are too expensive, are at-risk of theft, or that won't easily transfer on-site.
- Ill-conceived match lines. Use correct alignments and transfer informa-

tion accurately within documents. Incorrect bearings and erroneous or contorted split-sections that result from re-aligning details in order to fit landscaping onto different sized plan sheets or that are carried over to another consultant's work, can result in erroneous quantity take-offs and exaggerated design.

- Mucking out landscaping and topsoil. Don't plant or landscape ponds if you'll periodically muck them out. Why haul in topsoil if it's going to go out the door later mixed with sediment from storm runoff *along with* your investment in plantings? Performance or maintenance bonds associated with such plantings subject to excavation can create havoc for owners. Refuse to be a victim. Plant and bond once, and design ponds that clean out easily without removing growing medium and vegetation.
- Ignoring the maintenance department. Be cognizant of potential landscape maintenance issues even if permits have been issued that condone steeper than normal grassed slopes, difficult corners, or intricate water and landscaped features. Public works and maintenance departments responsible for handling deeded right-of-ways can comment or raise issues at any time that may impact an owner's budget and schedule. Many a change order has been triggered late in the game once grounds crews have gotten wind of something that they perceive will be a nuisance or difficult and overly-expensive to service, take care of, or manicure. It's smart to include street maintenance departments in the plan review and approval process.
- Planter information is missing. Design planter backfill below the elevation of final topsoil. The zone of planter material needed between subgrade and final grade must compensate for the varying thicknesses of road subbase, base course, pavement, structural fill, curbs, and sidewalks. More often than not, this area is missed in everyone's scope of work resulting in planter fills being a "no-man's land." Specify who is responsible for filling planters to elevations upon which landscapers expect to place screened or mixed topsoil. Specifying what type of materials are *not* acceptable for deep planter backfill can be a waste of good ink since planters are famous for collecting garbage and waste construction debris. Instead, in a contractor's scope of work, specify cleaning out garbage, backfilling with specified material, and compacting if necessary.

Techniques for improving plans and specifications

The degree in which an owner can impact the plans and specifications of a project depends upon the size and complexity of that project and the owner's innate ability and experience. Still, here are some ways to technically improve and add clarity to the site-construction documents:

- Never allow a footing drain to double as a French Drain.
- Stub all catch basins and most manholes to back of sidewalk.
- Include the names of adjacent landowners and detail their property corners and bordering structures.
- Create a list or table of acceptable substitute materials and have it placed in the plans and specifications.
- Tell design consultants early that a third party peer review entity will be employed to assess all construction documents upon completion. Then do it.
- Group all errors of tolerance per phase of construction or by trade and express them in the same system of measurement such as decimal, feet/inch, or metric.
- Insist that all wet-utility (sanitary, storm, and water) plans include both plan and profile views. Profiles reveal problems as well as opportunities and allow for more accurate estimating.
- List the names of key individuals to the project and utility service contacts including titles, addresses, phone numbers and e-mail addresses. Clearly note numbers for emergency, fire, police, and hazardous material contacts.
- Include all geotechnical work as part of the construction documents. Investigate contradictory work by other geotechnical firms and, if necessary, perform additional exploration to verify subsurface site conditions. Make all reports and studies available to contractors.
- Insist that scales remain consistent throughout plan sets. This promotes accurate quantity take-offs with less chance for error, and it ensures being able to overlay and compare plans and profiles on a true, one-to-one basis. This practice also provides non-technical people with a consistent point of reference for studying the plans.
- Consolidate information and details in schedules or tables for storm, sanitary, and utility manholes, catch basins, and vaults. Schedules should include structure number, type of structure, top elevation, horizontal coordinates, size, and direction of penetrations, invert elevations and any other related design information.
- Portray unfamiliar and new cutting-edge techniques or products in the

plans. Designers must do whatever homework is necessary to think out concepts and commit them to paper before asking a contractor to pull off something never seen before in the field. Drawings are also necessary for cost-estimating purposes. To do otherwise invites high costs and possible change orders, or a dispute.

- If the civil engineering work must be broken up, it makes logical sense to package erosion control, clearing and grading, storm facilities, roads, and storm lines into one contract and the sanitary sewer, water lines, and dry utility system design into a second package. (The same breakdown can be applied to contractors) It can be dangerous to breakdown design and engineering any further.
- Armed with potholing and field measurements, identify and solve conflicts between existing and proposed utilities on paper instead of in the field. Designate one consultant—usually the one who is developing the grading, road, and utility plans—to either take the lead with dry-utility coordination or be responsible for merging dry-utility design with project-wide engineered drawings. This helps to identify conflicts between wet and dry utilities and minimizes their collective impact on lot layout, easements, and dedicated right-of-ways.
- Some projects only provide infrastructure and sell parcels "as-is." Consider how these parcels will be developed before locating wet- and dry-utility stubs, curb returns and road entrances, street lighting, and landscaping. Run mock grading plans to determine possible retaining wall or cut-and-fill slope locations adjacent to the edge of right-of-way. With the possible exception of a retaining wall or yard drain stub, don't plan points of access or utility stubs where a retaining wall or slope appear likely. Provide a manhole to accept a future drain connection if walls will be likely. Also provide locations for stubs where it appears likely that a buyer will connect gravity storm and sanitary-sewer lines. This also helps to locate entrances.
- Objectively measure performance and completion of work. Loose and ambiguous language requiring that work be complete to the "satisfaction of the engineer or architect" may later breed controversy. Without objectivity, completeness lies in the eye of the beholder! Also purge ambiguous language defining material substitutions as "equal to or as approved by the engineer." When bidding lump-sum work, such phrases ring hollow. This is especially true for contractors responding on short notice or in fast-track bid situations. On a similar note, purchase and sale agreements that allow a buyer or builder wide and subjective authority to accept or reject an owner's (seller's) work based upon aesthetics, state of completion, or perceived quality, can lead to

long and heated meetings followed by grab-samples of aspirin and stout caffeine cocktails. If performance can't be measured against clear, objective, and easily identifiable standards, why measure it at all?

- Coordinate consultants and required studies. Projects normally follow a script where the owner assumes a passive role and elects to have an architect manage up-front coordination. With minimal owner influence, the architect then organizes subcontractors to perform environmental-impact reports, sensitive area analysis, geotechnical work, survey and boundary delineation, site engineering, and so forth. At some point in time, the project planning and design process gains momentum and can actually progress on its own. However, until a project materializes and begins to pass milestones, such as gaining initial financing, entitlements, and early approvals, it may be solely up to the owner to drive the process and keep it in gear. This mandates knowing when to employ the right consultants and studies needed to maintain schedule. The key element is allocating enough time to collect the elementary base data needed to support design, engineering, and permitting. Many times, this is best handled by the owner.
    - Clearly state which contractor (building or site-work) is to cross the line that separates the building envelope from the site in order to make utility connections. Relying on blanket statements that command contractors to "coordinate stub connections with each other" is like telling kids to behave in the back seat. If it's unknown which contractor will get to the stub location first, or, if the wrong contractor gets there first and out of sequence, then someone's probably going to be waiting around with a bucket in the air asking who's to make the connection. This issue can become especially sensitive when trade and union restrictions prohibit work by certain contractors within a specified distance of a building. Problems increase when connections on the architectural and civil drawings differ. Under these circumstances, it's almost guaranteed that lines coming into and out of a building will miss their connection point or be in conflict, which will then require flexible fittings and additional pipe work or even a complete utility re-installation. The solution, of course, is to coordinate the building to site utility connections.
    - Be explicitly clear when specifying and defining units of dirt. Don't label units of earth generically as "cubic yards" or as "CY" throughout the site-construction documents. Why? Because a cubic yard will change in size as its excavated, loaded, hauled, and used as fill. Who knows what a cubic yard is? Raise your hand. Wrong! It's not

that easy because there exists at least three ways to define a dirt cubic yard. Here they are:

1. Bcy–Bank cubic yard
2. Lcy–Loose cubic yard
3. Tcy–Truck cubic yard

Those who have read Book 1 in this Project Logic Series, the *Road Repair Handbook*, should already understand how cubic yard definitions differ from each other and what that means when working dirt. Project owners and designers need to realize that the old generic cubic yard (CY) definition common in site-construction documentation is a doorway to potential disputes, claims, and financial pain. Unless a cubic yard is clearly defined and used consistently in the same manner throughout the cost estimate, the schedule of values, and the construction documents, project owners remain exposed to having a contractor interpret the meaning of cubic yard for them, the result of which can lead to outrageous increases in dirt volume and cost. There's more on this subject in the next chapter on cost estimating.

The same applies to mixing tons and cubic yards. Besides having the potential for blowing a cost estimate right out of an accountant's briefcase, this dangerous combination can also leave gaping holes in the budget. It takes two factors to successfully convert tons to cubic yards. The first is defining cubic yard and the second is knowing the weight per unit of the material being hauled. Owners electing to proceed indefinitely with a fixed-tons-to-cubic-yard conversion factor are limited to importing material from a source or pit that offers product that is consistent in weight for that conversion factor. Cost savings that could be realized by changing pits or sources of material are reserved for owners willing to search out cheaper dirt and to send a geotechnical engineer to inspect it's quality, to run laboratory tests on it, and to review it's suitability for project construction. Recommendations for specifying how units of material should be labeled and measured is covered in the next chapter on cost estimating.

Improve your construction documents, improve your life

The benefits of proceeding with site-work (or any phase of construction for that matter) by using well-prepared and complete plans and specifications are many. Project owners should experience:

- Easier bidding
- Easier permitting
- Lower construction cost
- Reduced owner liability

- Fewer construction delays
- Attention of high caliber contractors
- Decreased risk of claims and change orders
- A positive reflection on owners and their project
- Improved chance of meeting construction schedule
- Increased consultant control and cost predictability
- More accurate and complete bids and cost estimates

Owners who manage the development of their construction documents proactively rather than reactionarily are more inclined to receive comprehensive and complete plans and specifications. They are also better prepared to address planning issues and to understand site-work as it affects sales and marketing. Here are ways to ensure receiving superior plans and specifications.

1. Use your contractor as a consultant.
2. Develop plans that remain market- and product-flexible.
3. Keep your vision for the project clear and don't deviate from it.
4. Use a flow chart to assign responsibility early on, and stick to it.
5. Keep everyone around you informed with the latest information.
6. Hold regular meetings, take detailed notes, and impose deadlines.
7. Share unique design elements with all consultants who should know about them.
8. Every question is a good question so ask them when you don't understand something.
9. Periodically obtain and file digital and reproducible copies of all plans and specifications.
10. Improvising is expensive—know when to abandon a design and start again with fresh ideas.
11. Designate one individual to coordinate and manage the development of documentation and plans.
12. Schedule prepermitting application meetings to ensure that government agencies are aware of project progress. This also gives agency personnel a chance to comment and to be involved in design and decision-making.
13. Request the inclusion of specific plan sheets for items such as base control and surveying, demolition, and erosion control. This gives you better-informed contractors, stronger bids, and an increased chance of proceeding with complete engineering and accurate details.
14. Whenever redlining the construction documents, resist handing them over and trusting that all changes will be integrated into a new set of

plans. Instead, hand them over only after procuring for yourself, a black line copy enabling you to later back check that all changes have been incorporated into the plans and specifications.

Remember that it takes less time and money to proceed with well-prepared, quality plans and specifications than it does to handle design through change orders and as-builts. Capitalize on doing it right the first time.

Tips for managing construction documents

Construction documents get physically mismanaged more than most owners would care to admit. Here are a few money-saving tips.

- Use half-scale drawings. Once you have approved and permitted sets of plans, run copies of them in half-scale. Half-scale drawings are sets of smaller-sized plan-sheets that are easy to carry in the field or manipulate in the cab of a pickup or piece of equipment. They are also cheaper to run and take up less space in the office.
- Don't pay for plan sheets and burdensome paper that you don't need. Some owners pay architectural and engineering firms based upon the number of sheets provided in a plan set. This can be a lazy and costly way of doing business. Besides the obvious and always viable temptation to increase the number of plan sheets solely as a vehicle for fattening billings, there exists no logical connection, no magical formula, that links project design or an architect or engineer's performance with the number of sheets included in a plan set. They're mutually exclusive. On the contrary, plan sets should contain the absolute *minimum* number of pages needed to do any phase of work. To save money and increase space on the plan rack, avoid adding redundant and unnecessary sheets to sets of plans.
- Know your printer. First, insist on using print and copy shops that possess enough resources to meet your needs and demands. Project owners cannot wait indefinitely for plans, documents, and reproductions. Second, periodically check the shops' pricing and customer service. Walk in and ask questions. Third, develop a feel for knowing when you or your consultants are ordering too many complete plan sets or when single sheets or partial sets will suffice. Until you're convinced that your consultants are making good business decisions and looking out for your best interest, examine whomever they use for your print and copy work. Afterall, print and copy work is a pass-through, marked-up charge imposed by consultants and sometimes, contractors. They order whatever quantities they feel are necessary to meet project objectives and owners ultimately pay for the service.

Scrutinizing documents is one of the easiest and cheapest ways to preserve profit. One way for owners to increase their return is to hire a third-party consultant to review documents and to, remove ambiguity, or suggest ways for improving the site-construction documents in advance of bidding or negotiating work. Inspecting documents regularly helps guarantee that they'll be what you expect when you expect them.

Chapter 13

SITE-WORK COST ESTIMATING

Luxury in site-work need not be illusive or expensive. However, it does require being prepared. Owners who engage in site construction with detailed, accurate, and informed cost estimates find pleasure and profit easier to achieve. In site-work, until there is a complete cost and quantity estimate, everything else is truly—make-believe.

When describing the components and attributes of a quality site-work cost estimate, words like detailed, thorough, and all encompassing should come to mind. Someone in the industry should be able to peruse a well-built site-work cost estimate and, without having reviewed the plans and specifications before, be able to accurately visualize the project and scope of work. Site-work cost estimates should be that good.

Options for estimating

Site work can be estimated in multiple ways. Estimates can be preliminary or rough in nature relative to where a project lies in design, financing, and entitlement. Cost estimates can also be based upon square unit area such as when estimating structures. Estimating costs per unit area may be useful when"rule-of-thumb" techniques will suffice for evaluating properties situated in like geographic areas where recent construction activity has produced a database of reliable unit costs for the same product type. Examples include large-scale business parks and high-rise construction. The key words when using area as a basis for estimating are *controlled conditions* and having a *database* of information. A third type of estimate involves detail and relies on having complete plans, specifications, documentation, and reports. In this chapter, we'll discuss uses and shortcomings of the first two type of estimates mentioned above but we'll focus primarily on the last type—the detailed, line-item site work estimate.

From worm to butterfly

Construction cost estimates drive budgets, and budgets support reports, summaries, and projections upon which proformas are based. Because these links are inextricable, it makes sense to create and utilize a well laid out cost-estimate format from the beginning. Because—like a worm—a bland and simple cost estimate can grow into, and underpin, a higher-level of documentation that staff will appreciate and depend upon.

Figure 13-1 illustrates how a well-built construction cost estimate can double as the base document for controlling contractors, costs, and schedule. Cost estimates that contain a wide unit-price schedule of values are useful in corralling and predefining change orders before they occur thereby helping set the tone for project accounting. The budgets and cost summaries linked to the base-construction cost estimate can seamlessly accept price and quantity changes and can automatically adjust the higher-level financial statements and reports. Therefore, according to changes made to the base site-work cost estimate, an in-house accounting system can track costs and tie those costs to budgets. Any alterations and forward comparisons linked to either the revenue, or the proforma, become straightforward.

Before settling on a final cost-estimate format, it's a good idea for the estimator to ask the accounting department how they plan to post costs. This provides a golden opportunity to discuss with accountants how the project will unfold and how that relates to organizing items sequentially in the cost estimate. The estimator can then format sections of the site-work estimate to best serve the accountants, so that the accountants gain a greater understanding of the estimate's structure as it relates to construction sequence and schedule. Accountants can also instantaneously tie budget changes to the changes in the base cost-estimate and better understand where money is being planned for spending. Both sides may not agree on everything, but they'll benefit from the exchange.

The potential for electronic entanglements makes it imperative for owners to settle on and use the right accounting software throughout the life of the project. Accounting software should be flexible enough to link with construction cost-estimating software and off-the-shelf spreadsheet packages. Compatible software that allows accounting to electronically link to the base cost-estimate eliminates the need to manually transfer loads of information between spreadsheets thus reducing the risk of error when conveying or interpreting information. Electronic linking also minimizes the opportunity for someone to indiscriminately alter construction cost and budget information. Compatible formats and spreadsheets enhance communication between field and office personnel and make it easier for financial lenders to review, correlate, and evaluate cash flow and cost reports. While a project is underway, it's a mistake to change an entire software system or to install new software as you may find out later that it doesn't merge properly with existing in-house packages. Begin projects with a software system that can go the distance.

The contractor pay request and accounting control package shown in *Figure 13-3* has been built on the chassis of the base cost-estimate spreadsheet shown in *Figure 13-2*. The Knights Bridge Office Park project site-work

A Well-Built Cost Estimate Can Serve Many Purposes

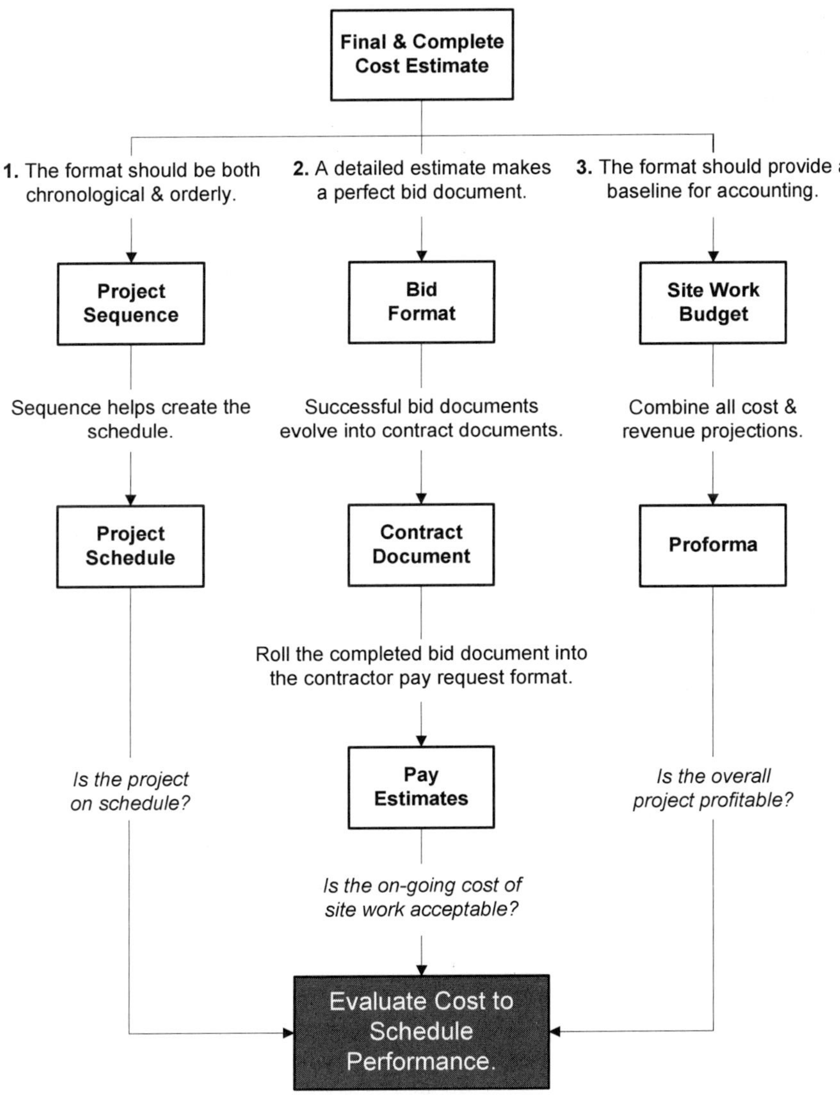

Figure 13-1

Site-Work Cost Estimate Example

Construction Activity Description	Quantity	Units	Unit Price	Bid Amount	Tax	Total Contract
	(1)	(2)	(3)	(4) (1)X(3)	(5) (4)X(Tax Rate)	(6) (4)+(5)
LOGGING & CLEARING						
Logging - (Net)	225.00	MBF	(275.00)	(61,875.00)		(61,875.00)
Clear, Grub, Chip, Haul off Chips	12.00	AC	4,400.00	52,800.00		52,800.00
Temporary Access Road Budget	1.00	LS	15,000.00	15,000.00		15,000.00
				5,925.00		5,925.00
TEMPORARY EROSION CONTROL						
Filter Fabric Fence	800.00	LF	4.95	3,960.00		3,960.00
Maintain Existing Fence	1.00	LS	10,000.00	10,000.00		10,000.00
Trenches & Ditching	3,000.00	LF	0.35	1,050.00		1,050.00
Rock Check Dams	12.00	TON	35.00	420.00		420.00
Riprap Basins	12.00	EA	750.00	9,000.00		9,000.00
Temporary Siltation Pond Facilities	1.00	EA	7,500.00	7,500.00		7,500.00
Temporary Fabric Over CBs	25.00	EA	35.00	875.00		875.00
8" Temporary Plastic Pipe	340.00	EA	13.00	4,420.00		4,420.00
12" Temporary Plastic Pipe	300.00	EA	16.00	4,800.00		4,800.00
Straw Mulch	25.00	EA	750.00	18,750.00		18,750.00
Hydroseeding	12.00	EA	975.00	11,700.00		11,700.00
Truck Wash	1.00	LS	21,000.00	21,000.00		21,000.00
				93,475.00		93,475.00
GRADING & EARTHWORK						
Strip & Haul Topsoil to Processor	15,000.00	BCY	2.90	43,500.00		43,500.00
Load & Haul Unsuitable to Waste Area	7,000.00	BCY	3.85	26,950.00		26,950.00
Mass Grade, Cut to Fill	8,500.00	BCY	2.65	22,525.00		22,525.00
Mass Grade Cut to Export	60,000.00	TCY	8.00	480,000.00		480,000.00
1.5:1 Slope Grading	6,500.00	SY	1.25	8,125.00		8,125.00
Import Structural Fill Budget	2,500.00	TCY	12.00	30,000.00		30,000.00
De-watering Budget	1.00	LS	12,000.00	12,000.00		12,000.00
				623,100.00		623,100.00
ROAD WORK						
Post-Hauling Subgrade Repair	1.00	LS	7,500.00	7,500.00		7,500.00
Import Structural Fill, Pit Run	4,500.00	BCY	12.00	54,000.00		54,000.00
Export Slop/ Cut	3,000.00	BCY	3.75	11,250.00		11,250.00
Roadway Fine Grading	22,000.00	SY	0.16	3,520.00		3,520.00
Grade, Compact, Backfill Curb & Gutter	8,900.00	LF	2.35	20,915.00		20,915.00
Backfill Planters with Raw Material	3,100.00	BCY	3.50	10,850.00		10,850.00
Road Fabric Budget	5,000.00	SY	1.00	5,000.00		5,000.00
				113,035.00		113,035.00
CURBS & SURFACING						
Vertical Curb & Gutter @ 18"	8,900.00	LF	9.00	80,100.00		80,100.00
Wheelchair Ramps	28.00	EA	240.00	6,720.00		6,720.00
6" Concrete Driveway Apron	32.00	EA	350.00	11,200.00		11,200.00
Sawcut @ 4"	200.00	LF	3.00	600.00		600.00
Asphalt, 2 inch Class B Overlay	21,600.00	SY	4.50	97,200.00		97,200.00
Asphalt, 4.5 inch Asphalt Treated Base	21,600.00	SY	7.00	151,200.00		151,200.00
Concrete Pattern Cross-walk	90.00	SY	85.00	7,650.00		7,650.00
				354,670.00		354,670.00
SIDEWALKS						
Fill, Grade, Compact Sidewalks w/ Native	7,100.00	SY	1.50	10,650.00		10,650.00
2" Pit Run/ Rock Base	7,100.00	SY	4.67	33,157.00		33,157.00
4 in Varied Width Concrete Sidewalk	7,100.00	SY	10.00	71,000.00		71,000.00
				114,807.00		114,807.00
ROADWAY LIGHTING, SIGNS						
Street Lighting Bases	64.00	EA	2,500.00	160,000.00		160,000.00
Signage Budget	1.00	LS	9,700.00	9,700.00		9,700.00
Street Markings Budget	1.00	LS	18,000.00	18,000.00		18,000.00
				187,700.00		187,700.00
LANDSCAPING						
Backfill Planters, Import Topsoil	6,000.00	TCY	10.75	64,500.00		64,500.00
Street Trees & Grates	80.00	EA	275.00	22,000.00		22,000.00
Landscape & Irrigate Shoulders	60,000.00	SF	4.00	240,000.00		240,000.00
				326,500.00		326,500.00
DRY UTILITIES						
Dry Utility Crossings	8.00	EA	1,500.00	12,000.00		12,000.00
Joint Trench Installation	3,500.00	LF	22.00	77,000.00		77,000.00
Data System Conduit	3,500.00	LF	5.00	17,500.00		17,500.00
Data System Vaults	14.00	EA	1,500.00	21,000.00		21,000.00
Adjust Existing Utilities	1.00	LS	12,000.00	12,000.00		12,000.00
				139,500.00		139,500.00
GENERAL SITE WORK						
Mobilization	1.00	LS	6,000.00	6,000.00		6,000.00
Install Irrigation Meters	2.00	EA	850.00	1,700.00		1,700.00
				7,700.00		7,700.00

Figure 13-2

	(1)	(2)	(3)	(4) (1)X(3)	(5) (4)X(Tax Rate)	(6) (4)+(5)
Construction Activity Description	Quantity	Units	Unit Price	Bid Amount	Tax	Total Contract
STREET & AREA MAINTENANCE						
Street Sweeper	240.00	Hours	90.00	21,600.00		21,600.00
Water Truck	240.00	Hours	90.00	21,600.00		21,600.00
				43,200.00		43,200.00
FIELD CONSULTANT BUDGET						
Geotechnical Inspection	1.00	LS	40,000.00	40,000.00		40,000.00
Survey Control, Layout, As-Builts	1.00	LS	48,000.00	48,000.00		48,000.00
				88,000.00		88,000.00
Wet Utilities						
SANITARY SEWER						
8" SDR 35 PVC 0' - 12' deep	240.00	LF	20.00	4,800.00		4,800.00
12" SDR 35 PVC Pipe 0' - 8' deep	860.00	LF	24.00	20,640.00		20,640.00
12" SDR 35 PVC Pipe 8' - 12' deep	440.00	LF	24.00	10,560.00		10,560.00
12" SDR 35 PVC Pipe 12' - 15' deep	30.00	LF	27.00	810.00		810.00
12" DI Cl 50, Pipe 15' - 18' deep	80.00	LF	47.00	3,760.00		3,760.00
48" Sanitary Manholes	9.00	EA	1,645.00	14,805.00		14,805.00
48" Sanitary Manholes, Extra V.F.	24.00	VF	216.00	5,184.00		5,184.00
60" Sanitary Manholes	2.00	EA	2,797.00	5,594.00		5,594.00
60" Sanitary Manholes, Extra V.F.	52.00	VF	297.00	15,444.00		15,444.00
Adjust Existing Utilities	1.00	LS	7,500.00	7,500.00		7,500.00
Cut Off Dams (If Required)	3.00	EA	190.00	570.00		570.00
Manhole Vacuum Testing	11.00	LS	475.00	5,225.00		5,225.00
Export Trench Displacement	900.00	BCY	8.50	7,650.00		7,650.00
Trench Import	425.00	BCY	9.50	4,037.50		4,037.50
Connect to Existing	2.00	EA	2,600.00	5,200.00		5,200.00
				111,779.50		111,779.50
STORM SEWER						
6" N-12 Pipe 0' - 8' Deep	230.00	LF	8.50	1,955.00		1,955.00
8" N-12 Pipe 0' - 8' Deep	175.00	LF	12.50	2,187.50		2,187.50
12" N-12 Pipe 0' - 8' Deep	600.00	LF	16.00	9,600.00		9,600.00
24" N-12 Pipe 15' - 18' Deep	130.00	LF	40.00	5,200.00		5,200.00
30" N-12 Pipe 0' - 8' Deep	1,650.00	LF	47.00	77,550.00		77,550.00
30" N-12 Pipe 8' - 12' Deep	220.00	LF	50.00	11,000.00		11,000.00
30" N-12 Pipe 12' - 15' Deep	248.00	LF	53.00	13,144.00		13,144.00
30" N-12 Pipe 15' - 18' Deep	55.00	LF	62.00	3,410.00		3,410.00
Type - 1 Catch Basin, Install Solid Covers	6.00	EA	125.00	750.00		750.00
Type - 1 Catch Basins & Grates	15.00	EA	680.00	10,200.00		10,200.00
Type - 2, 48" Manhole & Grates	12.00	EA	1,619.00	19,428.00		19,428.00
Type - 2, 48" Manhole Extra Vertical Feet	48.00	VF	216.00	10,368.00		10,368.00
Type - 2, 60" Mahhole & Grates	3.00	EA	2,797.00	8,391.00		8,391.00
Type - 2, 60" Manhole Extra Vertical Feet	56.00	VF	297.00	16,632.00		16,632.00
Adjust Structures to Grade	1.00	LS	6,500.00	6,500.00		6,500.00
Solid Locking Manhole Lids	15.00	EA	94.00	1,410.00		1,410.00
Cut Off Dams (If Required)	6.00	EA	190.00	1,140.00		1,140.00
Export Trench Displacement	1,400.00	BCY	8.50	11,900.00		11,900.00
Trench Import	750.00	BCY	9.50	7,125.00		7,125.00
Connect to Existing	1.00	EA	950.00	950.00		950.00
Miscellaneous Removals & Disconnects	1.00	LS	5,000.00	5,000.00		5,000.00
				204,898.00		204,898.00
WATER SYSTEMS - Trunkline & distribution						
6" Class 52 D.I. Waterline w/Fittings	1,100.00	LF	19.00	20,900.00		20,900.00
6" Gate Valves	4.00	EA	525.00	2,100.00		2,100.00
8" Class 52 D.I. Waterline w/Fittings	240.00	LF	21.00	5,040.00		5,040.00
8" Gate Valves	1.00	EA	621.00	621.00		621.00
12" Cl 52 D.I. Distribution System	2,550.00	LF	32.00	81,600.00		81,600.00
12" Cl 52 Restrained Joint Trunkline	2,370.00	LF	41.00	97,170.00		97,170.00
12" Gate Valves	5.00	EA	1,100.00	5,500.00		5,500.00
1-1/2" Irrigation Meter & Setter	2.00	EA	1,200.00	2,400.00		2,400.00
1-1/2" Irrigation Double Detector Check Valve	2.00	EA	1,700.00	3,400.00		3,400.00
Hydrants with Stortz Fittings	6.00	EA	2,400.00	14,400.00		14,400.00
2" Air Vac	2.00	EA	1,994.00	3,988.00		3,988.00
2" Blow Off	4.00	EA	1,100.00	4,400.00		4,400.00
Export Trench Displacement	1,650.00	BCY	8.50	14,025.00		14,025.00
Trench Import	450.00	BCY	9.50	4,275.00		4,275.00
Adjust Existing Utilities	1.00	LS	3,000.00	3,000.00		3,000.00
Connect to Existing	2.00	EA	2,300.00	4,600.00		4,600.00
Thrust Blocks	17.00	EA	200.00	3,400.00		3,400.00
Pressure Test, Chlorinate, Dispose of Flushed Water	1.00	EA	8,000.00	8,000.00		8,000.00
				278,819.00		278,819.00
Contract Subtotal				2,712,051.00		2,712,051.00
Contingency - at 12%				325,446.12		325,446.12
Total Contract before Bond				3,037,497.12		3,037,497.12
Bond Amount				0.00		0.00
Total Contract Amount				**3,037,497.12**	**0.00**	**3,037,497.12**

Figure 13-2 continued

contractor has seen this form at least twice. He first saw it as a blank bid document and again later as a spreadsheet for submitting electronic or hard-copy pay requests to the owner.

By using the base-construction cost-estimate format for other purposes, owners maximize their return on the time and expense spent building the estimate. Other benefits include streamlined paperwork, improved continuity between departments, and establishing a higher degree of trust between team members. Staff people and accountants are happiest when they can understand the site-construction cost estimate.

Challenges to estimating

If any phase of construction warrants the use of subjective reasoning, it's site-work. Unlike structures where known geometry and human design make things predictable, it's not uncommon for site-work to be riddled with variables that are tough to predict and tougher to estimate. While there are always man-made issues, such as grading, utilities, and pavement, to cope with, it's the intangibles—the hidden secrets belowground and the possible wrath of nature—that burden site-work estimators. Here's a partial list of elements that site-work estimators must contend with when projecting costs.

- Weather
- Temperature
- Subsurface water
- Variations in soil and rock
- Sensitive, critical, and contaminated areas
- Variations in topsoil thickness and root zones
- Inaccurate stripping- and earth-volume calculations
- The margin of error attributable to contour interval
- Overnight and unpredictable visits by endangered species

But it doesn't stop here. Identifying the innate differences between site-work and structures construction can lead one to surmise that succeeding at site-work is analogous to casting your fate to the wind. This is because errors in judgment and tactical mistakes preceding the site-work estimate can further compound problems for the estimator making it difficult, if not impossible, to produce an accurate and all-encompassing cost projection. Here's a palette of potential fabricated issues that always lurk in the shadows and that can heavily tax the site-work cost-estimating process.

- Working from inaccurate drawings
- Unknown or long-forgotten utilities

Example of a Cost Estimate Evolving Into a Pay Request Form

PHASE 1 - KNIGHTS BRIDGE OFFICE PARK

Black Mountain Contractors
Invoice:
Date:

		(1)	(2)	(3)	(4)	(5)	(6)	(7)	(8)	(9)		(10)	(11)	(12)	(13)	(14)
					(1)X(3)	(4)X(Rate)	(4)+(5)		(7)X10%			(9)X10%	(7)+(9)	(8)+(11)	(6)-(11)	
Cost Code	Construction Activity Description	Quantity	Units	Unit Price	Bid Amount	Tax	Total Contract	Previously Invoiced	Previous Retainage	Current Invoice	Current Tax	Current Retainage	Total Invoiced	Total Retainage	Contract Remaining	Retainage Paid
	LOGGING & CLEARING															
	Logging - (Net)	225.00	MBF	(275.00)	(61,875.00)		(61,875.00)	0.00	0.00	0.00		0.00	0.00	0.00	(61,875.00)	0.00
	Clear, Grub, Chip, Haul off Chips	12.00	AC	4,400.00	52,800.00		52,800.00	0.00	0.00	0.00		0.00	0.00	0.00	52,800.00	0.00
	Temporary Access Road Budget	1.00	LS	15,000.00	15,000.00		15,000.00	0.00	0.00	0.00		0.00	0.00	0.00	15,000.00	0.00
					5,925.00		5,925.00	0.00	0.00	0.00		0.00	0.00	0.00	5,925.00	0.00
	TEMP EROSION CONTROL															
	Filter Fabric Fence	800.00	LF	4.95	3,960.00		3,960.00	0.00	0.00	0.00		0.00	0.00	0.00	3,960.00	0.00
	Maintaining Existing Fence	1.00	LS	10,000.00	10,000.00		10,000.00	0.00	0.00	0.00		0.00	0.00	0.00	10,000.00	0.00
	Trenches & Ditching	3,000.00	LF	0.35	1,050.00		1,050.00	0.00	0.00	0.00		0.00	0.00	0.00	1,050.00	0.00
	Rock Check Dams	12.00	TON	35.00	420.00		420.00	0.00	0.00	0.00		0.00	0.00	0.00	420.00	0.00
	Riprap Basins	12.00	EA	750.00	9,000.00		9,000.00	0.00	0.00	0.00		0.00	0.00	0.00	9,000.00	0.00
	Temporary Pond Siltation Facilities	1.00	EA	7,500.00	7,500.00		7,500.00	0.00	0.00	0.00		0.00	0.00	0.00	7,500.00	0.00
	Temporary Fabric Over CBs	25.00	EA	35.00	875.00		875.00	0.00	0.00	0.00		0.00	0.00	0.00	875.00	0.00
	8" Temporary Plastic Pipe	340.00	EA	13.00	4,420.00		4,420.00	0.00	0.00	0.00		0.00	0.00	0.00	4,420.00	0.00
	12" Temporary Plastic Pipe	300.00	EA	16.00	4,800.00		4,800.00	0.00	0.00	0.00		0.00	0.00	0.00	4,800.00	0.00
	Straw Mulch	25.00	AC	750.00	18,750.00		18,750.00	0.00	0.00	0.00		0.00	0.00	0.00	18,750.00	0.00
	Hydroseeding	12.00	AC	975.00	11,700.00		11,700.00	0.00	0.00	0.00		0.00	0.00	0.00	11,700.00	0.00
	Truck Wash	1.00	LS	21,000.00	21,000.00		21,000.00	0.00	0.00	0.00		0.00	0.00	0.00	21,000.00	0.00
					93,475.00		93,475.00	0.00	0.00	0.00		0.00	0.00	0.00	93,475.00	0.00
	GRADING & EARTHWORK															
	Strip & Haul Topsoil to Processor	15,000.00	BCY	2.90	43,500.00		43,500.00	0.00	0.00	0.00		0.00	0.00	0.00	43,500.00	0.00
	Load & Haul Unsuitable to Waste Area	7,000.00	BCY	3.85	26,950.00		26,950.00	0.00	0.00	0.00		0.00	0.00	0.00	26,950.00	0.00
	Mass Grade, Cut to Fill	8,500.00	BCY	2.65	22,525.00		22,525.00	0.00	0.00	0.00		0.00	0.00	0.00	22,525.00	0.00
	Mass Grade Cut to Export	60,000.00	TCY	8.00	480,000.00		480,000.00	0.00	0.00	0.00		0.00	0.00	0.00	480,000.00	0.00
	1.5:1 Slope Grading	6,500.00	SY	1.25	8,125.00		8,125.00	0.00	0.00	0.00		0.00	0.00	0.00	8,125.00	0.00
	Import Structural Fill Budget	2,500.00	TCY	12.00	30,000.00		30,000.00	0.00	0.00	0.00		0.00	0.00	0.00	30,000.00	0.00
	De-watering Budget	1.00	LS	12,000.00	12,000.00		12,000.00	0.00	0.00	0.00		0.00	0.00	0.00	12,000.00	0.00
					623,100.00		623,100.00	0.00	0.00	0.00		0.00	0.00	0.00	623,100.00	0.00
	ROAD WORK															
	Subgrade Repair	1.00	LS	7,500.00	7,500.00		7,500.00	0.00	0.00	0.00		0.00	0.00	0.00	7,500.00	0.00
	Import Structural Fill, Pit Run	4,500.00	BCY	12.00	54,000.00		54,000.00	0.00	0.00	0.00		0.00	0.00	0.00	54,000.00	0.00
	Export Slop/ Cut	3,000.00	BCY	3.75	11,250.00		11,250.00	0.00	0.00	0.00		0.00	0.00	0.00	11,250.00	0.00
	Roadway Fine Grading	22,000.00	SY	0.16	3,520.00		3,520.00	0.00	0.00	0.00		0.00	0.00	0.00	3,520.00	0.00
	Grade, Compact, Backfill Curb & Gutter	8,900.00	LF	2.35	20,915.00		20,915.00	0.00	0.00	0.00		0.00	0.00	0.00	20,915.00	0.00
	Backfill Planters with Raw Material	3,100.00	BCY	3.50	10,850.00		10,850.00	0.00	0.00	0.00		0.00	0.00	0.00	10,850.00	0.00
	Road Fabric Budget	5,000.00	SY	1.00	5,000.00		5,000.00	0.00	0.00	0.00		0.00	0.00	0.00	5,000.00	0.00

Figure 13-3

- Reliance on poor or inadequate topography
- Discrepancies in how earthwork is calculated
- Proceeding with scant geotechnical information
- Multiyear build-out schedules on larger projects
- Misinterpreting the presence of rock or soil hardness
- Predicting timber volume from a sloppy or insufficient cruise
- Unexpected permit regulations or terms of a development agreement
- Off-site locations to dump export or sources for cheaper structural import material
- Conflicting designs between different utility service providers on the same project
- Being kept in the dark. If estimators aren't informed of upcoming changes or design modifications, they will miscalculate quantities and proceed with wrong assumptions.

The first issue mentioned, working from inaccurate drawings, also includes generating poor and erroneous surface-area calculations from cartoons and *crayon-lined* architectural drawings. Estimating from sketches leads to sketchy estimates. The same applies to estimating volume based on horizontal dimensions only. Horizontal and other plane dimensions provide only two-thirds of the required dimensions needed for estimating volume and mass. It takes three dimensions to estimate the bulk of site-work–length, breadth, and depth.

In light of all these challenges, what can an estimator do to better guarantee that he will build a credible and complete site-work cost estimate? Here are a few basic principles that estimators need to adhere to when estimating site-work.

- Learn from last year's estimate.
- Mark a set of plans as you estimate.
- Have a budget for controlling water.
- If a cost feels too light, it probably is.
- Pay attention to detail when estimating.
- Establish allowances for undefined quantities.
- Obtain more than one estimate for dirt volumes.
- Respect weather and how it can affect the project.
- Spend time visualizing how the job will be built out.
- Learn and apply the different ways of measuring dirt.
- Take the ongoing cost of erosion control very seriously.
- Stay consistent in the format of and your approach to estimating.
- Always assume timely survey and layout, and adherence to schedule when figuring base costs.
- Learn to read for, to expect, and to account for activity that is not

obvious in the construction documents.

- Break down construction cost centers into groups of similar work and, then, into elementary, easily understood, and easily measurable units.
- Don't underestimate the cost of dealing with small items, such as removing piles of unexplained dirt, protecting existing utilities during construction, garbage runs, and providing all-weather access for everybody, including marketing people in wing tips and stiletto heels.

The rudiments of site-work cost and quantity analysis

Cost estimating involves evaluating work and, then, matching quantities and price at some point in time. The time element is important because items such as site security, street cleaning, and erosion control remain active long after daily or seasonal construction has ceased. Other time-related cost centers not so obvious include re-establishing survey control and staking, and remobilizing. While most estimators account for inflation, it's much tougher to compensate for "shock-inflation" resulting from unexpected events, such as war, shortages of material, or disrupted oil supplies. Working year-round exposes owners to the risk that cold or wet weather will cause construction problems, and that the lack of daylight during winter will lengthen the construction schedule. Any of these situations can weaken a site-work cost estimate. Too many of them will render the estimate obsolete.

How work activities are broken down also affects the quality of the cost estimate. Size matters and detail is king. When cost estimates are broken done into manageable units, owner's benefit by:

1. better knowing where money is going;
2. easier tracking of contractor billings and charges for extra work;
3. being better equipped to evaluate multiple estimates that are similar in format; and,
4. obtaining information useful for building a comprehensive and detailed cost database.

Working with inherited construction documents (that you didn't control)

Estimating site-work from documents that the owner has had no hand in controlling may involve an inordinate level of interpretation and guesswork—that is, if they aren't stamped "Preliminary." Preliminary is synonymous with hope. Hope that further detailing, analysis, and missing pieces are being handled somewhere and will be included later in the final set of documents. Thus, estimates completed against "Preliminary" plans should be treated only for what they are—preliminary. They'll simply require updating at some point in the future.

More information on potential problems inherent in site-work documents is covered in Chapter 12, "Developing Plans and Specifications."

First, check inherited drawings for dates. Are they still relevant? Have environmental or market changes invalidated the design? Then, check for missing pages, incomplete sections, and be sure that dimensioning and stationing as described is correct and to scale. It's a supreme waste of time when estimators measure quantities only to learn later that the scale as advertised is wrong.

Framing the cost estimate

Estimating site-work often involves basing all or parts of the cost estimate on outside sources of information. Clarify the extent that third-party reports or backup data were used in building the cost estimate by documenting at least the following:

1. Note supporting information. Include with the estimate, copies of third-party backup or data used to build the estimate, such as earthwork volumes or an arborist's tree recommendations.
2. Remember who said what. Verbal direction and anecdotal information can be a valuable component of any estimate. Note all meetings, conversations, direction given, dates, and individuals who have had an impact on the estimate.
3. Cite all documents used in building the cost estimate, such as reports, agreements, test data, construction drawings, special details, and specifications. Note the construction standards referenced. For each document cited, include dates, page numbers, revision numbers, who approved them, and any other information pertinent to the estimate.
4. When estimating jobs that are purely speculative or that have no documents of any kind to reference, be lucid about the facts. Note whether or not there were field measurements taken or visual observations made, or if there was nothing tangible to work from. Otherwise, if things go sideways later, people who formerly understood that your estimate was purely "a stab in the dark" may develop a strange case of forgetfulness and not remember that there was absolutely no information available at the time and that the estimate was purely a seat-of-the-pants exercise lubricated by an urn of coffee.
5. List all the assumptions used in composing the estimate. If the assumptions aren't obvious, estimators should note in detail why they assumed what they did. This practice also helps estimators to remember their assumptions later. Assumptions include anything—methodology or formulas used for figuring quantities and costs, or subjective information concerning contractor production, potential working conditions, or schedule. Here are some examples of the type of

assumptions typically made when estimating site-work:

- Sawcutting existing pavement to a depth not shown on the plans
- Renting a crane instead of using an excavator to lift and set timbers and beams
- Scaling the architect's sketches in order to dimension and price entry rock amenities
- Applying standard in-city procedures for connecting to existing utilities in rural areas
- Deciphering the owner's schedule or phasing in order to adequately price construction
- Using average and accepted unit pricing, production rates, and construction practices for site-work in your area
- Calculating an increased volume of excavation, footings, or retaining wall area belowground that is not shown in the plans
- Using cost-per-gallon pricing for future storm-pond or reservoir construction in the absence of reliable topographic data, engineering details, or design
- Providing the aggregate cost for a retaining wall system that may include erosion control, wall material, excavation, moving (and possibly re-using) native material, grid or fabric, import fill, crushed rock, subbase ballast, landscaping, and piping the drainage system to the nearest usable catch basin or manhole.

Estimating the cost of construction with scant information happens frequently. A friend of mine in the industry has coined an acronym for this phenomenon calling it a "BEUCI". It stands for "Best Estimate Using Current Information." He applies this term when submitting preliminary or best-guess estimates. Using a unique term like this can help reinforce the point that the estimate is preliminary and subject to change. File it under damage control.

Setting up the spreadsheet template

Rows

The template or structure of the cost estimate structure is important. Build site work cost estimates in a fashion that emulates the construction schedule. For example, the section covering the costs of logging and clearing should precede the section covering mass grading. Next, construct the cost estimate so that it can double as a bid document and as a basis for contractor-progress pay requests. Finally, lay out the cost estimate in a fashion that is friendly to use project-wide. This means laying out rows of hard- and soft-cost items followed by quantities and pricing.

Site construction categories in land development are typically universal regardless of project, product, or region. Here's a broad list of categories (sometimes called hard costs) that may comprise a site-construction estimate:

Examples of hard-cost categories (in rough order)

- Mobilization
- Off-site improvements
- Erosion control
- Clearing and logging
- Storm facilities
- Demolition
- Grading
- Retaining walls
- Roadway preparation
- Sanitary sewer
- Storm sewer
- Water system
- Dry utilities
- Reservoirs and stations
- Infiltration system
- Hard surfacing
- Amenities
- Landscaping
- Trails
- General site-work
- Contingencies

Refer to the spreadsheet portrayed in *Figure 13-2.* Note that hard costs form the nucleus of the estimate and that they fall into headings and major categories that describe construction cost centers. More detail follows in the sub-rows below each major heading or category. Also, note that the item descriptions listed follow a chronological order that imitates the construction schedule.

Tabulate soft costs in similar fashion. Here are some examples of line-item soft costs that are usually associated with site work and are best estimated once the scope of construction is quantified.

Examples of soft-cost categories linked to site construction

- Survey control & layout
- Geotechnical inspection
- Materials inspection
- Concrete testing
- As-built surveying
- Utility connection fees
- Timber cruising
- Water-quality monitoring

The list is void of design, permitting, and management costs since these expenses aren't normally a direct consequence of site construction. Although design and management costs may be calculated as a percentage of gross construction costs, it's preferable to set up these budgets, or fees, as quoted or bid. Examples of design and management cost centers included in the overall project cost estimate may include any of the following:

Examples of soft costs pertinent to design

- Legal fees
- Signalization design
- Permitting fees
- Architectural and design fees
- Traffic analysis
- Storm-water analysis
- Civil and structural engineering
- Environmental assessments
- Geotechnical engineering
- Streetlight design
- Impact statements
- Landscape design
- Wildlife studies
- Critical-areas studies
- Plat applications and hearings
- Topographic, boundary, and aerial surveys

Columns

Refer to *Figures 13-2* and *13-3.* Notice that expanding the seven cost-estimating columns shown in *Figure 13-2* produces the contractor pay estimate form shown in *Figure 13-3.* Owners can choose whether to carry a full or an abbreviated cost-estimate format through to other accounting and finance project-reporting systems.

Whereas rows describe individual cost categories, columns manipulate costs and quantities for each cost category. The spreadsheet in *Figure 13-2* includes seven columns that are basic for detailed cost estimating. There isn't much flexibility in how these seven columns treat data since column numbers are fixed values or the product of column-to-column calculations. In order, they are:

1. Construction activity description
2. Quantity
3. Units
4. Unit price
5. Bid or estimated amount
6. State and local tax on the total bid amount
7. Total amount

In *Figure 13-3,* columns for cost code, retainage, invoicing, and summarized contract amounts are added to the spreadsheet producing a document that is useful for project control. Row descriptions remain constant throughout both spreadsheets.

Construction activity description

The level of detail used in the estimate drives item description. As detail increases, the number of construction activity descriptions and subdescriptions should also increase. Referring to *Figure 13-2,* construction activities and subdescriptions are both listed sequentially. Since activity descriptions provide a link between the estimator, contractor, and project staff people working in non-construction roles, descriptions must be clear, coherent, and correct.

Quantity

Quantities are best described and measured according to square area, volume, length, number of units, and weight. A quantity of one is a handy way to describe an integral construction package where components are installed at one complete price. Examples may include a bridge, retaining wall, or pump station. For a couple of reasons, assigning less than whole or decimal numbers to site-work quantities is impractical. First, unless the unit price is extraordinarily high, refining quantities by decimal typically makes little or no difference. Second, site-work quantities can be prone to error, approximation, or involve relative unknowns. Refining already rough or incorrect numbers doesn't make sense. An exception to this rule would be for clearing and grubbing where the price-per-unit is high and the units (acres) are large enough to justify fractional sizing. Remember—if quantities aren't verifiable, the cost estimate is close to being worthless. As a rule, detailed quantities are verifiable quantities.

Completion of work can verify quantities. How? If work is laid out per plan, inspected, and completed without changes, then, by default, it can be assumed that the correct quantities have been installed. Rest assured that if variances to the contract quantities occur during construction, most contractors would notice and contact the owner. Depending upon the contractor, the type of contract, or the amount of money at stake, an owner may not hear about minor quantity changes. Instead, the contractor may decide to absorb the additional cost. In other cases, a contractor may hold quantity changes until the bitter end and spring them on an unexpected owner as part of a composite claim. This one reason to define in site contracts, a limit that prohibits introducing a change order claim to an owner after a certain period of time has elapsed. Look for more on this subject in latter Project Logic Series books.

Site work isn't always verifiable as a function of completed work. Timber volumes measured by mill scale or weight are a good example. Timber is *cruised,* or subjectively measured on the stump, then cut, hauled, and *scaled,* or subjectively measured again at a yard or mill site. During the course of events, there's significant room for error or even theft. Unfortunately, there exists little or no opportunity for owners to return to the mill and squabble if the net log volume doesn't correlate with the original timber cruise. By the time owners receive payment for their wood, their logs are usually long gone—mixed into large piles, hauled elsewhere, or milled. Since most project owners aren't in the timber business, log compensation is one of those items that owners have to learn about, accept on faith, and then forget about. Owners can partially protect themselves by letting the logger have the logs as part or all of the logger's compensation, though this approach can still yield challenges related to accountability, determining accurate end-volumes, and ensuring that the site boundary doesn't get over-logged. Problems can also exist with weighed wood, as with pulp, but the opportunities for abuse or subjective errors decreases dramatically when scales are used to verify production.

Examples of other quantities that commonly require field measurements for verification include square area of retaining wall installed, depth and lineal footage of piles, and volume of extra overexcavation or fill performed during grading. The latter is where owners must rely on someone, other than the contractor, such as a geotechnical inspector to report on, to measure, and to verify change-of-conditions quantities.

There are a number of ways for owners to verify preconstruction earthwork quantities; and "cerebral triangulation" is a particularly good one. This involves having multiple entities, such as an engineer, a contractor, and a third-party consultant, calculate earthwork quantities. If a contractor completes work in accordance with dirt volumes estimated and verified by more than one independent study, it can be assumed that the work was completed per plan. Likewise, completing work as staked without incident verifies that the staking was probably correct and based upon accurate design calculations. In other words, everything checks. As a rule, don't proceed with earthwork if the third party's, engineer's, and contractor's calculations are in gross contradiction to each other. Instead, investigate and correct the differences and proceed once all parties can agree on correct volumes.

When earthwork estimates conflict or questions arise concerning a contractor's performance during mass grading, owners always have the option of determining net excavation and fill by cross sectioning and measuring, or by using instrumentation, software, and GPS technology, to check pre-excavation and post-excavation conditions. Elevation checks combined with daily geotechnical reports and an agreed upon volume of topsoil stripped provides a clear picture as to what has transpired.

Additionally, truck volume and weigh-scale tickets describing the delivery or export of topsoil, structural fill, crushed rock, concrete, and asphalt can be useful resources when investigating quantities. Lastly, contractor pay requests also work to verify, for good or for bad, the history of a project's earthwork.

Units

Units are segments and sections of completed construction. The trademark of an effective unit is that it is easily understood, measurable in the field, and verifiable in the plans. Unit prices are neatly packaged values that include all contractor costs to complete a unit of work including profit, risk, and overhead. Costs-per-item description are calculated when quantities are multiplied by unit price. If a problem arises with a unit-price contract, it only takes two people—the owner and the contractor—to measure work in the field and crosscheck it against plan quantities or the contract itself. Items that are verifiable by counting, or measuring lineal, or that are determined by square area, work best for unit pricing.

Simple and easily verified units protect owners in other ways. Work that's easily quantifiable reduces the need to proceed with nebulous lump-sum costing. Units reflect unit pricing and unit pricing creates a straight path to understanding and controlling site-construction costs. Owners who rely on simple to understand units are able to avoid more complex issues involving performance and activity estimating, projecting owning and operating costs, calculating margins, profit, risk, and overhead. These items are better left to contractors to slice, dice, and shoehorn into owner-specified unit prices.

The art of estimating site construction requires separating predictable and non-predictable costs into appropriate line items. Predictable line items can be used with confidence when estimating and bidding work. Unpredictable elements, such as items that the contractor has to subcontract out or items prone to variation, must be isolated from predictable elements and given their own line in the estimate. The reason—unpredictable items can contaminate predictable ones resulting in quantity busts, compound extra costs, and confusing change orders. Here are some examples:

- Estimate clearing and grubbing by large square-area units, such as by the acre or hectare. Cost per unit should account for organic density based on shrubbery, vegetation, and volume and type of timber. Predictable cost-per-area unit typically includes removing all vegetation, raking it up, hauling it to a chipper, and chipping all woody material down to a specified diameter or minimum-allowed-piece size. Unpredictability may involve hauling material if an owner remains undecided as to whether to use the material on-site, lose it on-site, sell it off-site, or haul it to a dump. In this case, estimate hauling as a separate lump-sum item.

- Estimate retaining walls in square-face-area units. Predictable unit pricing for walls normally consists of minor excavation, footing construction, ballast pads, drainage material, grid or fabric if required, modular block, crib, wire, rock or concrete installation, and backfilling. Depending upon the contractor and type of wall, pricing may not include full excavation or the furnishing of backfill. One reason for this is that some wall contractors may not want to deal with subcontracting out the excavation and trucking. Other wall contractors prefer that the hole be ready and waiting for them when they arrive. This exonerates them from blame should the hole end up in the wrong place, overexcavated, or changed in geometry and volume. Bigger walls usually require bigger holes and many contractors are reluctant to accept the increased liability. In this case, treat wall excavation, export of excavated materials, and import of backfill material as three separate line items. Export can mean to anywhere and is truly unpredictable. Another unpredictable item is tight-lining the wall-drainage piping to the storm system. This item also tends to be a subcontract item and should have its own separate line.

  Here are a couple of related wall tips:

  - Include stubs for wall-drain piping when designing the storm (and sometimes the sanitary) sewer system.
  - Wet-utility contractors are the logical choice for connecting wall drains to the storm system. Include this work in the wet utility contractor's scope of work.

- Measure and estimate by the lineal foot the wet and dry utilities according to pipe diameter, material, and depth. Predictable utility unit-price estimating in its simplest form includes the cost of excavation, movable trench protection, pipe and fittings, bedding, tracer wire, and native or select backfill as specified. It may also include wet- and dry-utility connections, pipe testing and sanitation. Using native material for trench backfill can be unpredictable since it may get wet, contaminated, and become unusable. Where this possibility exists, estimators should provide separate lump-sum line-item allowances to cover the import of select material and export of waste material. If native material is prohibited for use in trench backfill, it becomes predictable and, combined with import, can be rolled into the unit price for installing pipe. Items such as manholes, catch basins, valves, hydrants and trust blocks are all predictable and estimated as separate line items. Although predictable in quantity, when they are subcontract items, sheet piling, tunneling, and pipe jacking should be included as separate lump-sum line items. Upon completion of work, break down the *actual* total lump-sum costs into unit pricing. This strategy will prove handy for future estimating.

Any predictable changes in pipe material, regardless of diameter or depth, justifies using a new line item for installed pipe. Depending upon pipe size and ground conditions, utilities in land development typically remain constant to a depth of six or eight feet. From there, costs per lineal foot installed steadily increase. Beyond a certain depth of excavation or diameter of pipe (depending upon pipe material), material prices, equipment size, safety considerations, and slower production can cause wet utility costs to increase dramatically. For most land development work, pricing pipe per depth in three-foot increments works well. Reference the various line item pipe descriptions shown on *Figure 13-2.*

Utility conflicts that are predictable and clearly shown in the plans can be included in pricing pipe by the lineal foot. Unpredictability leading to extra pipe work or encounters with boulders, bedrock, and subsurface water usually ends up as expensive time-and-materials work. Work of this nature should be estimated as lump-sum. If encountering old pipe or impediments is a possibility, compensate by adding a separate line item for underground conflicts. Provide a line item for estimating trench displacement using truck-cubic-yard, or lump-sum pricing.

Physically sizing finished square area is open to debate. Should estimators use square feet (square centimeters) or square yards (square meters)? The answer is to use the largest size that you can get away with. Larger units absorb a certain amount of contingency and offset minor discrepancies or inaccuracies in the quantity take-off. Larger-sized units also command respect. Psychologically, they're tougher to indiscriminately inflate for billings or change orders. Smaller units, such as square feet, can more easily swell over the course of billings. The level of accuracy and the scales used in most site-work plans don't lend themselves to exact measurements thereby further discouraging the use of smaller unit measurements. Overall, it's good policy to stick with larger units.

Pricing

Site work enjoys a high degree of success when project owners possess:

1. Inclusive and accurate documents
2. Solid and comprehensive geotechnical work
3. A detailed, complete, and presumptive unit price schedule
4. Good weather and trusting relationships

Those estimating site-work rely on quantities and price. Of the two, pricing is more important for controlling budgets and construction. While many ways exist to price site-work, in this book, we'll only discuss two line-item systems: unit price and lump sum.

The virtues of unit pricing

During construction, quantities will vary but unit pricing remains stable. Unit pricing acts as a barometer to measure inflation and cost creep. A solitary unit-price schedule—also called a schedule of values—can apply to all on-site contractors. More importantly, unit pricing can help owners harness the cost of change-order work. It can't prevent cost overruns but when extra work does occur, owners who have a unit-price schedule in place are half way to containing the issue. All that remains is quantity verification, which underscores why accurately measuring quantities is paramount to project control. Fixed-unit pricing and accurate quantity measurements together remove much of the mystery in understanding the overall cost of site-work.

It's also reliable. Unit pricing is one of the few tools available to owners to judge the intent and direction of the same contractor over the course of a multiyear project. While unit price "creep" can occur due to differences in economies of scale for some construction activities, such as variations in price for installing 300 lineal feet of pipe in one location of a project versus installing 30 lineal feet of the same pipe in another location, control creep by entering bids and contracts with ample and diversified unit pricing that covers any foreseeable situation. Sudden increases in unit pricing may indicate that an owner is on the verge of being overcharged. Quantities and scope of work can meander, but unit pricing should only change through bidding and negotiations, or as needed to cover genuine cost increases. Unit pricing can also decrease over time. Decreases in unit pricing can reflect contractor error, lower material prices, or that the contractor has discovered a cheaper way to do something. It can also be construed as a sign of desperation for a contractor wanting to stay on a particular project, remain competitive during periods of economic downturn, or simply survive a scarcity of work. Finally, unit pricing provides owners with a record, history, and baseline for predicting future costs.

The lump-sum alternative

Line-item, lump-sum estimating works for owners looking for completed, turn-key, and finished work. It's also an acceptable substitute for unit pricing when the construction documents aren't complete or lack enough detail to support measuring quantities. Without quantities, unit pricing is meaningless. Lack of quantities or unit pricing results in estimators having to empirically derive lump-sum values based on known information and their experience and ability. Lump-sum pricing works well for one-shot subcontract work, in situations requiring minimal analysis, and when conducting work under simple contract terms. Examples include estimating or using bid numbers for specialized, arcane, and purchase-order type work, such as well drilling, logging, pile driving, and installing street lighting. When nobody but the subcontractor understands and knows the intricacies of the work involved, lump sum is a safe route when estimating costs.

Tax

There are two ways to handle taxes and retainage: include them as separate line items and calculate them against column totals, or provide a tax column and calculate taxes individually for each line item. Dedicated columns allow the estimator more flexibility to calculate row-by-row values for tax and accounting purposes. When calculated against entire contract amounts, retainage is included as a gross bottom-line item.

Pricing for weather

Too much moisture results in mud. Mud equates to loss of good material, erosion problems, and inefficient operations. Not enough moisture results in compaction difficulties, dust problems, and fire hazards. The art of estimating site-work often requires coping with a broad spectrum of unpredictable factors related to moisture, temperature, wind, and seasonal conditions. It's, therefore, good policy to plan using historical averages, but to prepare for the worst. To account for weather, it may be easier to apply a percent contingency against a whole-cost category or an entire project than to piecemeal contingencies against individual line items. Individual line-item contingencies work best when an estimator can positively identify individual portions of work that will be subject to extreme weather. When weather isn't accounted for, the estimator either expects that it isn't a factor, or is gambling that work will occur under perfect conditions without weather-related cost and schedule impact.

Evaluating soft costs

Estimators typically base soft costs on experience, calculate them as a percentage of the total hard construction cost, or simply use actual pricing as submitted from each consultant.

Percent projections may be the only way to estimate during the early stages of a project, however, once consultants come on-board, let them provide more accurate cost information. Calculating soft costs as a percentage of the total construction cost requires using caution since any errors in the construction estimate will be carried across on a percentage basis to the soft-cost projection. Soft costs based on capital intense construction can also end up skewed. An example is calculating soft costs as a percentage of construction on a project that involves moving enormous volumes of earth. Deep cuts and fills having little correlation with ongoing engineering and design will skew soft costs higher. Another example includes basing soft costs on the aggregate price of installing numerous similar-sized retaining walls or engineered slopes, which are all built from the same tables or set of plans. Here, the cost of engineering comprises a small percentage of total construction cost. Costs for other items, like pump stations, large-volume reservoirs, and elaborate entry monumentation, can easily eclipse proportionality with design and engineering. Correct this situation by not automatically basing soft costs solely as a percentage of bottom-line construction costs. The safest route is to apply detailed estimates from design and engineering firms to the estimate.

Quantity estimating–the art of the take-off

Most site-work estimators have their own style of measuring and tallying quantities. Estimating site-work appears quite simple but can be deceptively difficult. Good estimators tend to be detail-oriented, patient, and organized—and they're also a rarity.

Manual cost estimating requires measuring and counting construction items. To ensure clarity, it's good practice to codify utilities beforehand with felt markers—generally using blue for water, pink for storm sewer, green for sanitary sewer, etc. Colored pencils are useful for marking items already measured and tabulated. Estimators who use this technique rarely count something twice and typically miss nothing. The result is an accurate take-off and a record of what's included in the estimate.

Once a project is under construction, use one color, such as black, to mark work completed to date. In marking construction progress, note the date and limit of work completed. Tick marks work great for marking the limits of daily production. On both the plans and an adjacent calendar, note the weather daily. This system produces a record that is useful when reviewing contractor pay estimates. It also provides construction documentation in case an issue arises later concerning construction impact or the contractor's work site on any particular day. A record of construction and weather is useful in explaining why a week of work was missed, say, six months ago.

A quality and well-thought-out estimate will typically include line items for cost centers that don't appear anywhere in the plans or specifications. This is because good estimators anticipate circumstances that most project people fail to notice or forecast. Here is some hard-earned advice concerning site-work cost estimating:

1. If the odds favor something going wrong or happening unexpectedly, account for the potential cost.
2. Engineers are a reliable source for providing numbers for acres and earthwork volumes. Measure everything else off the plans or add items based on the estimator's experience.
3. Esoteric concepts, such as unique storm-water infiltration systems or wildly artistic community amenities, tend to get more complex as they evolve. Once under construction, complicated and intricate designs usually cost significantly more than anticipated. Be real and avoid emotion when estimating costs for uncommon work.
4. Quantities can change dramatically once construction begins. Experienced estimators play it safe by rounding up quantities to help cover small overruns or items overlooked by the designers. Having worked on many large public and private projects, I know that it's satisfying when someone mentions that they're glad there was enough money budgeted to cover the cost of minor increases in quantity.
5. Accept the fact that a number of line-item quantities in a site-construction estimate will probably be wrong. Maybe not by much, but they can be off for any number of reasons including last-minute tinkering, topographic error, or design change. One way to deal with uncertainties is to allow the contractor to supply the quantities. If his quantities appear accurate, include them as part of a lump-sum or unit-price contract. This works best when the owner establishes the cost-estimate format and the contactor follows suit. However, this maneuver poses other problems for owner, as we'll cover later in this chapter.

Another way to deal with variability in quantities is to add a "tolerance" to items in question. Likely candidates include items that are incomplete in design or that appear likely to increase during construction. Increasing quantities can also compensate for a degree of plan error and can cover minor quantity fluctuations that result during construction. Type of project and nature of the items under consideration help to determine the degree to which quantities should be increased. Here are some signals that may suggest the need to cushion certain quantities:

- When design remains in flux
- When working a previously developed site
- When it's likely that weather will impact work
- When grading extraordinary steep and uneven terrain

- When owners remain unsure or issue conflicting information
- When projects can't get off the ground due to civic challenges
- When deeper or larger than usual underground utilities are required
- When impending changes in permit requirements and standards appear likely
- When suspected but unverified pockets of rock, subsurface water, and poor soils exist
- When the estimator knows that the designer will need to upsize or correct something
- When it's apparent that a potential buyer may require significant grading or utility modifications
- When scope creep continues unabated for any type of work—especially—off-site improvements
- When design of items such as fills and compaction doesn't concur with the geotechnical recommendations
- When the estimator suspects that base topography is erroneous or when design is based on contour interval that is too wide for accurate engineering

When figuring quantities on steep terrain, estimators must remain cognizant of the difference between horizontal and slope distances. Estimators must also understand how horizontal road stationing relates to underground pipe lengths. In most cases, the key is to use horizontal plans for horizontal measurements and profiles for slope and vertical measurements.

Curbs, paving, and sidewalks are items prone to misestimation. At first, hardscape appears easy to measure since it's straight and predictable, but that all changes with road radius, intersections, hammerheads, and cul-de-sacs. Though geometrically simple, it's not unusual for paving, curb, and sidewalk contractors to submit higher square-area quantities than those estimated. Apply rules of thumb based on the relationship between circles and area when estimating non-linear hardscape.

The number of possible elements inherent in costing site-work is enormous. Every project is different and each one may demand a different set of cost activities and groupings. The following items highlight elements that appear regularly and apply to a wide range of land development projects.

Potholing

When working in-fill, existing sites, or off-site improvements, owners must budget for potholing in order to locate and identify underground pipe, rock, water, cable, conduit, anomalies, and unsuitable soils. When estimating potholing, include the cost of excavation, shoring, and backfilling. In structural areas, include costs for backfilling and compacting in specified lifts with clean inorganic material. Native material unusable as backfill will have to be moved on-site or exported which further compounds pothole costs. Also include the expense to survey and plot holes because potholing, especially in structural areas, should always be traceable. Include traffic control and street cleaning as applicable.

Potholing is accomplished manually by use of shovels or, more commonly, by use of heavy equipment such as backhoes and excavators. Pressurized air, water, and vacuum systems are other options for potholing. The time and money spent in potholing is insignificant when compared to the costs, legal exposure, and downtime spent in dealing with struck or conflicting utilities. Budget enough money to reduce the owner's risk.

Survey and control

Survey and control is most economical when it's provided and managed by the same firm performing civil engineering services. Property, topographic, and ALTA surveying can be lump-sum estimated per task or projected on an hourly crew-time basis. Base the construction staking according to units of actual construction, such as per lineal foot of pipe installed, or as a percentage of total hard construction cost. In addition to actual surveying and layout, remember that unit pricing must account for items such as multiple mobilizations; distance and condition of roads to the site; degree of bushwhacking required; thickness of vegetation; quality and completeness of the owner's documents; off-site information research; degree of accuracy and work needed to define existing infrastructure; and, coordinating with prior or existing consultants. Estimate the lot construction staking, the property-line surveying, and the plat work by the crew hour, by the acre, or by individual lot count.

When staking and layout are based on lineal units of work, then, if lengths change during construction, surveying charges change accordingly. For example, survey costs by the lineal foot or by the square unit measure can be projected or contracted according to the actual work that is bid or completed in the field. Pricing by the lineal foot only works for items that remain consistent in width like utilities, curbs, and strips such as easements. Conversely, pricing by the square-area measure makes more sense for items that are random in width. Topographic survey costs can be estimated by the acre or as a one-time lump-sum charge.

Construction activities that don't lend themselves to estimating the surveying costs by the unit price include items such as excavation and grading, logging, erosion control, storm ponds, and retaining walls. These types of items don't generate quantities on a one-to-one basis with survey effort. Building a schedule of values for surveying based upon units of construction activity is not automatic. It often requires bidding out work. Negotiating unit-price surveying is tough because owners don't know what's fair and what's not. For instance, with unit-price surveying, if staking gets destroyed, it is reasonable that owners may have to pay on a time-and-material basis to re-establish control. The possibility of extra time-and-material charges mandates attaching an hourly and crew rate schedule to the survey contract including the terms and conditions for items such as re-establishing control and layout.

Many estimators choose to estimate survey and layout by the crew-hour and then roll this number into a lump-sum line-item cost. Though most firms will gladly provide cost estimates for various stages of surveying and layout, estimators must grapple with redundancy. That is, they must calculate how many times the same construction item may require staking and then estimate accordingly. They must also factor in the cost of exploratory surveying required for ground truth, design, and engineering. Include appropriate costs for as-builts in any cost survey and control cost estimate.

Void of bid numbers or hourly pricing, the easiest way to predict survey and layout is to estimate it as a percentage of total construction. It may not be the most accurate method but this approach can cover survey and layout until owners get a chance to bid or negotiate work.

Erosion control

Erosion control may be the toughest cost center to predict. Costs can increase with moisture, out-of-season construction, poor soils, and growing regulations. However, here, too, quanities are more predictable when the appropriate pricing method is used. Unengineered temporary siltation ponds can be estimated by scaling the pond surface area off of the plans, assuming an average depth, and calculating the volume of excavation likely for construction. Estimate the piping in and out of each pond, ditch work, rock lining, and erosion-control fencing by lineal measure as installed. Price all ground protection including mulch, jute matting, and hydroseed by square area. Secured straw bales can be priced individually or, if blown, priced by square area per depth. Point-specific erosion-control work that protects catch basins and inlets, or that results in damming ditches or installing rock entrances and truck washes is estimated per unit installed. Most erosion-control items are temporary or require backfilling such as with ponds and swales. Estimates must include sufficient costs for mucking out and backfilling ponds, filling in swales, and removing fencing and other temporary devices. If structural backfill is required, price it accordingly. Consider the location of the project in relation to bodies of water. Costs for treating stormwater prior to off-site release may include anything from batch flocculation to a sand filter system with additives to remove suspended particles.

Erosion control can be a continual money-sink requiring time, equipment, and year-round material applications. Examples of items where costs can mount quickly include the cost of applying and re-applying hydroseed; the cost of renting volume pumps needed for pumping dirty water out of or between storm ponds; and, the cost of replacing stockpiles of straw mulch that haven't been protected from moisture. Include two separate line-items in the estimate for installing and removing erosion control fabric and boundary fencing. On high-risk projects, budget more than sufficient funds to handle erosion on a year-round basis.

Demolition

Few items in site-work can range as broadly in cost and require as much expertise as demolition. High-end jobs, such as figuring the cost to dismantle a steel-reinforced, cast-in–place, concrete power station are best estimated by bidding the work or by soliciting an estimate from a demolition expert or contractor. More common and less technical demolition includes removing septic fields, cesspool tanks, fencing, underground utilities, abandoned structures, and concrete or asphalt hardscape. Demolition work that presents a hazard or a possibility of contamination requires an environmental assessment; where as removing street utilities and infrastructure can usually proceed under normal permitting. If there exists environmental risk, add any additional remedial expenses to the base cost of removing, loading, and hauling off waste. Other site-related, demolition costs could include site security, erosion control, capping or filling utilities, dust abatement, site preparation and backfill, crew time, trucking, and dump fees. If excavation results in the need to backfill and compact a hole, include pricing for the import of structural material and geotechnical work. For most site-work jobs, it's easiest to estimate demolition by lump sum in one of two ways: in detail per line item, or as one all-inclusive work activity.

There exist many on-site uses for concrete and asphalt demolition waste. Concrete and asphalt can be ground or crushed and used to stabilize subgrade, and to build temporary access roads, or it can be included in fills. It is important, though, to investigate whether or not recycled materials can be used on-site since some jurisdictions may prohibit their use for construction. Correctly sized and clean demolition waste used on-site can save owners money in at least two ways. They spend less to export demolition waste and they save more by purchasing less import material. Chapter 16 provides more information on recycling.

Logging and timber

Foresters and loggers control this little corner of site-work. Logging timber involves two basic questions. What is the timber worth, and what is it going to cost to get logs from the stump to the mill? Finding out begins with performing a timber cruise to measure volume.

The timber cruise is a method of measuring stand volume before logging. Cruise information is used to predict overall value and provides owners with an approximate benchmark for checking possible timber theft and errant loads. Like construction estimating, timber cruising is an art. Timber cruising is based on area so foresters must know gross-project timbered acres (hectares) and gross-project areas destined to remain unlogged. Foresters also need a full legal description complete with records of corner monumentation, a topography map, any information related to access and roads, and a plan showing save areas, sensitive areas, buffers, and greenbelts. Aerial photographs are always useful.

Timber cruising involves measuring and grading trees in order to estimate merchantable log volume. Gross or net volume is typically expressed per thousand board feet per acre (mbf per acre) for lumber or in tons when logs are marketed for pulp. Fixed- and random-plot timber cruises are cost-effective methods when measuring volumes over larger parcels of land. On smaller parcels and right-of-ways where timber is high in variability or scattered across open areas, measuring and grading each tree is required. At minimum, a timber cruise will provide the following information required for estimating log value:

1. Average volume per acre
2. A listing of tree species by percent stem count or volume
3. Total net or gross volume in thousand board feet (mbf) and tons
4. The prevalence of disease, fungus, and mechanical damage inherent in the stand

Cruise reports may also provide other pertinent information such as:

1. Mill locations
2. Landing locations
3. Suggested method of logging
4. Method of loading and hauling
5. Options for alternative log markets
6. Possible local variations in log scaling

7. Residual volume expected after logging
8. Recommendations for manufacturing logs

Gross log value is assessed by matching timber cruise information to market pricing. Factors that affect whether or not trees are cruised per mbf or ton include the species, size, or condition of individual trees, the method of logging, and the regional mill demand. Weighed tons works well for pulp because the wood has no dimensional value. Geometrically measuring logs by cubic or board-foot dimension works for wood destined for the mill or for off-shore markets. Mills openly compete for and bid on logs depending upon geographic location, log quality, and forest-product demand. A smaller but growing industry also exists for specialty woods used to make furniture, woodsy décor, and musical instruments. Combined with the fact that weak markets may affect the salability of logs, it's worth having individual trees evaluated for unique or one-of-a-kind value. Most certified arborists, foresters, and loggers recognize high-value trees and know where alternative buyers exist.

Net log volume is the volume of logs available for shipment after incurring losses due to logging damage, defects such as rot, fungus, broken tops, and dead wood. Last but certainly not least, net log volume is a function of how each log is subjectively graded, meaning evaluated and measured, at the mill. Though not a problem with weighed wood, unless a project is heavy in high-quality straight logs, the net scaled value of timber can be shockingly low. Owners dealing with a general contractor who is subcontracting out the logging may find it easier to trade gross—not net—log volume against the cost of logging, clearing, and grubbing. Gross volume allows owners to negotiate with potentially more wood, leaving it up to the logger or contractor to manufacture as much value out of the timber as possible.

When dealing directly with loggers, owners may find them eager to trade logs for logging. This arrangement works well for owners because they don't have to deal with mills or log theft and it motivates loggers to maximize log value. Excess log value can pay for construction items that most loggers provide, such as establishing access, constructing haul roads, importing and placing crushed rock, clearing, grubbing and chipping, or falling hazard trees. This extra profit can flow back into paying for site work through the general site contractor, logger, or the owner depending upon how contracts are worded. Owners of poor-quality and low-volume timber may find that log value alone isn't sufficient to cover the logging and hauling costs. In this case, owners end up cutting a check to cover the balance.

For estimators, net volume multiplied by mill price per species per unit minus the cost of cruising, logging, and hauling the logs equals the owner's profit. Timber cruising is a soft cost usually charged on a lump-sum basis, although some foresters will cruise timber and manage logging in exchange for a percentage of log receipts. Estimate logging and hauling by bidding or by getting a logger to price the work. Expect pricing to be per mbf or ton as determined by local markets and mills. Estimate the expenses to handle and remove garbage trees and unmerchantable timber on a per-acre basis and budget it as a clearing cost. Include in the estimated logging price the minor amounts of construction required for logging, such as establishing skid trails, installing small culverts, and building landings, and then note that these items are included in the logger's scope of work. Since cruise volume, log purchase price, and logging and hauling costs are all typically expressed per mbf or by the ton, the last step in order to complete estimating is to perform simple mathematics.

Clearing and grubbing—and chipping

Regardless of the area's size or shape, the surest way to estimate clearing and grubbing of roads and developable area is by the acre or hectare. Convert small areas less than an acre to a decimal equivalent and hold the unit price constant for estimating. Consistent lengths of ground to be cleared for items such as easements and right-of-ways can be estimated by the lineal foot if these areas remain fixed in width. When clearing and grubbing is prohibited until the issuance of building permits, it may be necessary to estimate costs on a more expensive lot-by-lot basis.

Area measurements for clearing and grubbing are usually taken off the plans. Rarely are cleared and grubbed areas field-survey measured for payment. This works for most clearing and grubbing contractors as they'll base their contract amounts according to area plotted from plans or delineated from photographs. When bidding work, it is important that estimators price cleared and grubbed area from the same plans and photographs that the contractor uses. Once underway, if asked to supply chips for construction use, some chipping contractors, although saving money by having to haul less chips, may charge for the chips or at best, decline issuing any credits for waste material left on site. If your estimate is based upon using chips for items such as haul road construction or erosion control, budget money for purchasing chips or compose the chipping and grubbing contract in a way that guarantees owners the right to procure wood chips at no additional cost.

Include costs for an overnight watch and stationary or mobile vehicles that provide a source of water where burning is allowed. Generally, the cost of burning pales in comparison to the cost of hauling slash and stumps to a grinder, grinding or chipping the material, and exporting it off-site.

Contractors who specialize in performing clearing and grubbing use equipment built to pulverize and grind woody debris. Estimators may ponder the cost of piling, loading, and hauling woody debris to a chipper versus the cost of bringing the chipper to multiple piles of stumps and slash. This leads to analyzing the cost of using off-road trucks and a loader versus the cost of moving the chipper to scattered piles of material on-site plus the cost of providing project-wide access for highway-rated chip truck-and-trailer rigs. While such an analysis is noble, it's better to seek prices per unit acre or hectare from one or more clearing contractors and let *them* decide how they want to conduct work. Besides, a clearing and grubbing contractor may consider cost as second in importance to getting on-site, keeping the crew active, and making his payments. The key with clearing contractors is to get them on-site to grind or chip when you need them. Unfortunately, everyone wants his services at the beginning of the construction work season.

A note on specialty contractors

As a project matures, owners pick up additional responsibility for items such as securing the site, maintaining access, and posting signage. These items should be included somewhere when estimating the early phases of land development. Owners will find, however, that most specialty subcontractors are resilient. Drillers, pile drivers, demolition specialists, loggers, and retaining-wall subcontractors pride themselves on self-sufficiency. Since most of them have scheduled work lined up and waiting, they'll do whatever it takes to maintain production, do the job, and get on to the next project. Clearing and grubbing contractors are no different from these subcontractors. Most of them come to do a job, make adjustments on the fly, and get out as soon as possible without a lot of whining or nickel-and-dime change orders.

While this isn't a signal for estimators to lower cost projections or ignore related expenses when figuring clearing and grubbing, it's prudent to include a reasonable line-item lump sum-number to cover any additional costs that may occur during specialty work. Items to consider include constructing access, installing temporary gates, maintaining on-site roads, dewatering, flagging off areas, controlling dust and erosion, street cleaning, and stockpiling and hauling away excavated material.

Dirt, grading, and mass excavation

Land development is won and lost in the dirt. By the time an owner realizes that earthwork is over budget, it's usually too late to recover. Earthwork cost overruns often accumulate slowly over time beginning with stripping and ending with cut, fill, import, and export. Many owners and contractors remain unaware that earthwork is running over budget until the contractor's staff runs the numbers and bills the owner. Earthwork cost overruns tend to be more expensive than other site-work impacts. This is one reason that earthwork cost overruns are rarely recouped through savings elsewhere. Few owners can afford to allow their projects to act as "learning centers" for consultants and contractors. Quoting a client, "Tuition is too high." And he's right.

Earthwork and grading quantities are shear approximations as nobody can predict exactly how many bank cubic yards of material will be involved in developing a property. They are approximations drawn from grading plans and honed as a project progresses. Reliable topography, an understanding of the type of material that exists on-site, and depth of stripping are all prerequisites for calculating earthwork volumes. In order to price construction, estimators must combine gross cubic yardage and geotechnical information with knowledge of where quantities are to be cut and filled. This may involve conferring with the geotechnical engineer on matters pertaining to shrink-and-swell factors, cold- and wet-weather grading, compaction, and structural limitations of on-site soils. Estimators must also know if excavation or trenching is likely to encounter bedrock, expansive clays, or sandstone.

Geotechnical test pits and borings provide soils information but they may not reveal everything. Depending upon the frequency of testing, pockets of rock, subsurface water, or varying soils may remain undetected. Discovery of any one of these items can change conditions for earthwork and utility trenching. While it's true that cost estimates based on information contained in a geotechnical report are only as accurate and thorough as the report itself, a lack of information doesn't absolve an estimator from thinking. This is where experience kicks in and allows an estimator to make subjective, but reasonable, assumptions in pricing site-work. Estimators who suspect variations in subsurface conditions may compensate by adding a line item for such, or by inflating quantities proportionally to cover a potential cost overrun. Here are some reasons why an estimator may alter bank-cubic-yard volumes for earthwork, grading, or utility work:

- The topography plan is off.
- The contour interval is too wide.
- The grading plan is in limbo or is likely to change.
- Soils are prone to significant shrinking and swelling.
- The software used to calculate volumes is questionable.
- The assumed quantity or thickness of topsoil is suspect or wrong.
- Boulders, stumps, and rock displacement is expected to alter dirt volumes.
- Contractor, civil engineer, and third party dirt-volume estimates vary

significantly.

- Earthmoving volumes are extraordinarily high thereby increasing the potential for overruns.
- Poor weather or subsurface water is likely to impact moisture-sensitive soils resulting in increased import and export.
- Overly conservative specifications and compaction requirements may render marginal material unusable, therefore, effectively increasing the need for import and export.

Using third-party earthwork estimates

On projects with heavy earthwork, proceed with multiple dirt-quantity estimates. The site contractor and civil engineer normally supply quantity estimates but owners can further protect themselves by seeking one more estimate to verify estimates or to act as a tiebreaker. Commission third-party estimates from qualified firms unattached to the project and don't alert them beforehand to any estimated quantities that you have already received. All firms should figure earthwork quantities independently of each other. If all three estimates concur or are close, one option is to tie the third-party estimate to the construction contract as the official and neutral baseline for contractor payment.

Earthmoving quantities encompass cuts and fills. Here's a list of cut-and-fill possibilities or categories open to volume calculation and unit pricing:

- Topsoil stripping—cut to on-site or export off-site
- Mass excavation—cut to fill
- Mass excavation—cut to on-site or export off-site
- Import fill material
- Rock excavation

Study the above categories as any of them can apply to site-work. The trick for estimators is to recognize which of these categories should be included in the estimate. Then, develop a line-item description for each one as it applies to your specific project. But calculating earthmoving costs on larger projects involves more than number crunching. It requires understanding project strategy or, at least, the owner's objectives or philosophy for earthwork. On complicated multiyear projects with unlimited grading alternatives, this means understanding seasonal site-work phasing and how it relates to the overall construction schedule. The problem, of course, is that many owners don't have construction phasing and schedules figured out for the present, let alone next year. This leaves estimators no choice but to improvise and assume a logical path for completing work. Once modeled, they can predict likely quantities and calculate costs. Under these circumstances, the estimator assumes the role of an artist, prophet, and mathematician.

Smaller projects are often limited in the number of alternatives for earthwork or in the ways that the grading plan can be varied. The quantities are what they are, the owner is stuck with them, and the estimator applies unit pricing to volumes as calculated or reported. Fewer options mean less room for error elsewhere, so factors like weather and quality or usability of on-site material becomes proportionately more critical to the cost estimate.

The number and types of options for grading can lead to a couple of situations. The first involves an owner's desire for the project to take on a certain "look". In this situation, perception, layouts, or design may become more important than earthwork. In other words, quantities do matter, but unless the cost of grading becomes prohibitive, the quantities remain and the cost estimate stays as is. For all practical purposes, this alternative mimics the situation in the paragraph above, where grading alternatives don't exist, because in this case, they don't matter. It's the look that counts. Estimating simply requires applying costs to quantities, totaling the result, and submitting it to the owner.

Another and more likely alternative places the cost of earthwork on par with or above the layout and design. This option is more prevalent on projects where earthwork may be significant, where multiple grading alternatives exist, and where the owner is motivated to produce ground that is suitable for the highest number of builders at the lowest possible cost. Finding the preferred alternative is worth the time and effort but it doesn't always come easy. The analysis involves generating a series of road layout and grading alternatives aimed at producing one or more alternatives that will allow for maximum buildable area and an attractive but functional road layout. The optimal plan will accomplish both at the lowest overall price. And—it may even provide owners additional flexibility to grade in any number of ways.

Finding the optimal earthwork and grading plan typically involves an iterative process where all parties weigh land-use options against alternative designs and costs until a least-cost, maximum-value grading plan emerges. Owners gain added leverage by insisting that the final grading plan be flexible enough to serve a *multitude of different product types*. This allows owners to adjust quickly to shifts in the market and changes in buyer preference. The rewards for successfully pulling off an optimal flexible grading plan include an enhanced chance of meeting schedule, lowered construction costs, and the ability to make on-the-fly, last-minute adjustments.

At times, owners may choose to grade their project using only on-site material. An edict to limit or prohibit the importing and exporting of material or to balance cuts and fills means that grading can only continue as far as existing volumes allow. Grading schemes aimed at balancing or losing all material on-site often yield imperfect results where some areas remain too low or too high. When the objective is to make on-site-grading work, balancing dirt-heavy sites requires exploring every avenue for losing surplus and unsuitable material. Other techniques include "massaging" the grading plan to absorb more material or to uniformly generate more cut. It doesn't take much raising or lowering of site elevations to produce or to lose large amounts of material. Fine-tuning earthwork is often delegated to the civil engineer, however, an experienced construction individual, a seasoned heavy equipment operator, or an estimator who understands site-work may be a better choice for carrying out this task. Fine-tuning grading can also work well when the construction manager, civil engineer, earthwork contractor, and an architect or specialist versed in siting structures relative to adjacent lots and roads, work together to achieve a better and balanced lot or parcel.

Estimating the cost of a flexible grading plan that is likely to face future refinement should be done in using two estimates. The first estimate should cover the base cost of transforming raw land into a roughly graded product that is suitable for many product types. The second grading estimate follows later and is tailored to a specific product type or, if ground has to be modified, to satisfy a builder or buyer. The second estimate could be considered a sales expense or simply a cost of doing business.

Rather than looking for dirt, more often than not, dirt finds you. I'm speaking of material that regularly emerges unexpectedly in land development. Here are examples of material sources that estimators must account for depending upon the project:

- Erosion waste
- Builder-generated dirt
- Landscape displacement
- Pond excavation and swales
- Fine grading fronts of lots and parcels
- Retaining-wall and rockery excavation
- Fine grading lots, roads, curbs, and sidewalks
- Light, traffic, and other utility pole displacement
- Overexcavation of overly moist and unusable soil
- Inorganic material contaminated with organics and vice versa
- Excess fill material due to swell or minimal compaction specification
- Utility-trench displacement including manholes, vaults, and catch basins
- Foundation material for reservoirs and other small public works facilities

- Excavations for storm- and utility-vaults and other below-grade structures
- Additional material left within grading tolerance that will have to be taken to final "blue top" elevation
- Excess topsoil where it may not be feasible to excavate, haul, and use (or lose) significant quantities of organic material on-site.

Estimating earthwork quantities

Earthwork, mass excavation, and rough grading are three-dimensional cost items. These dimensions are comprised of a variable and often inexact vertical component surrounded by an imperfect perimeter. While engineers commonly describe volume in terms of X, Y, and Z, or length, width, and height, estimators should describe dirt volume in safe, clear, and sure terms, such as bank cubic yard (bcy). Never describe earthwork merely as cubic yards or in terms of square area since the first description is inexact and the second is two-thirds complete.

Topsoil can be an exception to the rule. It's *possible* to describe topsoil stripping in terms of square area if the vertical component or depth of topsoil remains consistent throughout earthwork. Depth cancels out allowing topsoil volume to be expressed on a per-acre or square-unit basis. When stripping is more expensive than structural cut to fill, strategically, it may be in the owner's best interest to estimate and contract that phase of earthwork out on a per-area or lump-sum basis. This leaves contractors less room to maneuver and removes cross sectioning or counting loads as a means of paying for stripping. Intuitively, consistent depth of topsoil also precludes the need to cross section after stripping and before structural grading. However, due to possible error in contour, it's still a good idea to survey check and verify post-stripping elevations. If elevations appear accurate, and the contractor agrees, cross section structural earthwork volumes again after cutting or filling is complete.

Cross sectioning yields volume expressed as bank cubic yards. With bank cubic yards, there are no guesses, conversions, or issues related to work complete because it's the most accurate method to calculate earthwork volume.

It's not always possible to describe earthwork in terms of bank cubic yards. Instead, it may be measured in tons, loose cubic yards, and truck cubic yards. Material measured differently on the same project is usually converted one way or another to the same baseline in order to measure and pay for work. At best, conversions are close approximations and, at worst (particularly for owners), approximations can result in heavy cost overruns. Converting in-place bank yards to loose yards can result in an average volume increase of 35 percent or more due to swelling. How? A bank cubic yard of material that is excavated out of its natural compressed state, then loaded, hauled, and dumped, breaks down and swells thereby increasing in volume. With rain or moisture, material can degrade even further. Bank-cubic-yard-to-truck-cubic-yard conversion increases can be even greater due to the dynamics of weight and the limitations of legal highway loading.

Estimating material volume by truck-cubic-yard measure works best when work is limited to on-site hauling or where weight restrictions aren't an issue. Still, failing to fill trucks to capacity or filling them inconsistently leads to problems. Inconsistent and light loads are common when contractors get their resource mix out of balance as when there are too many trucks or not enough excavators to perform an operation. Since contractors need to keep everything moving, trucks may end up partially loaded and well below capacity just to keep them rolling. When the volume being moved per fixed-unit price—such as by the hour, load, or truck cubic yard—falls short of, or doesn't match the volume being produced (cut or fill) on a one-to-one basis, owners pay more per unit and it's typically more than budgeted.

In addition to when trucks are loaded undercapacity, owners can also suffer when contractors use solo trucks instead of truck and trailer combinations to move dirt. Generally, the larger and lighter the box or the greater the number of axles, the cheaper it is to move dirt. Poor weather, restricted working conditions, or steep grades may leave contractors no choice but to utilize solo trucks. However, contractors who hire trucking on a daily basis *also* have to take what they can get. If that means using a subcontractor with a low-capacity solo truck without a trailer, then the cost per unit of material moved increases and someone—typically the owner—pays the difference. Estimators who suspect that solo trucking is inevitable must factor lower productivity and higher unit pricing into the cost estimate.

Another potential problem with truck-cubic-yard measurement involves keeping track of multiple trucks that vary in color, make, condition, number of axles, and box capacity. Add multiple on-site destinations for hauling, and owner control can quickly go out the door.

One way to cope with all of these issues is to revert to estimating by the bank-cubic-yard measure. Agree with the contractor on volume, negotiate a fair unit price, roll the unit cost and quantity into a unit-price adjustable lump-sum figure, and let the contractor have at it. This shifts the risk to the contractor, improves the odds of not incurring additional costs, and avoids problems with truck-cubic-yard measurement. Cross section the hole, fill, or the pile that was moved to verify its volume. Handle extra work and overexcavation on a case-by-case basis and pay for it using unit prices. Estimators expecting over excavation can develop a potential extra bank-cubic-yard line-item volume and match it with the same unit price used for estimating contract work. If importing fill material appears probable, estimate by the ton or truck cubic yard and covert it into bank cubic yards.

Estimating the cost of moving stockpiles of material is more complicated. Stockpiles consist of loose and possibly wet material. Soil type may vary or have organics mixed in with structural material. The solution is to cross section the pile and, depending on the material type and its condition, treat it as a pile of loose-cubic-yard material. Then, factor in the profile of the original ground and estimate the volume straight across to hauling. If a pile of fine-grained material has been sitting for an extended period of time, it may have consolidated back toward, but short of, bank density. It certainly will be heavier per unit if it has absorbed moisture. The solution may be to negotiate a fair conversion factor with the contractor and proceed on a per-unit basis. Another alternative is to measure the pile, establish volume, agree on a lump-sum cost to move it, and leave it to the contractor. It's always safe to estimate unusual earthmoving on the high side. Pricing the movement of stockpiles months in advance of the actual work being done, assumes that the pile volume will remain constant, that the haul distance won't change, and that the weather will remain consistent.

When pricing import, estimators must first consider where material will most likely originate. This in itself can answer the estimator's next question, which is whether to price delivered material by weight or by truck cubic yard. When estimating costs for importing on-site material, refer to that soil type as noted in the geotechnical report or material specification sheet, and use the correct bank to lose the cubic-yard conversion factor. Compaction can influence estimating since less compactive effort may yield excess material and thus, an increase in costs needed to handle it. Because of swelling and other factors, soils rarely compact exactly as they were excavated, which helps to explain why "balanced" sites may end up with unexpected quantities of excess material. If material is expected to be imported by weight, then convert tons to pounds and divide by the pit's weight per loose (net) cubic yard. Convert net imported cubic yards to bank cubic yards and price accordingly.

Unweighed or truck- cubic-yard material theoretically only concerns volume, however—weight is the issue. The estimator must begin by assuming that trucks hauling from public roads will not exceed legal weight limits. Begin by obtaining the import material's weight per loose- or bank-cubic measure. Then determine the approximate empty weight, or *tare*, of the trucks expected to perform the hauling. Subtracting the weight of an empty truck from the truck's legal maximum allowed gross weight results in the net allowable haul weight. Divide the weight per loose- or bank-cubic-yard of import material into the net allowable haul weight to determine the maximum number of loose or bank cubic yards that the truck can legally deliver according to weight. To calculate a conversion factor, take the calculated number of loose or bank cubic yards that a standard-axled truck can legally haul and divide it into that truck's advertised struck-box capacity without sideboards.

This formula results in a conversion factor relative to the number one. A conversion factor greater than one indicates that a fully-loaded truck of a certain material will weigh less than its legal limit. A conversion factor less than one indicates that the truck cannot legally deliver a full load of material of that weight. The inverse of the conversion factor for partially-loaded trucks is the *percent increase in cost* per unit that owner's have to pay when contractor's base their billing according to a truck's struck-rated capacity at maximum legal weight. Although he has legally hauled as much material as possible, the only sure way for a contractor to bill unweighed material is by that truck's rated capacity. The end-result is that owners pay more per unit of material delivered. This illustration further stresses the importance of estimating costs per bank cubic yard delivered and in-place. Let the contractor account for truck count, varying weights of material, and partial loads.

You may wonder—unless a truck is actually weighed or has an on-board scale, how does a driver know a truck's weight? Can a driver ascertain whether a truck is under, at, or above the gross legal weight limit? This answer is no—not exactly. Drivers wishing to avoid problems with the law typically underload their trucks by feel. Rarely will they risk a ticket by squeezing in a little more material that is billed by volume and not weight. Purposely underloading trucks that are already hauling below their advertised capacity further increases the cost per unit unless owners pay per measured ton or per bank cubic yard in place. Enough hauling on any project may justify installing a temporary on-site truck scale.

On-site scales or not, trucks hauling under capacity and under legal maximum weight means more trips than necessary resulting in extended haul schedules, increased risk of erosion control, more dust, prolonged street cleaning, and increased haul-route maintenance costs. Factor these costs and other peripheral impacts, such as possible off-site pavement repair, into the cost estimate.

The silver lining in all this is that unweighed and unclassified material originating from a non-commercial pit may still be cheap enough to offset losses due to partial loads or truck-cubic-yard measurement. If the price is right, it may even make sense to import and stockpile cheap quality material while it's available. Estimating costs for importing truck cubic yards requires predicting haul distances and assuming truck and trailer sizes. Further refinements can include number of axles, and accounting for lighter truck box weights. Estimators also have to decide whether to price pit material or, costs for locally available unclassified fill. It's easier to predict the cost of exporting surplus material to a known dumpsite than it is to estimate import costs. When estimating truck cubic yard costs, is wise to estimate high and inflate quantities in line with expected soil types.

Estimate fine grading, such as blade or screed work, by square surface area. Since fine grading follows mass excavation and utility work, estimates shouldn't reflect cutting or moving large volumes of material. Fine grading includes activity such as blading areas in readiness for fabric, bank or pit run; laying crushed rock; preparing hardscape for concrete and asphalt; and shaping buildable area. On large areas, consider estimating fine grading on a per-acre or hectare-unit basis. Where the infrastructure material thickness is set, estimating by square area protects owners by shifting the risk of providing quantities and area as specified to the contractor and the site inspector. Even the best estimators may overlook some earthwork items. Here are some tips for producing a more inclusive earthwork and grading estimate:

1. Budget costs for ongoing stockpile maintenance.
2. Account for backfilling curbs and gutters prior to paving.
3. On larger projects, include costs for rolling and maintaining roads.
4. On larger sites, always consider varying on-site haul distances when pricing earthwork.
5. Don't overestimate the value of surplus topsoil or structural material to on-site or off-site users.
6. When surcharging a site, account for the cost of site preparation and removing and hauling off surcharge material.
7. If encountering subsurface water is a possibility, include line-item dewatering costs to contain, channel, pump, and expel water.
8. Include heavy equipment expenses when estimating costs to stockpile material, maintain an on-site dumpsite, or spread recycled tailings.
9. When working rocky soils, include costs for picking out oversized rocks if the specifications restrict the size of material allowed in fills and embankments.
10. Add a line item for free-draining import material if it's possible that the native soil needed for structural fills may spoil due to moisture or organic contamination.
11. When estimating site-work on projects that provide infrastructure to large tracts or areas left natural, include the cost to clear and grade the zone between untouched ground and the outer edge of curb or sidewalk. This zone is necessary for installing utility stubs and as working room.

12. Not all material retains its shape and texture when exposed to weather, moisture, being hauled over, and excavation. This is especially true of silts, clays, and rock that can transmogrify into wet piles of liquid slop. Material in this condition is useless for fill and expensive to load, trucks cannot haul it in great volumes, and dump operators may charge a premium to accept it. Include alternative unit pricing to handle supersaturated and unsuitable waste material.
13. When possible, estimate in square units. This is possible when work depth remains consistent. Square-area measurement shifts risk to the contractor for providing accurate quantities and provides them incentive to prepare subgrade adequately so as to minimize installing quantities of more expensive material, such as crushed rock, concrete, or asphalt. Examples of surface work or materials specified by thickness that can usually be estimated in square units include:
    - Road fabric
    - Crushed rock pads
    - Subgrade preparation
    - Lot and area fine grading
    - Laying crushed rock or imported structural material for sidewalk and road base
    - Installing ballfield and park material including sand blankets, topsoil, and specified fill material
14. Estimate costs of filling planters to the elevation where landscape work begins. The site-work contractor can typically complete subplanter fills most economically. Why? Because deep-planter fills provide places to lose raw topsoil and excess material at no additional cost to the owner. A change order targeted at subplanter fill may put an owner at risk of receiving more expensive import material than necessary to support landscape stock. Growing stock is important, but the depth and quality of topsoil specified in the landscaper's scope of work should be adequate for supporting plants and trees. If not, the landscape topsoil specification is erroneous and in need of revision.
15. It makes economic sense to estimate the hauling of unweighed on-site material according to how much average-sized trucks can deliver pending grades and surface conditions. For projects in private ownership, unless bridges or other highway-rated structures have to be crossed, the legality of hauling with overweight isn't normally an issue. However, legal weight restrictions remain important. They have been established for a purpose including the need to protect infrastructure from damage. Preventing the damage of private roads or soon-to-be deeded public right-of-way should be foremost on every owner's mind. So, it's smart to maintain on-site weight and vehicle-

type restrictions to prevent utility, subgrade, and surface infrastructure damage. Avoid using infrastructure for haul routes by keeping off-road, tracked, and overweight vehicles on graded areas and by crossing infrastructure only at *select* and protected crossings. Estimate temporary haul roads accordingly.

16. Topsoil stripping can present several unique issues for estimators. For one, topsoil depth can vary widely. It may not be possible to identify limits of topsoil relative to root systems by only referencing the geotechnical test-pit data. Sites with a history of agriculture or landfills can make it especially tough to predict topsoil quantities. One solution is to perform additional exploration and potholing. Still, the only way to deal with highly variable sites may be to cross section or measure quantities as excavation proceeds. Another factor to consider when estimating topsoil stripping involves the contractor's style, ability, and equipment. It takes a keen operator, the right-sized piece of equipment, adequate staking or GIS control, and a tight machine to shave thin lifts of topsoil and not remove or disturb inorganic structural material below. How a contractor performs stripping greatly affects end volumes of material generated. Too much topsoil stripping on sites intended to balance may result in mixing structural and organic soil and effectively reducing quantities of both. This can strain cut-to-fill volumes resulting in quantity imbalances and increased earthmoving costs.

Rock

Many factors can influence the cost of excavating bedrock including hardness, type, on-site location, depth belowground, quantity to be removed, and applications for recycling or re-use on-site. Techniques for fracturing and removing rock can include the following:

- Ripping
- Sawcutting and rock trenching
- Drilling and blowing with controlled explosives
- Excavation and fracture by weight or hammer blow

Rock excavation and removal is expensive. Like earthmoving, estimate rock removal by bank cubic measure. Unit pricing for excavated rock that is unusable or that cannot be lost on-site will also have an export-and-dump fee component. Price export-and-dump fees separately since this item can change anytime. Fractured or blown rock doesn't compact well; therefore, estimate hauling costs by the ton, truck cubic yard, or lump sum. Because rock is impermeable, it can hold or channel water so it's wise to add a contingency item for dewatering and containment.

Retaining structures

Estimate rockeries, engineered slopes, and retaining structures by square-face measure in accordance with depth of excavation and layback requirements. When estimating volumes of excavation, overcompensate for laying back slopes. Some soils will stand vertical and others will fall like sand. Always anticipate removing enough material to be safe and legal and then add a little more.

Many contractors specialize only in walls and will avoid performing related work that they aren't prepared or equipped to perform. This may include excavating a hole for placing fabric or grid and providing suitable backfill and crushed rock. So, in addition to estimating walls by square-face measure, estimators should include separate line items for excavating, purchasing backfill material, handling excess cut, establishing truck access to wall locations, controlling erosion, pumping concrete, surveying and layout, fencing and guardrails, and connecting drains to street sewer systems. (For more information, review the retaining wall example discussed in the prior section, "Units").

Estimators must use intuition when pricing retaining walls beginning with careful evaluation of the topography plan in light of road design and grading. As with most everything in site-work, provided that the wall horizontal coordinates are staked correctly, the accuracy of base topography and how it interrelates with civil design can decrease or magnify wall length, height, and square-face area. Other cost consequences of topographic error can include laying back and landscaping more slope than anticipated, having to fall additional timber and clear unexpected areas above and alongside walls, encroaching and possibly dewatering wetlands and open space, or having to work in tighter quarters in dense urban areas. Too much topo error and its affect on walls can easily penalize net developable area or modify land use surrounding the wall. Estimate accordingly.

Walls planned in fill areas can cost significantly more than walls constructed in cut areas. Why? One reason concerns support and bearing strength. Cut areas, if excavated properly in good soils, offer a naturally compacted base for wall support. Fills may not be as reliable and thus require the geotechnical engineer to modify certain parameters and increase requirements, such as wider or thicker footings, a more elaborate drain system, overexcavation and backfill, more and longer lays of fabric and grid, and, sometimes, poured-in-place foundations. The estimator should defer to the geotechnical engineer and discuss the likelihood of extra wall costs in fill situations, or, hold off on cost estimating walls until designs are complete and approved.

Another factor that plays into estimating wall costs involves schedule and sequence for construction. Occasionally, retaining-wall installation has to occur simultaneous with lot grading, foundation building, or vertical construction. Note that estimated wall costs are only valid if wall construction remains synchronized with peripheral construction activity. Deviating from sequence will impact cost and schedule. Wall costs also increase when they have to be "shoe-horned" through another contractor's work or when they are placed in hard to reach locations. Issues regarding where to store wall material and stockpiles of excavated or backfill material can increase wall costs while possibly leading to claims of schedule impact from adjacent buyers or homebuilders.

Wet utilities

Estimating utility work can be as tough or as easy as the estimator chooses. Some estimators may price wet-utility work in as the same way a contractor would. That is, they estimate installation costs as a function of crews, equipment, and material expenses. This approach requires at minimum, knowing a contractor's real in-house hourly equipment-owning and operating costs and historic production rates. Unlike labor and materials, which are set or highly influenced by market forces, the expense structure for contractor equipment can vary widely depending upon whether it's owned or rented and according to it's size, type, age, condition, and make. The estimator must know or assume each eligible contractor's margins and minimal acceptable level of profit including overhead and risk and, know what that contractor's discount is on materials. Even if an estimator can accurately estimate owning and operating costs and production, things change once a contractor alters the equipment mix, adds or deletes personnel, changes materials, or deals with change orders that require yet another set of resources. Unless a project mandates that contractors use equipment that meets specified criteria, running hourly owning and operating costs bears little fruit for owners. Why guess at a guess? There's a better way.

Estimate utilities by lineal measure of pipe installed. Since utility unit pricing increases according to pipe material, diameter, thickness or gauge, and depth of installation, create separate line items in the estimate for every pipe and trench combination called out in the plans. To be safe, include a few more possible depth combinations in case trench work goes deeper than expected. Unit pricing should include trench protection, excavation, bedding, pipe, fittings, backfill, testing, and sanitizing. Unless specified throughout the project, special backfill, such as controlled density fill or concrete, is typically an extra cost and should have a separate line item. When backfilling with native material, the estimator must consider probable soil type, groundwater, and weather. Then, evaluate the potential for costs related to trench overexcavation, increased waste, dewatering, and importing material. If extras appear likely, add separate line items for each to the estimate. Other costs common to utility work include stockpiling or exporting trench displacement and running a video camera (TV'ing) through the system. Treat these also as separate line items.

Estimate sewer manholes, storm manholes, catch basins, and grates individually per unit. Depending upon locality, vacuum testing and sealing can be included in the unit price per manhole or added as a separate line item. Estimate extra depth of manholes per section or by the lineal unit. Handle connections to existing utilities as a lump-sum item per connection. If there is a possibility that urban stormwater overages may be diverted to the sanitary system, include costs for crossover piping and structures.

Elements of the water system that can be estimated per unit installed include butterfly and gate valves, fire hydrants, water services, meter boxes, blow offs, check valves, and thrust blocks. Lump-sum items include the cost to connect to existing utilities, to haul test water to remote locations, to dispose of flushed and chlorinated water, and the cost of pressure-reducing stations. It's easiest to include the cost of fittings for waterline in the installed price as opposed to performing a separate take-off per fitting or by weight. Who is checking weights anyway? When estimating the cost for water reservoirs and pump stations, include the cost of overflow, supply, and distribution lines serving these faculties. When looped water systems are required, add the additional piping to the line item covering that pipe of that size, material, and depth. If looped systems aren't required, but could be at any time, include the looped pipe segment as an add-on separate line item.

Topography affects how a project builds out and that can work to decrease utility size and cost. Here's how. If a project develops from dense to less dense areas, overall wet-utility costs should decrease per lineal foot even though the cost per unit served increases. Less demand should equate to smaller diameter systems, fewer connections, and thus, quicker production installing pipe. Similarly, as slope increases and density either decreases or remains static, the diameter of gravity-system pipes can also decrease since slope at a point, works to increase pressure, which helps to move the same volume of fluid through smaller pipes. It's the opposite of designing large diameter gravity systems to move fluid across flat grades. Decreased demand combined with smaller pipes should result in fewer numbers and smaller sizes of sewer and storm manholes, fewer catch basins, fewer connections, and smaller fittings. Smaller and less intricate systems equate to lower overall wet-utility costs.

Dry utilities

Estimators must first determine who is going to dig and backfill the trench before pricing underground dry utilities. If the service provider is going to control excavation and backfill, other matters concerning the owner's cost responsibility, easements, and meeting schedule materialize. Second, they must know which party is going to supply the materials: the project owner or the service provider. If this issue remains unresolved at the time of estimating, assume that the owner is going to supply all materials and then price accordingly. This isn't all bad because allocating funds for dry-utility pipe may come in handy later if a service provider fails to deliver pipe as scheduled. Plastic pipe is cheap. Its price is far less than the cost of losing schedule, paying for crew downtime, or dealing with open trench and windrowed backfill degradation due to weather. So, estimators who suspect that pipe may not be available for installation should include its price in dry-utility costs. Later, owners will have plenty of time to argue about reimbursement. The goal is to get the pipe in the ground, inspected, backfilled, and to move on to completing infrastructure.

Consolidate dry-utility design

Dense urban-type communities that combine narrow roads, restricted right-of-ways, and tightly spaced housing with state of the art communication services, such as cable, fiber, and wire, can create estimating problems. To begin with, it costs more per unit to place more of anything into less area. But, estimators may also have to follow a mosaic of unrelated and freely designed dry-utility schematics prepared by any number of firms and by any number of people from planners, to buyers, to individual service providers. The solution, as we'll discuss in more detail later, is to depict dry-utility lines, vaults, and easements against already designed road, storm, sanitary, and water plans. Provided that the dry utility plans are complete and true in both the horizontal and vertical planes, an estimator can then measure quantities, locate pipe conflicts, and adjust existing unit pricing according to depth.

Unfortunately, not every owner invests in combining each service provider's dry- utility layout and design onto one master sheet or plan set. This can put overall engineering at a disadvantage and create havoc for estimating unless the civil engineer has already investigated dry utility depths and conduit as it relates to possible utility conflicts with wet utility lot stubs. Difficulties, errors, and oversights can also occur when estimators have to wrestle with a collection of drawings each at a different scale, on different sized-paper, or drawn with no regard to exiting wet utilities, other dry utilities, or plat layout. A primitive way to solve this dilemma is to estimate high and hope that someone later figures out the cheapest routes for installing pipe and structures before commencing work. As a last resort, it's a mistake to expect contractors to "make the pipe fit" as they go because installed pipe is committed pipe that follows a trajectory that may sooner or later collide into an expensive utility conflict. Still, issues may linger concerning trench depth, stacking pipe, lot stubs, connections to existing systems, and how to route pipe in and out of right-of-ways to vaults and structures. Estimating from one comprehensive document avoids many of these issues. It also leads to more predictable, accurate, and complete cost estimating.

Pricing considerations

Estimators may also have to price deep utility installation by boring, jacking, or horizontal-directional drilling (HDD), all of which are specialized and can be expensive. Unless the estimator is well-versed in these methods of work, it's best to solicit a contractor's price or to pre-bid the installation and apply that figure to the estimate. Divide lump-sum pricing into the total lineal feet of pipe to produce a price per lineal foot that will be handy for future estimating.

More traditional methods of deeper pipe installation involve dragging a trench box or cutting back and terracing the trench walls. Pulling and setting a trench box requires a trench wide enough to accommodate a box and generally requires larger and more powerful equipment. Economical and smaller equipment used when laying slopes back must have ample reach, the ability to lift pipe and manholes, and possess enough power and weight to perform excavation. Laying back slopes also requires room to move, windrow excavation, and backfilling of additional volumes of material. Production also plays a role. Up to a point, work progresses faster with larger equipment and a capable crew. Regardless of trenching, workers may have to "hand-tunnel" under existing cross utilities in order to install wet or dry utilities. Estimators who know of potential conflicts must price the work accordingly, however, unexpected tunneling and reroping or modifying pipe is time consuming and expensive change-order work.

Still, no amount of on-site production can negate the effects of having service providers not show up as scheduled to install their system or inspect the work of others so that a trench can be backfilled. Production suffers, downtime charges accumulate, trenches can fill with water (or vehicles), and backfill is at risk of spoiling until work can commence and be completed. For owners, the result is higher costs and longer schedules. Make no mistake about it; it only takes one errant dry-utility service provider to hold a project hostage.

Estimators can remove some uncertainty by either bidding or selecting a construction firm specializing in dry-utility installation to provide pricing, however, the contractor still needs what the estimator needs and that's uniform and complete plans and specifications. A potential drawback to using any independent dry-utility contractor is that some service providers may only approve certain contractors to install their systems. Thus, their permission or recommendation may be required before committing to a particular contractor. Additional scrutiny is okay because owners also want the trust and cooperation of service providers and they want dry utilities installed right.

Create a cost center to handle bringing utilities *to* a project. It's not uncommon for owners to believe that all dry utilities are waiting and stubbed at the project boundary. While this may be the case, dry utilities aren't typically available and ready for easy connection. If they are, it's unlikely that they'd be of high enough capacity to serve a large project at full build-out. Check with service providers on capacity. Low or antiquated utilities may need to be upgraded at the time of connection. Due to street congestion and service demands, off-site-work on active utilities may have to transpire at night and on weekends bringing higher overtime and off-hour pricing. Estimates should include costs for traffic control, pavement restoration, and any vaults needed for connection. If vaults appear likely, note that a dedicated easement or land may be required for aboveground or underground boxes. If vaults or boxes are required at the project's main access road, smart owners will factor vaults in with entry design.

Construction sequencing can bear heavy on cost estimating and few activities are more restrictive than the window available for installing dry utilities. When pricing dry utilities, the estimator must interpret the status of dry utilities relative to other site construction. For instance, in roadways, dry utility systems almost always lay above or fall in between already installed wet utilities. With the exception of crossings, they may go in after curbs but before fine grading for paving, sidewalks, or landscaping. Therefore, the estimate should reflect required post-dry utility work involving road preparation to handle uneven trench displacement, rough grading, shaping, installation of missed sleeves, possible curb and sidewalk damage, and final subgrade work.

Except for excavating and setting vaults, dry utilities are best estimated by the lineal foot either as a group of conduit per service provider or in aggregate for all dry utility piping installed and backfilled together in the same trench. Estimating the cost of installing sleeves or "crossings" at intersections based upon curb-to-curb road widths can be misleading. Instead, scale crossings at intersections long enough to include the additional length needed to compensate for corner radius, curbs, and sidewalks. Estimate the cost of sleeves on a lump-sum-per-crossing basis. For example, a four-way intersection with a group of sleeves crossing each of the four roads would cost estimate as four lump-sum items according to length.

Estimators must also account for stubbing and marking the locations of pipes into each lot or parcel. Although stub lengths can vary per project or municipality, never estimate them to be less than five feet beyond lot line. Similar to wet utilities, it should be clear which contractor, the seller's or buyer's, is going to make dry-utility connections to structures that lie on the edge of right-of-way. This problem can be especially acute where land is restricted such as in dense communities or in-fill projects.

Service providers usually provide off-the-shelf items such as precast vaults. Owners may end up paying for them but service providers specify the size and type of structures required to meet their standards. Examples of above- and belowground structures may include hand holds, "J" boxes, pedestals, specialized vaults, transformers, and switch gear.

When owners are required to provide concrete pads, retaining walls, or excavations for dry-utility structures, estimators must consequently price this work. Belowground installations require a hole, a level bottom, and backfilling. Depending upon the size and depth of the structure, estimates should include whatever it takes to pick the structure, place it in the hole, adjust it, and backfill. Costs may mandate renting a crane, large excavator, or boom truck that is capable of safely lifting a large concrete or steel structure off of a flatbed delivery truck and setting it in place.

Larger and deeper vaults may require laying back the slopes of the excavation for safety reasons or for gaining additional working room between the box and the excavation. Vaults located a distance back from curb will require an access road, so include the price of one even if it's not shown on the plans. Access roads sneak up later after the field service guys complain that they need reliable truck access to service the vault. Check to be sure that there is an access apron. Required fencing for aboveground boxes not supplied by the service provider will be another cost to the owner. Estimate accordingly.

Another cost center often overlooked is surveying. More often than not, owners are required to provide layout for centerline and depth of pipe, vault excavations, stub locations, access road layout, and, yes, as-builts. Don't miss this one.

If an owner chooses to install private community-wide conduit or to just include pipe for future use, it's usually the owner's problem to install it whatever it will fit. In this case, an estimator may opt to compare pricing of, say, one four-inch (10 cm) conduit versus four one-inch (25 mm) individual conduits bundled or single. The trick with estimating an owner's dry-utility system is to quantify the number and *locations* of "J" boxes and structures required to serve groups of homes or individual buildings. Beyond that, it's estimating by the lineal foot installed one-way, or possibly two-ways, if conduit is slated to enter and exit on-site data centers to lots and parcels. See *Figure 14-4.*

Consider when and where dry utilities are going in. Is there a chance at that at some point in time, dry-utility work may potentially undermine a retaining wall or something else? Or, is conduit installation planned under future walls, foundations, or rock ornamentation? If so, first try to move the utility or change the plan in terms of what's going above it. If everything is fixed, include costs for block outs or the cost to protect utility lines with ductile iron or steel sleeves, or concrete backfill.

Plan for schedule disruption and lower daily production when dealing with layers of plastic pipe owned by varying service providers. Multiple pipes dart in and out of the roads into structures, lots and parcels, and into side streets. Not only does this work require more than usual trench excavation and backfill, it can result in a myriad of pipe conflicts requiring deeper trenching and increased crew time. Estimators should discuss potential impacts with whoever is responsible for scheduling work.

Here are some general tips to consider when estimating utility work:

1. Budget for hand, machine, or vacuum potholing. You may need it.
2. Controlled density fills, lean concrete backfill, and other means of protecting data lines, power, and communications are increasing in popularity. If there is a chance that maximum utility protection will be required—perhaps by a high-tech firm, government office, or permit agency, include costs for backfilling trenches with structural soup and hauling off the additional trench displacement.
3. Research and contact wet- and dry-utility service providers having a legal right to be in your project. Involve them in overall utility planning or else risk having them come in later and disrupt the project by cutting finished pavement and landscaping, and possibly removing curb and sidewalk sections in order to install their system. Owners may end up paying all costs beyond those that service providers would have normally incurred had they installed their system in sequence after grading.
4. Honor trench displacement. It's amazing how many estimators miss or aren't aware of the activity and costs associated with trench displacement. They must figure that after the placement of pipe and bedding, everything that comes out of the trench conveniently fits back in. That is until piles of dirt remain long after the utility contractor has left. The fact is, trench displacement happens and it can be costly. At best, trench displacement is comprised of surplus native material displaced by pipe, pipe backfill, and structures. At worst, it may include wet slop and organic material such as limbs and stumps. Estimators must consider the cost of dealing with trench displacement by pricing it for on-site disposal or to off-site export.
5. Native material isn't always reusable. Smaller piles of excavated material are liable to spoil. The more fine-grained or expansive the soil, the higher the risk of it being unusable as structural trench

backfill. There just isn't enough critical mass in long windrowed material to absorb rain and construction activity. Unless the estimator is sure that usable soils can be promptly backfilled with good weather, it may be prudent to include pricing for exporting native trench material and importing select backfill. The same thinking applies when working with shallow trenches, less than four feet deep (120 cm), in moisture sensitive fine-grained soils. Excavated clays and silts exposed to overnight moisture may not be reusable as trench backfill without having to be dried or amended with select material. Estimate trench costs with one eye on soil type and the other eye on volume, but fix both eyes on expected weather and subsurface moisture conditions.

Hardscape

In land development, asphalt and concrete typically define hardscape. However, hardscape can also include pavers, pavements that infiltrate, or honeycombed cells that support highway loading and grass. Most applications include roads, sidewalks, trails, and curbs.

Being approximately six percent oil, asphalt pricing reacts to spikes in crude oil prices. Although the cost differential between crude and asphalt-in-place isn't one to one, the relationship is direct and can make estimating asphalt paving both difficult and precarious. The same is true when pricing other products dependent on oil or oil derivatives such as PVC plastic pipe, sealants, lubricants, and certain fabrics. Oil shortages increase costs for diesel fuel that affect nearly every aspect of site construction from running trucks and heavy equipment to drying aggregate. If it appears that politics may adversely affect oil prices, factor costs for petroleum inflation into unit pricing.

Asphalt can be unit-estimated by tons or square area. For whatever reason—company tradition, reliance on other people's numbers, or just a total misunderstanding of how to protect an owner—some estimators succumb to the "asphalt-tons" trap. From an owner's perspective, always estimate *and think of* asphalt paving in terms of square area and depth. Estimating and budgeting asphalt by any other measure including by the ton is risky. Here's why:

- It can portray a lack of understanding. What does "tons" mean? Nothing really. Tonnage draws an image of mass—not area, shape, or volume. Tonnage speaks nothing of asphalt thickness, number of lifts, challenges in paving corners and tight radiuses, or matching irregular borders and subgrade. Unless used by a paving contractor, communicating in "tons" is meaningless for describing what an owner is or isn't getting in terms of three-dimensional pavement. Permit authorities inspect and approve work based on geometry—not weight. Reliance on the word tons when estimating asphalt can evoke deeper questions concerning the estimator or manager's understanding of how-to control infrastructure construction.

- Conversions. In order to use tonnage as a means for measurement, one has to convert area measurement to volume measurement or vice versa. Why tie cost items to conversion factors when square area and depth are all that matters? Another issue concerns the fact that asphalt delivered by the ton compresses when rolled and compacted in-place. How do owners know what they are getting? Why accept the risk? Converting loose tons to compressed square area can be tricky and is better left to contractors. Instead, estimators should concentrate on obtaining accurate area measurements and applying square-yard-per-depth pricing as appropriate. Afterall, it's the properly compacted three-dimensional product that counts.
- What a deal! Estimating and pricing by the ton usually leads to bidding and contracting by the ton. When providing tons is the objective, it can overshadow a contractor's responsibility to provide adequate coverage and depth per design. Contractors also have less incentive to grade and roll subgrade or underlying lifts of material when they can charge by the ton to fill in dips and gouges with more expensive asphalt. In its extreme, this can lead to a circus of sloppy work and accumulated cost overruns. Pricing asphalt work by weight does nothing for the owner except further pressure quantity and quality control. Estimate and pay for fixed thicknesses of crushed rock and asphalt paving by square area and not by weight.
- You get what you pay for. If an asphalt quantity dispute arises, owners may find it difficult to prove that they've been shorted or overcharged solely on the basis of tons. One reason is that the documentation supporting gross tons delivered is irrelevant—load count and weigh tickets measure performance. From an owner's perspective, it's far easier to prove that you haven't received adequate coverage and, more importantly depth, than it is to argue over tonnage. Plus, it can be futile trying to equate tonnage to plan geometry without considering subgrade anomalies, variations in paved thickness, compaction coefficients, wasted and spilled product, and tons-to-volume conversion factors. With these considerations in mind, on what would an owner base claims of being overcharged or receiving insufficient quantity?
- Who determined the tonnage? Estimating by the ton reveals that the estimator probably hasn't measured the square area to be paved and is instead relying on a paving company or a contractor's estimate for pricing and quantities. It may also suggest a streak of laziness, inexperience, or lack of knowledge on the estimator's behalf. Estimating by the ton requires that somebody convert tonnage to a geometric equivalent

and then deliver per plan and specification. In the owner's best interest, the party assuming this risk should be the supplier or the contractor. The same holds true when pricing road materials under pavement. Once an estimator knows the square area for paving, other quantities in the same footprint automatically drop out. As long as thickness is specified, square area is the common denominator for everything down to the subgrade including fine grading, subbase, base rock, fabrics and geotextiles, lifts of structural material, and, ultimately, pavement. One calculation is applicable to many items with zero conversions.

Measure, bid, and control hardscape according to in-place, square-area units per thickness. Estimate other items pertaining to hardscape as follows:

1. If applicable, include traffic control.
2. Estimate wheelchair ramps per unit installed.
3. Estimate driveway aprons per unit or by square area installed.
4. Estimate sawcutting per lineal measure and adjust the cost according to variations in depth. When asphalt depth is unknown, price work high.
5. Estimate gravel driveways per square-unit measure installed. Include subgrade treatment, fabric, rock thickness, and type of gravel in one unit price.
6. Asphalt sidewalks and trails are more expensive per unit to install than pavement placed using a highway-sized screed. Estimate accordingly by square-area measurement.
7. Estimate curbs by lineal measure and include separate line-item costs for curb grading, adding a base or leveling course, and backfilling. If a precast or a thickened edge is an option to cast-in-place curbs, include them with alternative pricing.
8. Tight quarters and hard-to-reach sections often require handwork resulting in uneven subgrade surfaces. Inconsistent subgrades can require more fill material than uniform surfaces, so consider increasing fill- or crushed-rock volumes proportionately in these areas.
9. Estimate sidewalks uniform in width per lineal measure per depth and width. If width varies, then estimate sidewalks by square-area measure per depth. The same holds true for any patios or broad areas requiring concrete flatwork. Consider extra line-item costs for cleanup, coloring, texturing, backfilling edges, and concrete pumping.
10. Most items carefully constructed in units are demolished in complete chunks. One method is slow and methodical and the other is quick and decisive. This makes lump-sum estimating a good choice for

concrete demolition. Estimate asphalt grinding and cold planing by lump sum, or possibly per square yard depending upon the situation.

11. Hand-placed asphalt in restricted parking areas, around tight radius points, and along irregular planters is more expensive than pavement placed using a screed. The same applies to hand-formed concrete. Examples may include matching sidewalk grades around existing utility vaults and planters, or placing poured-in-place curbing along the top of a retaining wall. Unit price these items significantly higher than if they are placed by a machine.

Storm Ponds

Storm-pond design may not always reflect the degree of construction that is required to build them as envisioned. This happens in part because ponds often require overlying designs by a civil engineer, a geotechnical engineer, a landscape designer, and a biologist. So, when the estimate is performed, all of the plans may not be available. Ponds can also lack detail concerning items such as storm lines in and out, outfalls and spillways, fountain configurations and power, how fill embankments knit with native subgrade, or how multiple ponds or vaults are to be constructed sequential to each other.

Quantities and limits of excavation are contingent on the accuracy of the topography plan. Calculate bank-cubic-yard dirt volumes in similar fashion to mass excavation, and estimate work in accordance with working area, degree of slope in and out of the hole, cost of export, and to what extent cut can be lost in adjacent embankments. Details, such as construction access, fencing, signage, perimeter trails, construing where plantings and topsoil meet high-water mark, and handling interim runoff, leave the estimator much latitude when calculating quantities and costs.

Storm facilities frequently are mini-projects where multiple ponds occupy the same surrounding acreage and are connected by piping, landscaping, roads, hiking trails, common berms, and flow diversion structures. Estimate site-preparation and utility costs similar to the rest of the project with one exception. Unit costs for excavating in restricted areas, unconventionally laying pipe or installing short-lengths of pipe, and placing specialized structures must all be priced higher than in normal site-work. Pond utility work commonly includes energy dissipaters, outlet structures, grates, filters, and control structures. Estimate these items individually per unit. Price liners in-place by the square foot and rely on multiple measurements to check pond bowl-area calculations. Estimate the cost of permanent-gravel access roads into and around the pond no differently from the cost of the same work elsewhere—per finished square area. Price short paved roads, which may or may not be surveyed, into the pond per square area *or* lump sum, and price longer paved roads per square-area units. Estimate associated roads costs for subgrade preparation, fabric, crushed rock, and surfacing individually. Assume that any permanent roads planned into ponds will be concrete or asphalt pavement and estimate accordingly. Estimate the cost to stake the location of, and fence around ponds on a lineal foot basis. Estimating units for landscaping varies. Price individual trees per unit per caliper diameter, and estimate ground cover, small shrubs, and irrigation by square area. In case it has to be replaced later, estimate expensive landscape stock on a per-unit basis. Ponds initially fill with turbid construction runoff or become murky immediately following construction, so releasing water is usually contingent on water turbidity. Therefore, include costs for ongoing water-quality monitoring.

If the project site contractors complain when asked to honor the same-unit pricing for storm ponds as for site-work, then it may present owners with the opportunity to bid out pond work. Depending upon how soft the housing market is, or how hungry contractors are, bidding can produce a whole set of lower unit prices. As a rule though, inordinately low pricing never lasts. Owners thinking that they're getting a good deal in one location typically end up paying more somewhere else, or they set the table for heavy change orders and future price inflation. It's usually safe to negotiate pond pricing with an existing on-site contractor.

Predicting storm-pond work is more complex if the number or size of ponds fluctuates with lot count, density, impermeable surface regulations, or final developed acreage. One solution is to develop a storm-pond unit-price schedule that can apply to future pond work and adjust it for inflation.

Wetlands

Few items in land development are more variable and unpredictable than constructing new or enhancing existing wetlands, and then trying to get all plantings to survive as designed. Without plans and specific details, estimating the cost of replacing or mitigating wetlands hinges on experience and a touch of creative thinking. With wetland design, many people are usually involved, the initial base topography may not be accurate or even close to reality, and the scope of work can change dramatically once contractors get underway.

When wetlands are designed by an individual who is better versed in plantings and hydric soil than in construction, owners are likely to receive a collection of stock illustrations and ordinance standards that serve as backup to a cartoon titled "Wetland Rehabilitation", but that are lacking in-depth construction detailing. This is a double-edge sword. It frees estimators to price work as they see fit, but it also restricts the amount of information that they have to work with. Either scenario can lead to inaccurate pricing.

In short, wetlands are tough to estimate. Typical cost centers for wetland work include surveying, clearing, grading, building berms, importing specified soil mixes, installing a litany of new plants— and possibly replacing stock— and monitoring on a multiyear basis. Additional costs related to irrigation, spillways, riprap, water piping in and out, storm facilities, water circulation and aeration, ditch building, and fencing might also be included. Depending upon soil type, basins may require bentonite, welded fabrics, or other methods for sealing and holding water. Another item typical of wetland construction can involve surveying and recording costs for boundary line adjustments and for composing sensitive area legal descriptions. Of course, it's preferable to budget these costs as plat work and to not have them occur later as unanticipated add-on expenses.

Since wetland construction is largely comprised of the same elements as site-work, estimate the costs on a unit basis in line with the recommendations offered throughout this chapter. In a worst-case scenario, pricing wetland work without a design usually means estimating on a per-acre basis using rule-of-thumb parameters. Short of excavation, pipe work, and access, it's also possible to price wetland work according to square area. Landscapers who gravitate to wetland work typically supply costs for site preparation and uniform plantings on a square-area basis. In all estimates, include fees for watering and ongoing maintenance before, and possibly after, the contractor's warranty period.

Miscellaneous items

Here are suggestions for estimating miscellaneous site-work items:

1. Estimate bollards per unit or lump sum as a group installed.
2. Estimate signalization, street lighting, and roadside furniture per unit installed.
3. Estimate street and traffic signage per unit and then combine all costs into a lump-sum number.
4. Estimate dry-utility pipe and trenching costs by the lineal foot plus any additional costs for street rehabilitation.
5. Estimate striping and stop bars per lineal measure. Price special or individual street markings per unit after receiving quotes from suppliers.
6. Depending upon length and number, estimate culverts by material per length installed. Estimate bridges per square unit, but bid them as lump-sum items.

7. Price trees, large shrubs, and irrigation meters per unit installed. Uniform seeding, bark, ground cover, small plantings, and irrigation can be estimated on a per-square-foot basis.
8. Estimate costs for large dry-utility vaults and switch gear per unit installed. Costs for smaller transformers, J-boxes, and handholds can be rolled into the overall system price per lineal foot.
9. Estimate crushed-rock construction entrances by square area installed or on a lump-sum basis per entrance. Include alternative costs for placing pads over fabric to preserve rock and provide additional structural support.
10. Surprises and complications expected from relocating utilities justify initially pricing work based on hourly production rates, hourly crew costs, and material costs. After calculating an hourly cost for work, roll this number into a lump-sum line-item figure.
11. Estimate French drains per lineal measurement installed. French-drain price per foot includes costs of surveying and layout, excavation, pipe, drain rock, tying the pipe into the storm system or to daylight, abating water during construction, backfill, and fabric if required.
12. Gauge mobilization in terms of how often the contractor will move in and out of a project. Factor in the size and number of pieces of equipment expected for use in construction. Since contractors charge whatever they feel is appropriate, estimate mobilization as a lump-sum item.
13. A number of expensive items may appear towards the end of a project. These items may not be well-defined or they may rely on specialized design or undefined government specifications. Signalization, ornamental rockwork, and entry monumentation fall into this category. Quantify and price those items not yet designed on a lump-sum basis.
14. Estimate wet-utility pipe according to material type, soils, pipe diameter, backfill requirements, and depth per lineal measure beginning at a minimal depth of four feet (120 cm). Increase pipe installation costs in increments commensurate with soils, rock, and pipe diameter. For most land development, three-foot increments work well.

When the plans and specifications are lacking, there is no substitute for experience. Knowledgeable estimators familiar with land development can assign ball-park numbers to items until more information is available. If they can't, they usually know where to find reliable figures that work.

Applying costs to someone's dreams

When estimating, how do you cope with incomplete construction documents or details and specifications that cry for the impossible? One path is to change or amend the documents but that move isn't always welcome or feasible, especially just before bidding. Another alternative is to cost and quantify the dream as it exists. What do I mean by dreams? Dreams include anything half-baked, poorly thought-out, or woefully lacking in technical details. Here are four examples of how to apply costs to someone's dreams:

1. Removing, preserving, and replacing fragile items. The lighter the gauge and larger the panels, the less likely a chain link fence will move well. The fix: ignore the specification and include a contingency cost to demolish and replace the fence in its entirety. The same holds true for galvanized culvert, concrete pipe, and thin-walled and bell-and-spigot plastic pipe.
2. Heavy earthwork and no details that describe how to treat construction runoff. Most, if not all, projects that involve excavation and grading should have an erosion-control and storm-water treatment plan. If one doesn't exist, create one, develop the costs, and then include it as a line-item budget contingency. Include your plan or sketch when submitting costs to handle storm and water quality facilities.
3. Preserving individual trees that are in areas of cut and fill. Field-examining the condition of individual save trees in question isn't always possible or practical especially if tree flagging or staking doesn't exist. Instead, find out what tree species are present and rely on an arborist for answers. Most arborists know the level of disturbance that particular tree species can handle and still survive. According to an arborist's advise, include a cost contingency to purchase and install three substitute trees for every native tree that is likely to perish during development.
4. Amphitheaters, gathering places, kiosks, intricate entryways, and spiritually inspired architecture are arguably the most difficult examples of site-work to quantify and price. Somewhere in space—between the mind of the artist and the blade of a bulldozer—there exists a world, a vision, that must be interpreted and translated into design and engineering. Arcane, one-of-a-kind work usually entails custom detailing, hand-wrought materials, and unique fabrication by individuals who abide by no man's schedule. Once work is underway and the vision evolves into wood, stone, and metal images, artists often find it hard to leave things alone. Instead, they may tinker, adjust, and perform changes as construction proceeds. The resulting scope creep may continue if the artist is intent on pushing the job

closer towards a realm reserved only for true masterpieces. The safest way to estimate costs for these types of jobs is to create a not-to-exceed ceiling and hope that ongoing value engineering results in one of two things; more amenity for the same money, or the same amenity for less money. Estimating unusual work is tedious and should be as detailed as possible so that everyone, including the artist, knows where money is, where it isn't, and what the limits are.

Items commonly overlooked in site-work estimates may include:

- As-builts
- Potholing
- Soil aeration
- French drains
- Dust abatement
- Trench displacement
- Site-security measures
- Landscape pest control
- Multiple water services
- Importing structural soil
- Water and irrigation meters
- Weed control prior to paving
- Backfilling temporary swales
- Pond fountains and pump systems
- Treating, mixing, and aerating soil
- Removing snow and tree blowdown
- Taking and testing concrete cylinders
- Electrical work and field connections
- Off-site export of unsuitable trench soils
- Repairing and sawcutting existing streets
- Removal and backfill of temporary pipes
- Ongoing street maintenance and cleaning
- Relocation of fences, signage or structures
- Hydrating soil in order to gain compaction
- Temporary pavement marking and signage
- Connecting new utilities to existing utilities
- Pressure reducing stations for water systems
- Costs to adjust and raise utilities with paving
- Installing monumentation and permanent control
- Overexcavation required for placing imported topsoil
- Dual water systems where differing pressure zones meet
- Area preparation and utilities for temporary on-site offices

- Researching and decommissioning wells and boring holes
- Removal of erosion control, protective fencing, and sand bags
- Hauling in and disposing of water for remote waterline testing
- Dual storm systems that separate water for release or treatment
- Repairing builder damage to curbs, sidewalks, and landscaping
- Cleaning existing wet-utility pipes and structures prior to construction
- Protecting existing storm catch basins and manholes during construction
- Performance and payment bonds as a percentage of total hard-construction cost
- Premium costs attributed to meeting ultra-tight grading, pipe, and paving tolerances
- Collecting, loading, and exporting garbage and contractor waste to an off-site dump
- Importing material for pipe trench backfill in lieu of using unsuitable native material
- Costs for an on-site nursery to store and care for landscape stock purchased in quantity
- Costs of removing piping and structures, and of backfilling temporary erosion control ponds
- Maintenance and unexpected surfacing of access and on-site haul roads due to wear, inclement weather, or control erosion
- Cost impact of site-work crews having to work around and among building contractors and their staging areas and equipment
- Costs to collect, load, haul, and dump organic debris, such as slash, limbs, brush, and stumps, during extremely dry weather and no-burn work periods
- The cost/premium ratio of using small screed machinery or hand-placing asphalt in restricted areas typical of dense urban developments, parks, trails, and narrow accessways
- The costs of permanent slope treatment. Compacted inorganic material works great for building embankments but it's lousy for growing anything short of weeds. The cost of preparing slopes, called *scarification*, of placing a uniform lift of topsoil, of, perhaps, installing biodegradable woven mats, and of landscaping are items that are usually underfunded.
- When there is a desire to balance grading on in-fill projects with pre-existing hardscape, then account for the volume displacement caused by removing asphalt, concrete, underground fuel and septic tanks, underground utilities, and small building slabs and foundations. The cost difference between material-mass removed and replacement fill added can be considerable.

Contingencies and exclusions

What determines adding contingency funds to an estimate? Or, when should an item be included as a contingency rather than be excluded from the estimate? A proven route is to base the cost estimate on preconstruction reports and to account for worse-case-possible conditions such as poor weather, difficulties in gaining permits, variable soils, and subsurface water conditions as part of other, or as individual, line items. How well an estimator translates "potential problem" information into costs affects everything from the budget to the proforma and it says a great deal about the estimator's ability to interpret conditions and to create accurate and complete cost estimates.

But, questions can linger regarding how an estimator separates contingencies from exclusions. When is a site-work estimate too conservative or woefully inadequate? In site construction, the difference between overrunning by 3 percent and by 20 percent in cost can sometimes be traced to earlier decisions regarding contingencies and exclusions.

Here's an exercise for estimators. Peruse this list of potential cost centers and ask yourself under what conditions you would treat these items as exclusions or contingencies. Could any of these potential budget busters emerge as a surprise extra cost on your next project?

- Requirements to encase pipe
- The unexpected need for well points
- Need to raise or move existing power lines
- Unexpected wetland rehabilitation or creation
- Requirements to transplant native trees on-site
- Working during poor weather or out of season
- Requirements to add a sound attenuation fence
- Sudden requirement for utility-trench dewatering
- Costs attributed to possible owner-induced delays
- Watering and maintenance of erosion-control grass
- Cost to mitigate work or disruption on public streets
- Year-round maintenance of erosion-control facilities
- Handling of unexpected contaminated soil or asbestos
- Special requests from the sales and marketing departments
- Schedule impacts and delays resulting in the need for paid overtime
- Being responsible for making utility connections to buildings or homes
- Subgrade repairs while restoring pre-existing curbs, paving and sidewalks
- A request for exotic erosion-control facilities that exceed current regulations

- Trench-excavation costs resulting from unexpected rock, boulders, or sandstone
- Adding a post-construction contract contingency for follow-up finishing touches
- Overexcavation beyond limit of subgrade due to unsuitable soil, moisture, or rock
- Soil treatment and conditioning of on-site soils in order to make them all-weather usable
- Loading, hauling, and dumping boulders greater than one cubic yard or cubic meter in size
- Schedule delays caused by government agencies or due to lack of having sufficient permits
- Encountering varying and deeper depths of topsoil due to prior dumping or farm activity
- Drought-induced or inordinately high water costs to establish and maintain landscape plantings
- Requirements to paint the inside walls of utility vaults or to provide floor drain connections to the storm system
- Installing utilities by laying back slopes or installing utilities because other methods for shoring without use of trench box are required

Contingencies

Contingencies are one of the most misunderstood and abused tools in construction. They're also one of the most effective ways to control project costs. Contingencies are included in cost estimates to cover unquantified but possible expense. For example, during the course of developing land, any or all of the following issues can happen, and when they do, they tax a project financially.

- Incomplete design
- Delayed permitting
- Rapid price inflation
- Overtime and off-hour work premiums
- Unexpected wet, cold, and poor weather
- Impact resulting from errors in judgment
- Unexpected inspection or survey expenses
- Newly imposed regulations and restrictions
- Changes in design standards, market, or product type
- Underestimation of quantities, costs, and schedule requirements
- And—the one fate that some people believe is just pure myth—bad luck

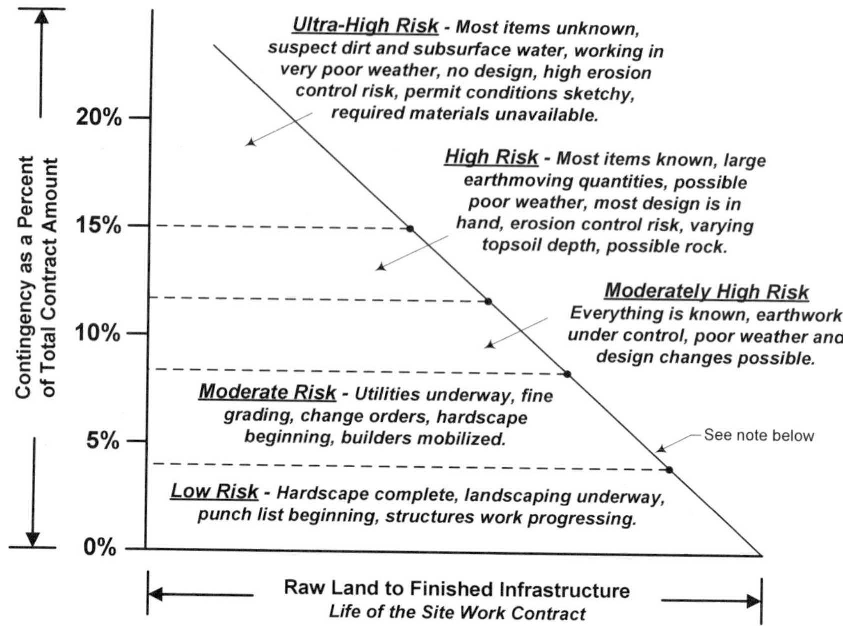

Figure 13-4

Uncertainty makes budgeting contingency funds a mandatory operating procedure. Contingency funding, however, can mean something different to each individual on the team. To the estimator, it can cushion errors, omissions, and poor assumptions. For the construction manager, contingencies can leave room for meeting budget in spite of people problems, operational delays, or complications in logistics. Architects and engineers may view contingency as a means of compensating for shortfalls due to fast-track design and insufficient funding or, for just not knowing what something will cost. The financial team may view contingency funding as money that someday may be available to help improve the internal rate of return. Owners may view contingency as a sort of latent profit center, a fund not to be touched if things go as planned. To a certain extent, all of these people are correct since contingency funds can take many shapes and support the owner's team in multiple ways provided that the rest of the job is set up right from the beginning.

The estimator must factor in contingencies to help ensure that not only the estimate is adequate to perform the scope of work, but that budgets based on the estimate are also beatable. However, some owners may look at contingencies as funds available for completing routine work. Or, they may be willing to allow the scope of work to expand to meet contingency levels. This, of course, is a mistake. Rather than being thought of as an expendable well of money, contingency funding should be thought of as untouchable—a last resort when no other option exists for financing work.

Figure 13-4 portrays contingency as a function of risk. In land development, risk declines as work progresses. The percent risk portrayed on *Figure 13-4* is a calculation based on total construction cost. As risk comes under control, contingencies should decline. On many projects, risk diminishes: with completion of mass excavation; by averting the wet season; when crews work from complete plans; and, with timely permitting. Contingency amounts are subjective but there are reasonable limitations. For example, logging through heavy grading on steep, broken terrain may constitute a 12- to 20-percent contingency. As these activities evolve into utility work and fine grading, a contingency of seven to 12 percent may be acceptable. Hardscape, street lighting, and landscaping may warrant a three- to seven-percent contingency. As a site gains predictability, contingencies should decline.

One strategy is to assign contingencies to line items. This task can be arduous and it questions the practicality of calculating estimates based on an estimate. Afterall, it's just as easy to assign a master contingency to the bottom line. A smarter strategy is to list the line-item contingencies for highly suspect or impossible-to-predict items in the same cost category as the item in question. Typical candidates for this method include tough to quantify items that are expected to overrun, such as installing dry utilities, handling subsurface water, providing erosion control, or importing truck cubic yards of material from an unknown source. Structure the contract and contingency pricing together per cost category and avoid the use of sum total, bottom-line blended numbers.

Estimators experienced and skilled in evaluating production and risk can also add contingency by "imbedding" into select line items, additional quantities or *margins* of material beyond those measured per plan. Typical candidates for inflating quantities in this manner include high-risk activities that are expected to overrun but that are verifiable, such as overexcavation, utility pipe, and street trees.

When bidding work, or when the owner-contractor relationship remains distant, don't reveal in-house, inflated budget quantities to the contractor. However, when relationships are trusting, or when negotiating unit-price work, inflating or imbedding the contractor's contract quantities can be beneficial to both parties. First, the contractor continues working through change of conditions without disruption to schedule or downtime. Next, there is no paperwork or time lost in processing the change order. Last, verifiable quantities protect the owner. But, the bonus is this; owners pay unit pricing for extra work. The prerequisites for imbedding are knowing how much to inflate quantities, keeping unit pricing constant, inflating quantities that can be clearly verified, and working with a willing contractor. Imbedded quantities allow owners to skip the traditional change order process and, instead, apply quantities against pricing as work continues.

Imbedding margins also works to shield others from meddling with and adjusting costs when translating or manipulating cost-estimate values. Imbedded margins are a fair, honest, and efficient way to deal with undeniable changes in quantities. It's a cost-savings technique that is useful in controlling damage before it happens—with less overhead.

Use any combination of contingencies to suit a particular project. However, don't allow multiple contingencies to compound and skew the budget and proforma. The problem is that everybody on the manager's team who is responsible for cost control and accounting can inadvertently inject burdensome and redundant contingencies into the cost structure. The story may unfold something like this:

1. The estimator increases quantities by imbedding a five-percent margin into select line items.
2. The same estimator adjusts the bottom line of the construction cost estimate with a 10-percent across-the-board contingency.
3. The person responsible for consultant contracts adds another seven-percent contingency to the estimate in order to lock in expenses for surveying and control, geotechnical work, and lab testing.
4. Along comes the person in charge of financing and he's going to secure the deal by adding in a hefty 10 percent everywhere, just in case. Afterall, he's responsible for procuring the working capital. If this thing goes sideways, he'll be the one explaining to the bank and board of directors what went wrong.
5. Finally, the head of sales adds three-percent to all soft- and hard-costs within an individual parcel because a potential buyer is "floating" requests for the seller to perform open-trench surgery, add a new retaining wall, and regrade in order to complete the sale.

The above vignette—although fantasy—can and does happen when parties with unrestrained electronic access to project records act without communicating. Total in-house construction contingencies can mount in convoluted ways and needlessly exceed 25 percent or more. Not only does this move result in excessive costs for borrowed money, but artificial cost projections can alter an owner's perspective and cause him to make poor decisions about issues such as planning and layout, degree of acceptable grading, or product type. Too much contingency can turn a protective shield into a crutch or, for the timid, an expensive pair of stilts.

Unfortunately, high inflation and skyrocketing interest rates can dampen the ability to reduce contingency during construction. If everything is under control, review the contingency amounts across the board and reduce them where possible. Gauge contingency as a project matures.

On a side note, land development projects that are carrying accumulated contingency *and* are experiencing out-of-control construction spending may have a serious money-sink somewhere that is in need of plugging.

Exclusions

If contingency pricing can be described as subjective, exclusions can be regarded as objective. That is, as opposed to contingency pricing, which is based on unknowns, exclusions typically encompass work activities that reasonable evidence suggests may transpire during construction. Many reasons exist for estimators to exclude items from cost estimates. An estimator may choose to omit something if the estimate already appears too high. Or, an estimator may be directed by a consultant or the owner to "forget it" after inquiring about an item. Provided that the item never materializes, this is good advice. If however, an omitted item does come to life, additional costs and schedule may hang in the balance. A party other than someone on the owner's staff is more likely to exclude items in their estimate or bid. For instance, a fixed-fee contractor or consultant hired to estimate work may be inclined to shove uncertain items into a box titled exclusions and move on rather than take the time to research and provide for them, line-item values or at least, an "informed cost contingency". Exclusions can include anything tangible and considered a remote possibility, or a key construction item that will most-likely be required but is missing from the plans. Unfortunately, multiple exclusions that compound cost estimates can represent inadequate budgets and insufficient time devoted to earlier planning, design, and coordination. This can be an indicator for owners that more wasted time and higher-than-expected construction costs my soon follow. Cost items prone to exclusion include:

- As-built drawings
- Temporary culverts
- Surfacing haul roads
- Hand-placed rock headwalls
- Storm catch-basin protection
- Dewatering and French drains
- Accounting for prevailing wage
- Select backfill in utility trenches
- Pricing vehicle-resistant barricades
- Cleaning ponds of accumulated silt
- Installing ladders in deep manholes
- Estimating pipe of higher than usual strength
- Providing temporary utilities to contractor job shacks
- Pricing imported blended topsoil in lieu of raw native topsoil
- If burning is prohibited, the cost premium for grinding and exporting woody debris
- Greater than forty-hour workweeks excluding possible overtime and weekend work
- A cost premium for export when it's assumed that all topsoil strippings and excess cut will be lost on-site
- Costs to seismically construct footings and foundations that support unoccupied structures such as sculptures, rockwork, and ornamental amenities.

For estimators, the objective should be to balance exclusions and contingencies in order to produce a reliable and comprehensive site-work cost estimate regardless of incomplete, missing, or impossible-to-know information.

Preliminary cost estimates

Limit the shelf life of preliminary site-work cost estimates. Rough order of magnitude and range estimates can accentuate potential risks and bloat various cost centers while, at the same time, minimize or completely miss others. They can overdrive a proforma resulting in unrealistic expectations, undo pressure on sales performance, and mislead owners trying to determine whether a project "pencils" financially.

One way around this issue is to *model* the project. Modeling is a method used to determine rough development costs without paying for full-blown design and engineering services. With modeling, the owner commissions a consultant to perform a quick road layout and grading plan that is buildable, mimics local standards, and fits the terrain. The consultant adds utilities, infrastructure, landscaping and the like, performs a quantity take-off, applies pricing, and builds an informal cost estimate. This approach bases the estimate on something tangible instead of estimating by feel, per acre, square foot, or unit, or by assigning arbitrary and blended lump-sum numbers to multiple categories of construction. With modeling, owners not only obtain an economical and quick technique for evaluating projects, they advance with a baseline concept that can be measured against, extrapolated upon, or used again to predict other future costs.

The better the cost estimate, the better the budget. Owners that enter construction with budgets based upon preliminary and incomplete plans or poorly constructed estimates are at risk of overrunning in cost. *Figure 13-5* portrays the relationship between budget, actual construction cost, savings, and cost overruns. This example profiles a typical site-work cost curve over the life of a project against thresholds defined by three different budgets. Construction costs break even where the total cost curve intersects each budget amount. Unfortunately, the break-even points occur well before site construction is completed. Note that the highest and most accurate budget results in cost savings. Two other budgets, though, fall short of meeting construction costs resulting in overruns. Of the more profitable of the two, poor estimating is offset by good construction documents but still results in a subpar budget. Concerning the least profitable alternative, poor construction documents combined with inadequate estimating leads the project deeper into cost overruns.

This graph also illustrates, to nobody's surprise, that high construction costs combined with low budgets are a formula for significant cost overruns. Conversely, low construction costs and adequate budgets can result in cost savings. More specifically, this graph stresses the importance of proceeding with budgets that correlate with cost estimates based on accurate and all-inclusive construction documents. Because owners can control budgets more than they can control the cost of site construction, proceed with estimates that are accurate, complete, and not dated or preliminary.

Estimating and managing work

As alluded to earlier, owners shouldn't allow anyone to change the format of a logical and well-crafted cost estimate especially if the person who is going to manage and be responsible for construction has built the estimate. Experienced estimators build cost estimates according to how they envision a project to build-out. Provided that the scope of work doesn't change, it's unrealistic to expect a construction manager who didn't estimate a project to meet or beat an original estimator's numbers. Without speaking to the original estimator and comparing notes, it's unlikely that a construction manager will know exactly what the estimator was thinking when pricing the complicated activities such as erosion control, grading, and storm-pond work. Afterall, site-work doesn't follow

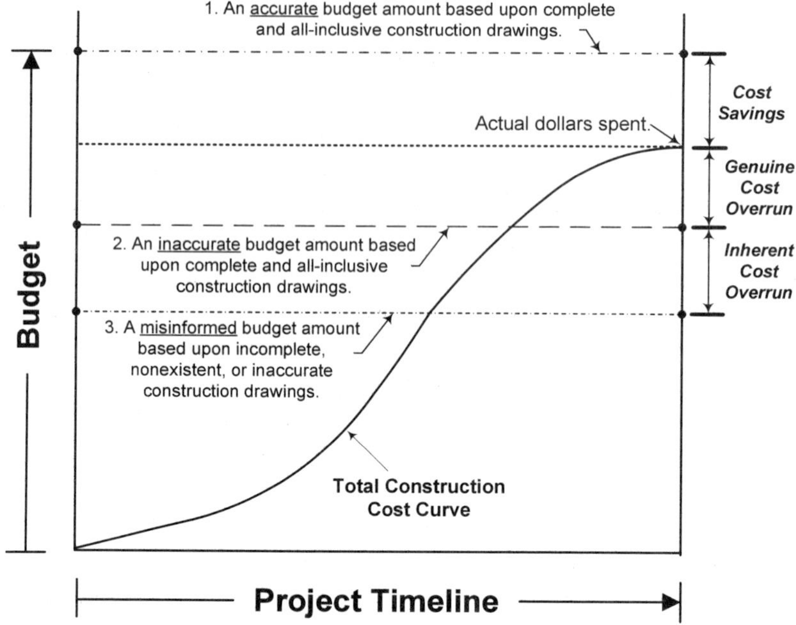

Figure 13-5

cookie-cutter patterns—there are too many ways to perform work and to mine savings. Strategy matters. This reinforces the doctrine that requires construction managers to do their own estimating.

Value engineering

Owners who are serious about enhancing their project, meeting objectives, and collecting better bids invest in value engineering throughout the life of their projects. At minimum, value engineering is a technique useful in helping to ensure that all the options have been considered for maximizing profit, minimizing construction cost, expediting schedule, producing more product for the same money, and reducing the risk of disputes and litigation.

Owners may sense resistance to third-party value engineering from some design consultants on their team. Designers—whether it's planners, architects, or engineers—naturally feel that their work is "good-to-go" and that owners are already reaping the benefits of ongoing design checks, thorough in-house review, and, of course, sharp design. And, they may be right. Some firms routinely save owners money but, sadly, some firms don't. There is usually too much involved in site-work to allow one design firm total independence without some form of oversight or third-party peer review.

High-risk work provides fertile grounds for value engineering, and site-work is no exception. Vulnerability to cost overruns increases with project size but so do opportunities to save money. The basic objectives of value engineering site construction are three-fold:

1. Reduce costs across the board.
2. Shorten the construction schedule.
3. Reduce the overall requirements (scope) for construction.

Value engineering is most effective when instituted early in a project. Go back to *Figures 3-1* and *3-2* since they both apply here. The closer a project gets to bidding or breaking ground, the less opportunity exists for increasing value and cutting costs. Though value engineering can still be useful anytime throughout a project, opportunities rapidly vanish once owners commit to grading, retaining walls, storm ponds, utilities, and locating roads.

Once higher-risk items are brought under control, value-engineering opportunities largely involve finding ways to accelerate schedule and to get builders in quicker. With the exception of building cheaper road sections, opportunities to save significant money day-to-day in site-work through product and material substitution are limited. Sure, changing utility-pipe material, pond linings, streetlight poles, or landscape stock can produce handsome savings, but the greatest values usually result from manipulating dirt—both physically and in content. For instance, trading structural fill for crushed rock and crushed rock for asphalt or concrete can bring owners hefty dividends. Reservoirs, pump stations, bridges, and storm vaults can also provide large-scale opportunities for savings through material substitution provided that they're value-engineered early in the process. Retaining walls can be ripe for value engineering but with walls, cheaper may not always be better if it results in an ugly wall. More often than not, the big dollars are saved by manipulating site design, by altering earthwork volumes, building smaller and more efficient storm-water treatment facilities, minimizing or eliminating walls, raising utilities and shortening pipe lengths, or building less infrastructure.

Different reviews for different objectives

Value engineering, constructability reviews, and document review services act like a three-legged stool. They all support each other. Refer to *Figure 13-6*. Value engineering is useful to help reduce cost, make money, and, if necessary, challenge a designer's work. Constructability reviews are used to determine whether a project can be built per design and plan. The reviewer traces the course of site-work to determine conflicts, to identify missing information, and to check for technical completeness. A key element is ensuring that match lines work. Can the contractor build the project as advertised? If not—why? The intent of document review services is to scrutinize the plans and specifications by identifying and correcting wording and detailing that are in conflict, are canned and useless, or are nebulous enough to lead a contractor astray. Document reviews reduce an owner's liability and risk during the site construction and help to ensure a better, more accurate construction bid. Eliminating vague language, conflicting information, and inaccuracies greatly narrows an owner's exposure to cost overruns, unintended change orders, and potential disputes. It provides owners with an edge by effectively reducing a contractor's room to maneuver. In construction documents, more potential problems usually lie in words and wording than in pictures.

Moving on to *Figure 13-7,* in this true-case example, the civil engineer who was charged with value engineering a roadway submitted a quickly prepared pavement section that was unacceptable to the permit agency. Following through the diagram, a final and acceptable paving and base section finally materialized after multiple design efforts, meetings, and redesign activity had occurred—all at the owner's expense. It didn't have to happen this way. What could have happened differently to reduce the time and effort spent in seeking the variance? Here are a few ideas. First, the geotechnical engineer would have been more effective as the lead consultant. Second, the civil engineer could have included alternative road sections at the time he submitted the plans for approval. Third, the owner should have sought value engineering earlier in the process.

Keeping in mind the above example, now refer to *Figure 13-8.* This spreadsheet portrays a smorgasbord of road-section alternatives each structurally suitable for the same standard of road. This chart of structural equivalents for base course and pavement was created after analyzing road-section alternatives based upon native soil strength and available road materials. *Figure 13-8* works because the geotechnical consultant performed the pavement analysis well before the civil engineer submitted the road sections for permitting. So, there was no need to switch plans at city hall or to ask that a new cost-savings concept be reviewed. Instead, when options are submitted and approved as part of the

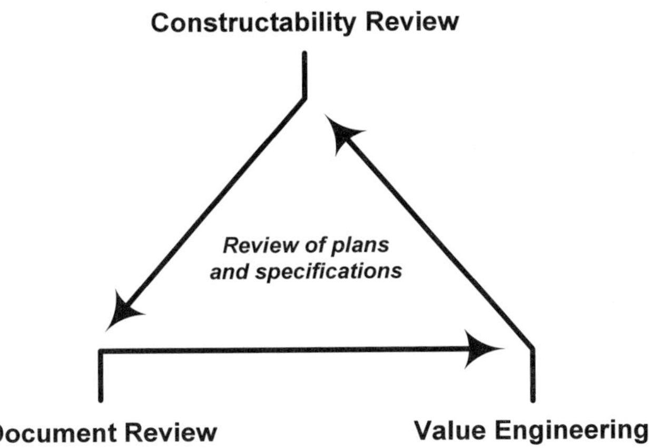

Figure 13-6

road plans, all an owner has to do is consider cost and weather and then pursue the cheapest and most workable road section during that time of the construction season.

Referring to *Figure 13-8*, the cheapest road section per square yard (not including earthwork) is alternative four. However, if your permitting agency insists on laying an asphalt base, alternative three may be the best option. If wet weather is expected, owners may opt for the cheapest alternative that also provides a paved running surface to protect subgrade and allow easy access. Taking *Figure 13-8* one step further, consider the total section thickness column. Where crushed rock is cheap and abundant, it pays to consider using sections of thicker rock. A thicker rock section may also be preferable if deeper grading results in a cut that generates additional fill material that is needed elsewhere on-site.

Conversely, fill sections requiring imported material may benefit from a deeper road section where the combined thickness of crushed rock and asphalt results in requiring less imported material. Dirt-heavy projects may benefit from thinner road sections that require less cut and, therefore, produce less material to deal with. When comparing alternatives and value engineering roads, owners must carefully monitor the total cost of earthwork in combination with the cost of pavement and crushed rock. This same logic applies to other categories of site-work. By using the matrix in *Figure 13-8,* the chart "manages" the work.

When owners want alternatives ready and waiting for the construction manager or contractor, they shouldn't wait for construction to begin before

An Example of Value Engineering Following the Approval Process

(Italicized words tell the "behind-the-scene" story)

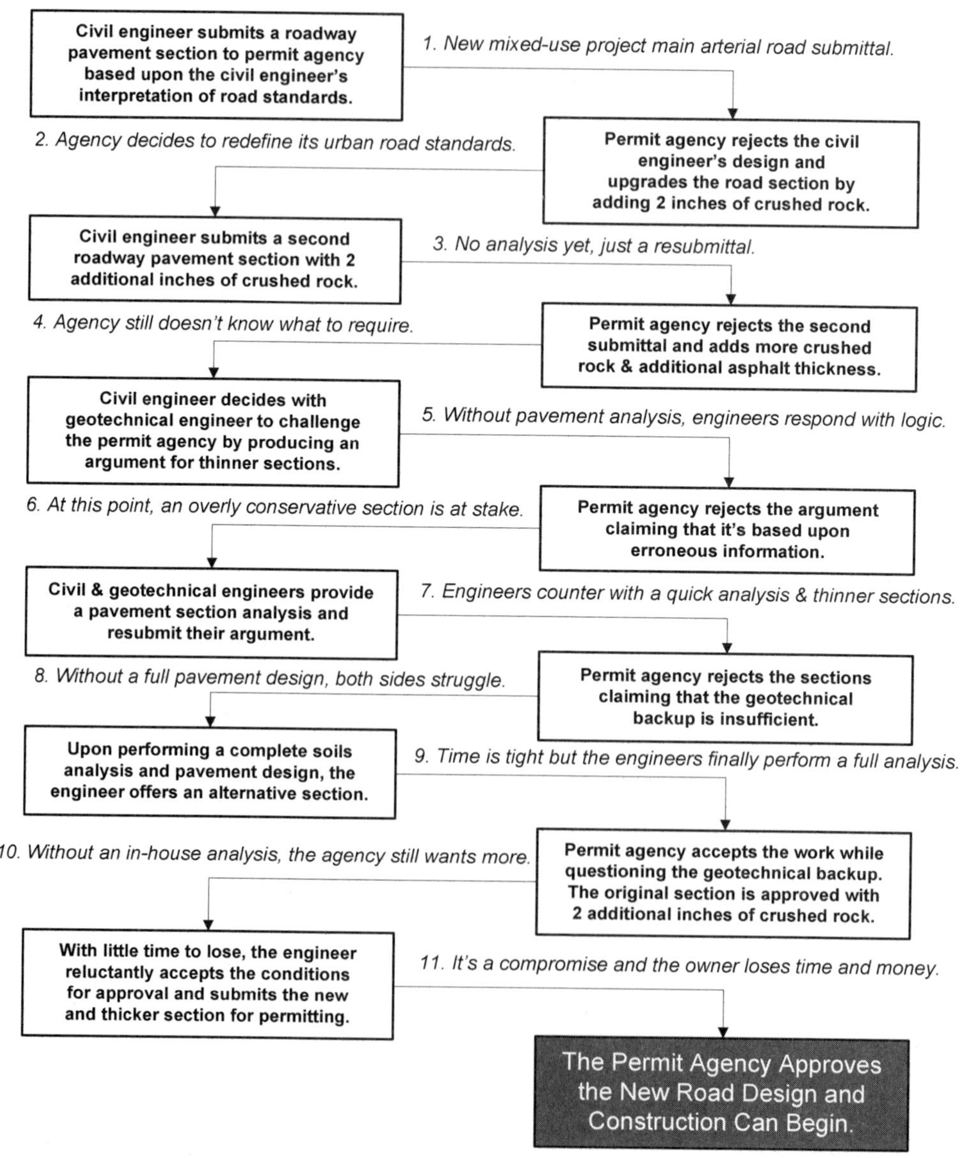

Figure 13-7

Road Section Alternatives Matrix

(To be Designed and Submitted Early in a Project)

	Material Choice (1)	*Material Choice* (2)	*Material Choice* (3)	*Material Choice* (4)				
Cost per SY per inch	**$2.30**	**$1.70**	**$0.36**	**$0.42**				
Road Pavement Section Alternatives Inches	**Asphalt Surface Course (ACP) Inches**	**Hot-mix Asphalt Base Course Inches**	**Crushed Rock Base Course Inches**	**Crushed Rock Subbase Course Inches**	**Total Road Section Cost Square Yard**	**Cost Differential to Standard Section Square Yard**	**Total Section Thickness Inches**	**Comments**
(Current Standard)	6	6	0	0	$24.00	$0.00	12	2 lifts (1), 1 lift (2).
2	6	3	7	0	$21.42	$2.58	16	2 lifts (1), 1 lift (2), 2 lifts (3).
3	4	3	14	0	$19.16	$4.84	21	2 lifts (1), 1 lift (2), 3 lifts (3).
4	6	0	15	0	$19.02	$4.98	21	2 lifts (1), 3 lifts (3).
5	6	3	0	14	$24.78	($0.78)	23	2 lifts (1), 1 lift (3), 2 lifts (4).
6	6	0	4	21	$24.06	($0.06)	31	2 lifts (1), 1 lift (3), 2 lifts (4).
7	4	4	11	0	$19.96	$4.04	19	2 lifts (1), 2 lifts (2), 3 lifts (3).

Figure 13-8

value engineering the design. Perform value-engineering work well in advance by collecting and costing out each alternative, listing them in table format in the plans, and getting them approved. Then, the construction manager is then free to choose whatever alternative best suits the situation.

Other prime hunting grounds for value engineering site-work may include:

- Manipulating lot grades
- Optimizing cuts and fills
- Fine-tuning erosion control
- Scrutinizing retaining walls
- Processing or losing topsoil
- Minimizing throwaway items
- Optimizing storm-facility design
- Maximizing existing timber value
- Developing smart strategies for access
- Pushing the limit of road layout and grading
- Creatively losing storm water overland and on-site
- Testing whether the owner's build-out schedule is realistic
- Placing utilities in non-structural areas in lieu of structural areas
- Improving or eradicating wrong and redundant boilerplate details
- Utilizing slash, wood chips and grindings for on-site construction use
- Eliminating errors, omissions, and poor design buried in the documents
- Mining holes for generating quality material and for losing unsuitable material
- Ensuring that special projects are buildable with available off-the-shelf materials
- Minimizing haul-and-dump fees by recycling on-site material in productive ways
- Challenging standard pavement sections with thinner designs that perform as well
- Using reasonable compaction specifications that better match final loading and use
- Retaining as much native vegetation as possible in lieu of landscaping and irrigation
- Minimizing the number of catch basins and using minimally- but correctly-sized manholes
- Manipulating the relationship between roadway vertical curves, lot design, and utility design
- Minimize import and export by grinding on-site asphalt in-place and using it as base material

- Eliminating as many acute angled areas of hardscape as possible to save money, reduce cracking, and improve ease of construction
- Including cement-treated-base (CTB) and kiln-dust treatments as legitimate bid alternative items should you need to stabilize or strengthen on-site soils
- On poor soils, using on-site mined material to surcharge building pads and then replace the surcharge material with import select fill or treat and stabilize the ground with kiln dust.
- Lower the risk of inheriting overdesigned culverts engineered solely from hydrographs. Use data from existing culverts, such as upstream high-water marks and basin radius measurements, to provide cheap historic flow data and verify, or at least influence, culvert redesign or sizing.
- When estimating the costs of surcharging a site with piled material, consider the weight per cubic measure that the engineering calculations are based upon. By searching out heavier material per cubic weight per bank cubic yard, it may be possible to reduce the total volume of material needed to accomplish the same surcharge. When choosing alternative material for surcharging, other considerations include the price per unit, whether or not the material can be used on-site after surcharging, and the cost consequences of not being able to legally haul as much heavy material per truck.

Encourage your contractors to perform value engineering. Tap into their knowledge and experience. More often than not, some of the brightest ideas come from the trades' people. Even the slightest field adjustments can at times yield tremendous savings and improved efficiency. Many cost-saving ideas can also come from on-site subcontractors. As specialists, subcontractors are closer to the work, more intimate with ways to shave time and money, and are experienced with solving or working around problems. Many also know how to work around flaws in stock design. On a side note: Good ideas and cost-saving suggestions flow quicker when dealing with individual field personnel than when dealing with groups. Group settings can be distracting, dampen creativity, and discourage participation.

Here are some off-the-shelf examples common to value engineering site-work that may be immediately applicable to your project and bear on cost estimating:

- Do you really need five-inch thick sidewalks, or will four-inch ones work?
- Install fabric under any and all crushed rock pads to keep the rock from sinking into the mud.
- Design utility trunklines as close to catch basins and manholes as possible in order to reduce lateral piping.

- Investigate selling wood chips and organic grinding debris off-site to local utilities for use in power generation.
- When faced with oil contaminated structural soils, explore the option of losing this material on-site under paved areas.
- On steeper terrain, when projects gain elevation and density decreases, the diameter of gravity wet-utility piping may also decrease.
- In an excavated state, hydric soils are valuable. Stockpile and cover it for use in enhancing or building on-site wetlands, or sell it off-site.
- Utilize one contract for site work and another for the general building contractor. Why? An immediate benefit is that it saves the owner markup fees.
- Explore changing curb sizes and types. For instance, is it possible to reduce vertical curbs from 24 to 18 inches wide? This move results in cheaper curbs. Changing from a curb and gutter to a vertical curb or thickened edge can further save on curb costs.
- Before laying down a thick lift of expensive imported landscape bark or wood chips, consider first laying down a base of hog fuel grindings. This effectively reduces the volume of expensive imported material while aiding in the disposal of recycled wood waste.
- Where required, install colorful plastic protective fencing in half sheets. This saves the cost of purchasing and installing full-width temporary fencing. Native growth area protection requires a clearly marked line of demarcation, not an impenetrable barrier that can affect wildlife.
- Order bulk landscape stock early in order to guarantee that you will have it when you want it. Many nurseries will hold and care for purchased stock until it's ready to move. Alternatively, an owner can purchase smaller stock in bulk and allow it to acclimate and grow on-site. This requires having suitable land, sunlight, irrigation, growing medium, and pest (deer) control.
- Lose large rocks and boulders on the edges of greenbelts, at the toe of fills, into terraced embankments, and in holes under unstructural areas, such as parks. Doing this before the wet season allows the fill over the boulders to water-settle. Overexcavate storm ponds to develop material for projects that are short of fill. Then plate the pond bottoms with boulders. Or, excavate vertical shafts in pond bottoms and fill them with boulders. These strategies help save on initial pond clean-out costs since silt will bed between the boulders. Losing boulders in ponds can also work with pond liners. Overexcavate good material, bed the boulders *below* grade, and backfill with vibratory compaction or water-settle material. Export the usable material, fine grade the pond bottom, and install the liner. Even dirt-heavy projects may benefit from this strategy, as it can be easier to lose structural dirt than it is to lose large and hard-to-handle boulders. Use boulders as a means of channeling water, armoring streams, and controlling wetland levels.

Another way to lose large boulders is to set them around greenbelts, wetlands, and ponds for visual and habitat purposes.

Lastly, always perform a "backcheck" to make sure that third party value engineering, constructability, and document review comments have been at least read and considered by the primary planner, architect, or civil engineer. Useful ideas and suggestions for improving the documents should have been integrated into the plans, specifications, and bid documents. Backcheck each comment and verify with the design consultant that this has happened.

Here are some principals and global tips to remember when estimating

- Always remember that an estimate is not a bid.
- Don't intermingle time and materials pricing with unit or lump-sum pricing.
- Include sketches or drawings that have been composed to assist in estimating.
- The *owner* or a competent agent should design the schedule-of-values template.
- Avoid using or accepting "blended numbers" in a cost estimate. Simply put, they are dangerous.
- Before finalizing the estimate, verify that costs related to potential schedule impact have been included.
- Account for extra utility line lineal footage needed for looped systems, such as water-service lines and most dry-utility systems.
- Expect to pay upfront monies to contractors for material on-hand, insurance, mobilization, and other costs necessary for beginning work.
- Never paint yourself into a corner by committing to completing a quality estimate in too short a period of time. Something always remains unresolved as questions linger or further clarification is required. Solid estimating takes time.
- When allocating dirt-moving costs between parcels, divisions, or tracts, charge the cost of excavation and hauling to the ground being graded or cut. Only charge the cost of placing (blading and compacting) the material to the parcel receiving the fill material.
- When packaging elements into a single price, such as lump sum or by the unit, specify what's included in the price. For instance, when entertaining a unit cost for importing and placing dirt per BCY, find out what's included in the price. Beyond purchase, does the price include:
    1. rolling and compacting;
    2. placing it in specified lifts;

3. hauling it anywhere on-site;
4. preparing native subgrade for the fill;
5. a guarantee of consistent quality material;
6. importing quantities no greater than agreed;
7. delivery only as required or asked for by the owner;
8. providing a push bulldozer should the owner or shipper elect to import and temporarily stockpile large amounts of material;
9. hauling off material that the owner's geotech deems to be too wet, uncompactable, undesirable, contaminated with organics, or that is spoiled due to shipper negligence?

- Be cognizant of the relationship between cost and density. Dense urban-oriented projects tend to have more expensive site-work. Here's a partial list of reasons why.

1. Tighter curbs and parking result in a greater likelihood for hand-placed asphalt and curb work.
2. Interconnected roadway grids may increase hardscape costs and decrease land available for development.
3. Alleys and woonerfs result in higher total hardscape costs, higher costs per square area of vehicle use, and increased loss of land.
4. Wider sidewalks, bike lanes, and pedestrian friendly hardscape typically equate to narrower roads and higher design and infrastructure costs.
5. Increased numbers of shorter streetlights that are required for uniform and subtle lighting may not be the optimal choice in terms of material cost and installation.
6. Combining minimal or no building setbacks with smaller lots on steeper terrain is likely to result in retaining walls, increased grading costs, and product restrictions.
7. Owners wanting to balance less garage space and shorter driveways with typical needs for parking may have to build wider streets resulting in higher infrastructure costs and less net buildable area.
8. Narrow setbacks and tighter tolerances mean more dry-utility appurtenances installed in front yards or adjacent to sidewalks. This can mean higher costs to move vaults and boxes out of view or to pay for landscape screening.
9. Shorter curb radii meant to reduce the distance pedestrians must negotiate when crossing intersections may result in continual curb or corner-pole damage from vehicles. It also may require design of alternative routes needed to accommodate long and cumbersome truck and trailer rigs such as moving vans or fire trucks.

- Inadvertently mixing materials during site construction is a fact of life. Layers of usable material placed over, or kneaded into, unusable material results in cross-contamination and piles of spoiled waste. If this material has to go to export or has to be lost on-site, costs increase. If replacement structural import material is required, costs can easily double. Areas prime for generating volumes of mixed material can include zones where materials interface during:
    1. clearing or stripping where woody debris can mix with topsoil;
    2. stripping where topsoil can mix with native inorganic material;
    3. trench excavation where dry material can mix with wet slop or mud;
    4. grading if free-draining pit material mixes with fine-grained silts or clay;
    5. construction where usable soil can mix with hazardous material or garbage; and
    6. landscaping when delivered topsoil is dumped over ungraded structural material.

Computers and the virtues of manual estimating

Years ago while renting a bachelor pad in northwest Montana, I accepted my landlady's offer to join her for dinner. After dinner, she asked if I would stay while she met with an insurance salesman. I agreed and sat through his presentation, which centered on life insurance. As the meeting progressed, the salesman reeled off information into input figures into a handheld computer eventually producing a printout that just delighted my landlady. After the salesman left, she asked me what I thought. I told her it was too expensive. She responded that the price must be a good one because it came out of a computer. Whether or not she moved on the policy I'll never know, but what struck me was how easily she accepted a price merely because it came out of a computer. Relying too heavily on automated site-work estimating can create a similar illusion that looks and feels right but later proves to be in error.

The positive attributes and capabilities of estimating software are undeniable. Choosing software and setting up a database depends upon many factors including work style, price, existing hardware, available support, type of construction, and level of detail and management required. Some estimators or managers however, hide behind automation and lean too hard on flashy graphics, pretty output, and quick totals without understanding what's happening in the field. Automation can be a beautiful trap.

Academically speaking, software cannot supercede or replace the benefits gained from thinking through the process, visualizing the actual ground where work will transpire, and picturing potential pitfalls. Estimating is meticulous work. It expands intuition, better prepares one for the unexpected, and builds an individual's managerial capability. The confidence and understanding gained is recognized by other professionals and fosters mutual respect and productive chemistry. It isn't until someone gets to manage what they've estimated that they gain a complete picture of how site construction works and where they made mistakes in the estimate. Technically speaking, if an estimator cannot perform a calculation by hand and rationalize the mechanics of how dimensions, volumes, and logistics interplay against unit pricing and the cost of construction, that estimator shouldn't be relying on an automated system to produce results upon which the estimator has no genuine understanding. It's expensive folly.

Who should compose your cost estimates?

Ask yourself who is going to be responsible for construction budgets, scheduling, approving pay requests, managing consultants and surveyors, bidding, formulating contracts, and overseeing work. Whoever that person is, that's the person who should perform the cost estimate. On most projects, this task falls on the construction manager. Managers must know the job and work within whatever construction schedule or estimate that they've created. Understanding the job leads to searching out opportunities for value engineering, contemplating which contractors are best equipped and experienced to complete work, and focusing on the goal of getting land ready for building structures.

The problem is, not all construction managers possess the quantitative skills, knowledge, or experience to estimate site construction. Unsurety may lead to insecurity or to a level of doubt causing some managers to request that somebody—anybody— be it the contractor or a consultant, perform cost estimates for them. For managers who are aware that they're in over their head, this can be a stroke of genius. For those that don't, it can spell tragedy.

Letting others perform cost estimates for you of course, leads to that estimator—be it a contractor or consultant—to know the job better than the owner's manager. If other unrelated problems develop between the manager and a contentious contractor or an unsavory consultant who has, coincidentally, *also* performed the cost estimate, guess who has the edge? Managers who farm out their estimating also suffer because they have no secrets—no margins to play with. If modifications occur or something unexpectedly happens causing quantities to change across multiple categories, not only is the manager likely to be ignorant as to what transpired, the estimate might become invalid and, therefore, releasing whoever built it—the contractor or the consultant—from ongoing responsibility. And, responsibility to the cost estimate is important. If the contractor or a consultant provides a woefully inaccurate or incomplete base estimate, how does an owner evaluate responsibility for going over budget? Who is responsible? Well—to put it mildly—owners have only themselves to blame. Estimating the work and managing the work go hand-in-hand and project owners who realize this fact stay ahead of problems and prosper. Construction managers who don't are liable to be forever "out of synch" with their project.

Estimates by architectural and engineering (A&E) design firms

Another issue relates to the de-facto or convenient practice of having engineers and designers perform quantity estimates in concert with developing site plans and specifications. If a consultant can provide a complete, automated, and reasonably priced quantity take-off electronically as a by-product of their design work and software, the owner benefits. The more information that an owner has, the better the chances are of building a solid estimate. However, not every design firm is capable of producing cheap quantity take-offs while designing work. Hand-measured quantity estimating is always best and usually is an extra and expensive charge.

A side benefit of having designers or engineers perform quantity take-offs is that it gives them a chance to quantify what they're creating. Engineers and architects can better understand an owner's costs if for instance, they calculate earthwork volumes or measure depth and lineal footage of pipe as design proceeds. It's true—owners shouldn't provide schooling, but here, the cost of education may be cheaper than overhauling an entire design and engineering team or paying exorbitant construction costs later. Ultimately, if this exercise doesn't improve a designer or engineer's interest or strength in value engineering, an owner should consider changing consulting firms.

Beyond quantities, owners must also question pricing. Where is a design firm going to get unit pricing? The most likely sources are from it's own database or last job. Are the prices bid or negotiated, or are they more applicable to another region or perhaps another state? A firm with a limited unit-price database can easily pass along bad information. What about format? Is the consultant going to provide a chronological estimate with logical categories and adequate detail? Will the estimate be structured in a way useful for other in-house purposes? What do the accountants think *(see Figure 13-1)*? Type of firm also matters. A firm specializing in public works projects may price work higher than a firm that primarily does private work and vice-versa. In some regions of the country, the differences in labor and equipment cost between public and private work can be huge easily spelling doom or tragedy if misapplied. Then there's prevailing wage. Does a private developer want or expect to pay union scale?

Some A&E design firms that know their limits may discourage owners from having them perform sole and final cost estimates. For them, providing rough order of magnitude pricing may be as far as they're willing to go, and that's okay. Afterall, unless they manage the work, some consultants feel that their survival may be at stake if their estimate ends up bearing no resemblance to actual construction costs. For them, abstinence may be the safest policy.

Owners also have to ask about pricing site construction. Who knows the most—the architect or the civil engineer? The answer may be either for owners that lack the services of a site contractor or construction manager. They may have no choice but to use a design consultant for cost estimating. The obvious solution is to obtain estimates from consulting firms, contractors, or individuals well-practiced in site-work.

Other concerns in cost estimating

Validation by bid is another trap that owners must work to avoid. In this scenario, after collecting contractor bids, owners find out that they are over budget. This practice wastes both valuable time and resources and may require frantic redesign and repermitting, and result in schedule impact. Also remember that bids, the second time around, are more apt to rise than decline. Contractors may resent educating an owner especially if it comes with hard bidding work that they have no guarantee of procuring.

Cost estimates must also be right for financing purposes. Owners borrowing for land development don't want to pay for more money than they need and they certainly don't want to go into a project undercapitalized. Owners only need to borrow what they need based on sound cost estimates. This alone justifies the expense of moving forward with a complete, detailed, and comprehensive site-work cost estimate influenced by value engineering. Besides, those that submit for financing with detailed and complete backup information project an image of professionalism and success.

Finally, owners who solicit cost estimates from a designer, engineer, or contractor should collect more than a one-page, roll-up, bottom-line estimate. Owners must insist on obtaining all back-up numbers, quantity calculations, and cost assumptions in a basic and elementary format. Assume nothing.

Paying for estimates

As an owner, have you contemplated seeking a cost estimate just to see what something will cost or because you need an estimate as a basis for financing? Site-work contractors are a logical choice for providing quick and informed estimates, but owners need to be careful. Contractors may be willing to provide cost estimates as a way of getting into a project or of improving their chances of winning work later but nothing comes free. Everything comes at a cost, including "free" estimates. Besides possibly receiving an estimate that is lacking detail, is void of unit pricing, or is otherwise nebulous and lump-sum-worthless, a contractor may feel that an owner is indebted for accepting no-charge estimating services.

Also, be aware that contractors may at times offer free estimates but overprice the work or add onerous contingencies that are not clarified to owners. In a devious way, a contractor may be indirectly trying to set budget or possibly trying to raise the bar high enough to where expensive change-order work later may appear palatable to an owner. Or, a contractor may compensate for inadequacies in the construction documents, or even pad the estimate upfront, if they believe that the project is risky. Today's padding may be tomorrow's value-engineering opportunity, so owners beware.

When asking contractors to provide cost estimates, the best policy is first, to specify the format, and second, to pay for services. Even if the contractor is a strong candidate for performing the construction, it's beneficial to maintain an unbound relationship in case, later, an owner wants to hire a different contractor. Owners who pay for their estimates can also expect to receive something worth paying for. Compensating a contractor helps to ensure that the contractor will invest time and a higher degree of care in formulating the estimate, as well as accepting a certain degree of responsibility for its numbers. Paying for an estimate makes it less awkward for owners to dictate the format and later question the results. Do you want to know where weaknesses exist in the documents, or where you're exposed? Ask—because you're paying for it. So, owners can enjoy a no-strings-attached cost estimate rich in detail with a list of exclusions and assumptions when they specify format and pay for the work.

Owners may find it easier to obtain contractors' quantities than their unit prices. Contractors reluctant to offer unit pricing may feel that such information is too revealing. They may counter by stating that they're a full-service contractor and that their lump-sum numbers are all inclusive and safe for budgeting—and they may be right. But, that argument does nothing for an owner determined to use line-item unit pricing for estimating and bid purposes. Owners may feel comfortable negotiating work only if they have a grasp on unit pricing. A preliminary estimate lacking detail and described broadly in lump-sum values may be convenient for the contractor but it may prove useless for an owner wanting to know "what's in the box."

The need to obtain unit pricing is a good reason to bid the first couple of site-work jobs on a project. From there, owners can either continue bidding or evolve into negotiated contracts based on original bid unit pricing.

Negotiations and trust can work well with contractors as many of them appreciate being part of the team and are worth collaborating with.

How much?

How much should an owner expect to pay for an estimate? It depends on the level of detail, who is doing the estimate, the number of reports and information requiring evaluation, how quick the owner needs the estimate, and the project complexity. Obviously, a rough, one-page collection of broadly based lump sums will cost less than a multipaged, detailed and comprehensive estimate. The best policy for owners is to pay once and get the best cost estimate that they can afford. In fact, while they are at it, they should pay twice and get two estimates. It's that critical. How expensive can it be? Certainly not more than the cost of dealing with construction problems later that could have been avoided by basing work on a more comprehensive estimate.

After paying for an estimate, owners need to address the longevity of the estimates provided by the contractor or consultant. How long will a contractor cover his pricing if, later, the contractor has a chance of negotiating or bidding the same work? Adhering to these principles will lower an owner's chances of paying for one-time, throwaway, site-work cost estimates.

Multiple cost estimates

Third-party estimates can provide owners with a more complete picture of project costs, scope of work, and can verify that an owner is using adequate cost and quantity data for bidding or negotiating work. Owners who use multiple firms for estimating costs should define what firm or individual is to handle various portions of the estimate. It can be a mistake to allow one firm to estimate all facets of a project ranging from logging and earthwork through signalization and landscaping. One solution is to use a point person as the lead estimator or a firm to coordinate other consultants who are qualified to review and evaluate specific and specialized site-work items. This act is more important than it appears because estimates are commitments. Estimates leave a fingerprint describing a firm's ability, effort, and experience.

Multiple estimates by a contractor or designer may be biased or contradict each other, but an independent third-party estimator is less likely to offer something half-baked or incomplete. Since they don't have a stake in the project, they have nothing to protect. Third-party estimators are free to search out cost savings and design errors. Other benefits of third-party estimates include testing the accuracy, completeness, and relevance of the construction documents against the schedule and existing budget. Outside consultants are eager to please and typically want to impress owners with thorough work. Once a project becomes predictable, the need for multiple estimates should diminish.

Here are guidelines to follow in order when pursuing multiple cost estimates.

1. Again—the owner provides the estimate format. Include lines for adding items not included in the owner's original estimate format.
2. All estimators must use the same plans, documents, and revisions. They must reference the same road network and grading plan alternatives. Copies of all backup documentation must be transmitted with the estimate. Allow every estimator to figure his own quantities. Provide nothing except item descriptions and units. This is a fact-finding mission.
3. When all estimates come in, build one spreadsheet, list all submitted values, and compare all costs on a row-by-row basis.
4. Options for utilizing multiple estimates include:
   - averaging all estimates together into one budget amount;
   - gauging the estimates and adjusting project design or scope of work commensurate with overall cost;
   - formulating a conservative cost projection by accepting only the highest values from each estimate; and,
   - isolating troublesome, widely priced items to determine the reason for confusion or lack of information.

Adjusting costs as a project matures

As a project matures, owners should assess their estimates in comparison to actual cost-to-date. Herein lies a path to understanding site-work with the side benefit of critiquing the accuracy and strength of the site-work estimates. How did the estimates fare? Are they low, high, or on the mark? Low-ball estimates may portray too much optimism, an estimator's lack of knowledge and ability, or the likelihood that estimator worked off preliminary or incomplete construction drawings. High estimates may indicate pessimism, or exorbitant quantity or unit-price inflation. Either one of these problems can work to distort management decisions and penalize profits.

Chapter 14

DEALING WITH DRY UTILITIES

By definition, dry utilities are systems delivered in a waterless state. Examples include power, gas, phone, cable, and fiber. As extensions of existing mainline and distribution networks, newer systems are commonly located underground in backbone and primary road infrastructure and in dedicated easements and right-of-ways. However, some services are still delivered above ground on poles. In sequence, underground dry utilities usually follow rough grading and wet utilities, but precede fine grading and hardscape. This narrow window can make scheduling and installing underground dry utilities a challenge.

In the long term, service providers are dependent on owners and the projects that they build, but in the short term, it's the opposite—project owners are dependent on service providers for design and construction support. This reliance carries risk and worsens when individual utility companies suffer from take-overs, mergers, downsizing, or change driven by new technologies and consumer demand. When service providers experience turmoil, it typically translates into design and construction problems for owners. Unfortunately, the quicker that an owner needs action to be taken, the higher the risk.

For instance, you may begin with service-provider contact person A. After months of calculating loadings, tweaking product, and messaging vault locations, person A assumes a different role in the company and hands you off to person B who is new and knows nothing about your situation. So, you explain the project to person B. After the initial, and only, meeting with person B, person B goes AWOL for weeks and returns no calls. Then, person B leaves a message stating that your case has being routed to person C and leaves a new phone number with an unrecognizable area code. So, you call person C and explain everything again. Person C is in the business development department and has no idea how you got their number so, Person C will have to call you back. You never again hear from Person C but, shortly thereafter, person D calls from some head office located on a different ocean. Person D needs more information. After explaining everything again, person D recommends that you to contact person E in regional engineering and around and around it goes.

In this vignette, the person who finally handles your project may not know anyone that you dealt with prior, let alone have access to the trail of your plans, files, and transmittals left behind. Besides accomplishing nothing, the result is lost time, higher cost, and frustration. Complications related to franchise rights and government regulations can further retard getting dry utilities designed, scheduled, and installed in a timely manner.

Unfortunately, with service providers, it can be hit and miss. However, owners can facilitate this process by planning utility and vault locations in advance. Design wet utilities with dry utilities in mind—especially in narrow and restricted roadways typical of dense urban style developments. The same forethought applies when laying out lot lines, structures, and plats since these elements must also accommodate the placement of above- and belowground utility cabinets, vaults, and boxes.

Owners must also build relationships with utility service providers. A good start is to prepare site plans that accommodate dry-utility design and engineering. And, don't expect service providers to get excited when contacting them at the eleventh hour with approved preplat drawings and wet-utility plans that fail to accommodate dry utilities. This approach can suggest to them that they're not being taken seriously enough to be included as part of planning and design. Worse yet—it can illustrate naivety and a lack of management skills on the part of the owner. Nurture relationships with service providers as soon as plat design and initial road and utility layout begin. If granting easements and right-of-way land appears imminent, or if construction of facilities such as trunklines, substations, odorizer stations, switchgear, or transmission lines are likely, build your relationships as early as possible. Which brings to us sequence.

Coordinating plans and design

Most service providers prefer to design their own system. This is fine if the utility is the only one going into the ground, or if their system can fit without conflicting with existing wet utilities or other dry utilities. Typically, this isn't the case.

Part of the solution to ensuring that dry utilities don't clash with wet utilities, and vice versa, lies in coordinating all utility plans. Check pipe conflicts and unacceptable tolerances between lines by horizontally and vertically overlaying and comparing wet- and dry-utility systems. This task is easier when all plan and profiles are drawn to the same scale and when one entity, such as the consultant designing roads and wet utilities, leads dry-utility coordination. This party would also be responsible for merging dry-utility designs with road and wet-utility engineering.

Figure 14-1 illustrates in flow-chart form issues relating dry-utility design and layout to road right-of-way, plat layout, and wet utilities. Coordination is crucial when working through a process that tracks similar to *Figure 14-1*. Here, any number of iterations may be necessary to produce a set of plans that accommodates the design and installation of wet and dry utilities.

The balance of the solution lies in *joint trench*. Joint trenching means excavating one trench to handle conduit for multiple dry-utility service providers. The dry-utility companies collaborate and produce a joint-trench plan for installing their collective conduit and vaults. Here, the key is having utility system designers use project base maps such as wet utility and road plans for joint-trench design. Show dry-utility piping and structures in bold against already-designed or existing wet utilities and roads shaded in the background. The owner then reviews the joint-trench plan and provides comments or changes as necessary. If the utilities don't conflict, and if vault and appurtenance locations are acceptable, the owner and service providers can apply for permits and schedule work. Note that not all jurisdictions require permits for dry-utility installation.

A joint-trench plan integrated with infrastructure design also reveals potential conflicts between wet and dry utilities in relation to lots, easements, and intersections. Combining utilities exposes the less obvious conflicts at water valve clusters, manholes, and with existing crossings. It also exposes where the sewer and water laterals, which reach to catch basins and lots, have to cross under or between layers of dry-utility conduit. Consolidating design provides contractors with a point of reference when subjectively locating lot stubs. It prevents them from translating on-the-fly or working from a quilt of various utility plans, each one varying in scale, detail, and paper size. For this reason alone, it pays to horizontally and vertically distill a separate dry utility plan from the road design and wet utility plans.

Density and dry utilities

As mentioned earlier, density and utility conflicts go hand-in-hand. Dense urban-type projects can present a set of conditions that strain utility design. *Figure 14-2* exhibits conditions common when installing dry utilities in tight and narrow urban-style road right-of-ways. Here are some reasons why density can make designing and installing dry utilities difficult:

- The number of streetlight poles may increase.
- Narrower lots and parcels effectively limit options for locating stubs.
- Room for vault farms, large transformers, and switch gear is restricted or non-existent.
- Minimal setback lots may have to trade lawn or landscaping for aboveground utility boxes.
- Planters may not exist and the lot lines may coincide with the outer edge of sidewalk forcing all utility vaults and structures onto private lots.
- When alleys are common and roads are narrower, right-of-way is often more restrictive, therefore, leaving fewer options for cost-effective conduit installation.

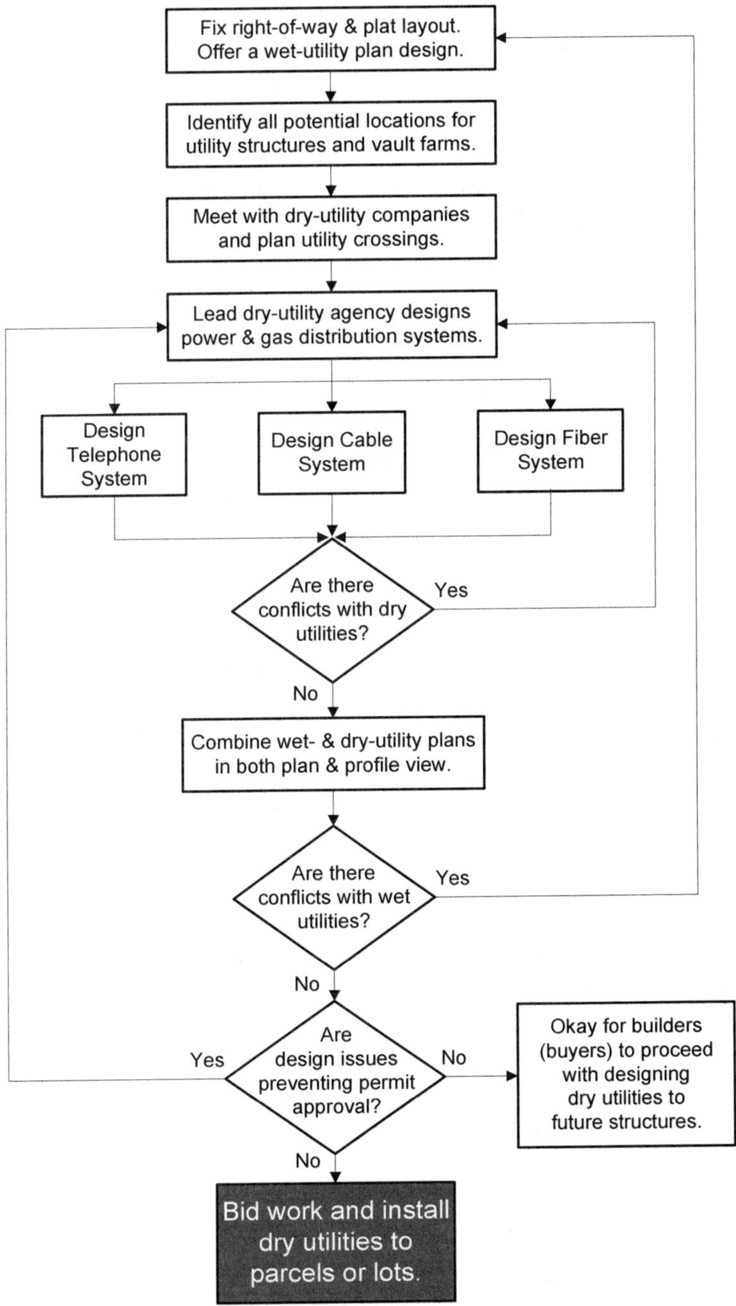
Dry-Utility Design Process
Fix right-of-way & plat layout. Offer a wet-utility plan design.
Identify all potential locations for utility structures and vault farms.
Meet with dry-utility companies and plan utility crossings.
Lead dry-utility agency designs power & gas distribution systems.
Design Telephone System
Design Cable System
Design Fiber System
Are there conflicts with dry utilities?
Yes
No
Combine wet- & dry-utility plans in both plan & profile view.
Are there conflicts with wet utilities?
Yes
No
Are design issues preventing permit approval?
Yes
No
No
Okay for builders (buyers) to proceed with designing dry utilities to future structures.
Bid work and install dry utilities to parcels or lots.

Figure 14-1

- Zero setbacks can leave little or no room for conduit sweeps and utility vaults and structures. This forces dry-utility conduit away from lot lines towards road or alley centerline and potentially into conflict with other utilities. Vaults and other structures typically have to be flush-mount-located underground in right-of-way, which leads to more expensive and complicated design and installation.

Meanwhile:

- Wet-utility demands increase with density often equating to larger pipes and manholes.
- Service providers may install multiple and redundant conduit to meet future utility demands or to be leased or sold as excess capacity.

Plus:

- Multiple zone-water or fire-flow systems may require the installation of more than one waterline.
- Storm-water regulations may require the installation of large-sized trunk lines or separate storm lines to carry clean water in one pipe and runoff requiring treatment in the other.

The result often leads to installing or stacking dry utilities vertically with layers of pipe laid as deep as necessary to fit. Installing pipe in this fashion makes routing conduit in and out of vaults and structures difficult and expensive. Due to increased trench depth, trench-wall layback requirements, and the need to dig their way out just to stub into and out of vaults and appurtenances, crews typically end up performing additional excavation. Backfilling layers of pipe with stubs running every direction also makes backfilling, filling voids, and gaining compaction difficult. Inadequate backfilling can cause settling problems and eventual pavement or utility-line failure. Hardly an inspiring sight for eyes set on completing work and moving on.

Other issues can include having to overcut or lay back slopes on private property in order to install conduit in sloped right-of-ways with little working room. Consider the retaining wall and slope cut illustrated in *Figure 14-3*. Here, the construction of a retaining wall facilitates the installation an underground vault. The option of *not* adequately addressing dry-utility design in a timely fashion can lead owners down a path to:

1. permit delays
2. potential for continual conflicts with other pipe and structures
3. crisis design and grading, retaining wall, or fill slope construction
4. paying a premium to have service providers or contractors work around

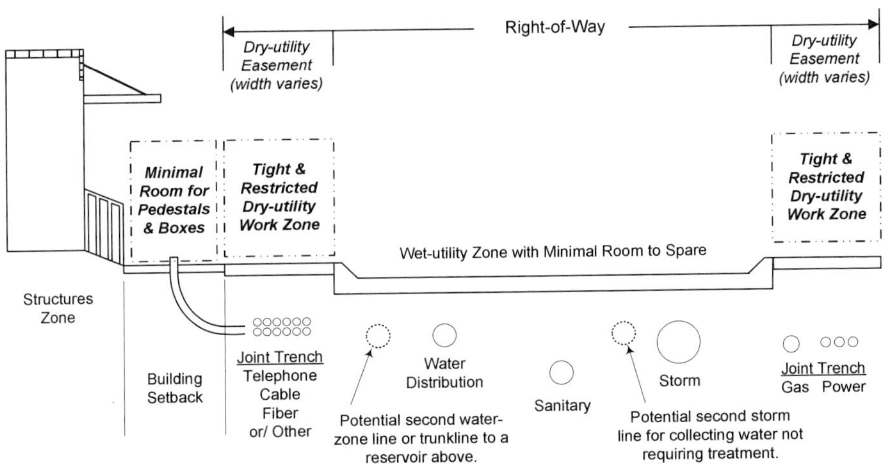

Figure 14-2

the clock installing systems in order fix problems or meet schedule

5. unaesthetic vault farms and aboveground boxes going wherever there is room, such as in front yards, nestled into parks, or pigeonholed into corners of landscaping
6. the inability to pave or finish work until the dry utilities go in which may result in working through the wettest part of the season, experiencing acute erosion control issues, compromising road subgrade, and incurring other unnecessary extra costs.

Figure 14-3 also illustrates a common problem where the sidewalk doesn't match the top of a belowground flush-mount utility vault. Vaults that don't match future final elevations of sidewalks and roads can collect water, ice over, or become impediments to drivers and pedestrians. Additional surveying may be required to guarantee that belowground vaults are set exactly in line with future final grades. Surveying, though, doesn't solve everything. Vaults that don't settle at the same rate as surrounding asphalt or subgrade may rise or drop, producing a lip or low area. Settling around vaults can also cause stressing and fracturing of the conduit stubbed into and out of the vault. Slope also matters.

Flush-mount vaults must match final slope or super elevation. Matching final surfacing to vaults can compensate for minor grade differences between the tops of flush-mount vaults and roads or sidewalks, but extreme variances usually require lowering or replacing the lid. The solution for smaller, unadjustable monolithic structures may be total re-installation or simply living with it as is.

Alleys and woonerfs are a common characteristic of dense urban-type developments. These features tend to be narrow and sometimes steep. Before agreeing to place utility appurtenances down alleys and woonerfs, check to make sure that there's ample room for utility vaults and boxes. Tight quarters may mean jamming vaults and boxes into unserviceable corners or placing them in areas shielded by parked cars. In a worst-case situation, there may be nowhere to place them. And, remember, service providers resist locating aboveground vaults or structures where they can become easy targets for vehicular damage. This further discourages placing dry utility structures in narrow accessways.

Alleys and woonerfs can pose other issues. Non-essential portions of infrastructure, which can include alleys and woonerfs, are typically built later to satisfy sales or to complete project build-out. When mainline or first-phase utilities fall into non-essential infrastructure, it forces owners to accelerate design and construction ahead of schedule. In fact, secondary alleys and woonerfs destined for project-dependent utilities may land on the critical path. At the very least, advancing alley and woonerf design may be required so that contractors can stub conduit and vaults correctly in location and at the proper depth and elevation in preparation for hardscape. Non-essential elements pulled into construction on a buyer's property can result in *accelerating* that buyer's schedule causing premature investments in design and construction. Unless disclosed in the purchase and sale agreement, the buyer may seek a credit from the seller.

Gas lines

Gas providers prefer separate trenches for their lines. Depending upon diameter, gas lines can be unyielding and relatively inflexible. This, combined with the need to lay gas lines at a minimum distance from other utilities, mandates they be installed before other dry utilities. The gas company also likes to weld pipe. This can mean having to string long lengths of large diameter pipe along roadways that are still being graded or where utilities are being installed. As a result, dry-utility planning and design often hinges on first, locating gas service, and second, meeting the overall site-construction schedule.

Large high-pressure gas trunklines pose a separate set of issues for land developers. When developing land around high-pressure gas lines, begin the

Challenges of Installing Dry Utilities in Restricted Areas

(Not to Scale)

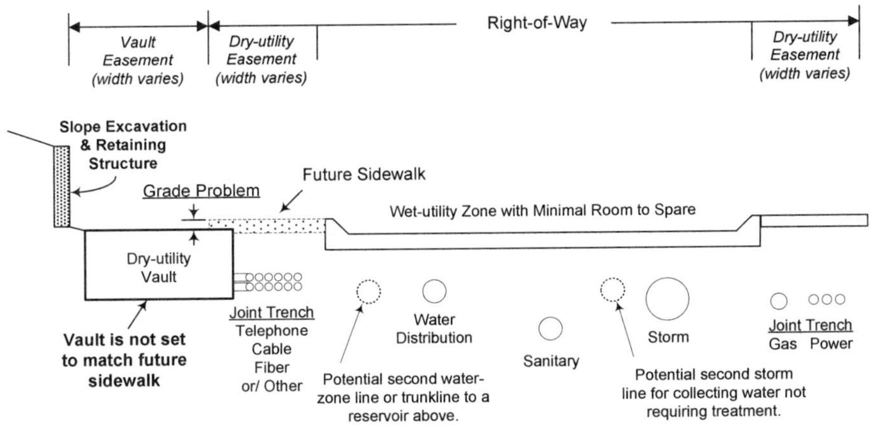

Vault Installation with Slope Excavation, a Wall, and Grade Problems

Figure 14-3

design process as *early* as possible. Work in advance with the gas company in order to understand their concerns and requirements, and to schedule work. Owners who aren't attentive to coordinating work around high-pressure gas lines set themselves up for cost and schedule impact.

Building infrastructure over gas lines creates a barrier to future inspections. Before construction, the gas company may want to excavate and expose sections of pipe to inspect it's condition, to install corrosion-control systems, such as cathodic protection, to apply fresh pipe coating, or to install a sleeve allowing the sealed section to be removed later if necessary. When high-pressure gas lines are exposed, the owner's schedule is secondary in importance to safety, pipe preparation, and proper trench backfilling. No exceptions. Projects that cross existing high-pressure gas pipelines with paved road infrastructure, utilities, and grading must view these lines as control points. Treat them with respect. Coordinate with gas companies throughout the design, permitting, and construction process.

Common problems with dry-utility plans

Here are examples of common mistakes when planning dry utilities. Work to avoid the following:

- Placing dry utilities or conduit under wet utilities
- Conflicts with irrigation, signalization, and streetlight conduit
- Underground vaults expected to drain where there's no outlet or sewer connection
- The absence of conduit sleeves under hardscape such as curbs, sidewalks, and flat work
- Designs calling out for only one small-diameter conduit stub to lots which are intended to accept multiple lines including phone, cable, and fiber
- Situations where there's not enough room to handle large-diameter conduit stubs sweeping in and out of vaults or utility boxes resulting in having to place conduit deeper in order to gain enough vertical and horizontal sweep radius.
- Poor global sequencing

An example would include bringing dry utility conduit to lots under construction, which are dependent on hot-connections to items such as vaults, transformers, or switch gear that are not yet installed within the project. The placement of vaults, boxes, hand holds, junction boxes, transformers, switch gear, or pedestals should coincide with planning and layout and must precede the installation of service lines when buyers or builders expect "juice". Otherwise, dry-utility design can influence on-site planning for the wrong reasons. Stay ahead of easements and vault farms or else risk having to plan around them later.

Before committing dry-utility stubs to a buyer or builder's site, check with their architect to ensure that stub locations are correct. Here are some ways to avoid dry-utility problems when dealing with residential, retail, or commercial buyers and builders:

- Locate places for gas-line vents or risk having their contractor place them in the wrong place.
- When placing dry-utility vaults and transformers for buyers, ensure at least, that the vaults and transformers don't block or impede future perimeter fire access around their site.
- If the architectural plans call out for utility meters to be camped together at one location, there must also be adequate room for installing, maintaining, and monitoring the meters.

- Make sure that the buyer's architect doesn't plan a transformer location based solely on the location of wall footings. Overhead canopies limit utility-truck access required for lifting, installing, and servicing transformers.
- If the buyer's architect plans for all utilities to converge at one location, the joint trench must be large enough for the installation of multiple pipes into and out of the central building area. This includes providing ample room for pulling lines and cable to the facility.
- Make sure that the buyer's architect designs vault space that is adjacent to buildings to be sufficient in size to handle installing and maintaining utilities. Trucks need access to move or replace equipment, feed cable, and pull lines. This means not installing planters, and providing aprons and thickened sidewalks where service or delivery trucks are likely to travel.

Vault Farms

No one but the utility companies wants vault farms. For most project owners, vault farms represent a loss of developable area, an aesthetic challenge, and, on dense urban developments, tough choices concerning where they *can* be placed. Echoing prior advice given in this book, project owners need to meet with service providers early in the planning process in order to discuss expectations and requirements for vault farms, easements, and land set-asides.

Vault farms are designated areas within a project that contain a number of utility structures usually owned by different service providers including, but not limited to, vaults, transformers, and switch gear. Some utility providers discourage sharing space in vault farms. Instead, they would rather locate their gear in their own separate area divorced from other providers. Depending upon your project and available area, this may not be possible, practical, or cost-effective.

When discussing vault farms and land set-asides, terms are often negotiable. There usually exists more flexibility in locating appurtenances than may be initially stated by the service provider. Many providers overshoot requirements with the hope that, in the end, they'll at least get what they need. Owners can get ahead of this game by first presenting to service providers a plan noting all potential on-site areas eligible for vaults and appurtenances. To be successful, present dry-utility companies with options. If this isn't possible, find out what the dry-utility companies require, increase their total area request by 30 percent, and begin locating suitable land. Thirty-percent more land gives owners latitude when committing areas for large utility structures and vault farms. Working with an abundance of vault-farm possibilities makes it easier to locate land for providers who require oddly shaped or specifically sized areas. Knowledge of having multiple vault farm sites can also benefit a seller when negotiating with buyers or builders that would prefer to not have such utility appurtenances located at their future entryway. Options can provide leverage or at least room to maneuver when pricing lots.

Because vault farms typically require screening and result in loss of developable area, they impact land value. Vault farms can also lead to other surprise cost centers including unanticipated excavation, curb cuts, apron construction, retaining walls, fencing, and thick sidewalk sections capable of supporting utility service trucks. Because they present visual problems, vault farms become streetscape control points sometimes requiring owners to install rocks, signage, trees, and landscaping to hide them. Most people, particularly those in sales and marketing, find them unattractive and want them hidden. Few people can envision what a vault farm is going to look like until it's too late. By then, allocating additional money for screening design and construction may impact budget.

Utility companies have basic requirements for accepting easements and right-of-way land for housing their equipment. At minimum, they need land of a certain square-footage and shape. And, they're usually concerned with setback and safety guidelines as well as guaranteeing truck access. Setting an underground box the size of a baseball dugout isn't an exact science. Vaults usually go in the ground before conduit because it's easier to stub pipe to block outs (predetermined vault penetration points) than it is to dig a hole and place a vault and block outs in lead with existing stubbed conduit. Aboveground vaults and boxes install easily over existing conduit stubs. When vault farms are laid out for general use, some service providers may race to get there first in order to install their equipment in the most convenient and accessible locations. To avoid problems, configure each farm beforehand and get each service provider to agree to where their equipment will go. Schedule them in accordingly.

Here are tips to consider when negotiating long-term set-asides or land deals with utility companies.

1. Limit scope creep by defining early what the service provider is asking for and what the project owner is willing to provide. Get all the details upfront and when satisfied with the arrangement, immediately stamp and sign the papers. Do not let it sit.
2. Vault farms must be located within a certain distance or splay of each other. Other appurtenances including individual vaults, boxes, hand holds, junction boxes, transformers, switch gear, and pedestals must also be set within a certain range of each other in accordance with

load and demand. Find out what these distances and requirements are early enough to correctly-locate dedicated areas for utility equipment. Also, push for full-project dry-utility layout at least in plan view. This may mean laying out density and product earlier than anticipated. Try not to settle for year-to-year designs. Getting the dry-utility land set-aside and easement wildcard figured out project-wide is a major step towards project control.

3. Be mindful that over time, regulations increase, utility capacity requirements, such as kilowatt per square foot, change, and items such as power transformers tend to grow in number. The demands from high-technology customers (buyers and builders) for luxuries, including power redundancy, immunity from power outages, and maximum poured-in-place trench protection, can end up costing project owners significantly more money than will wiring a simple residential or retail project. And, it can worsen if a project owner (seller) agrees to perform something for a buyer in the distant future and unexpectantly falls prey to an increased scope of work, inflation, higher costs, and even more regulations. Project owners must enter into utility agreements with both buyers and service providers with eyes wide open.
4. If an owner agrees to perform land improvements for a utility company, the owner should clearly spell out both the owner's and the utility's role by defining items such as maximum-not-to-exceed-dollars per task, who directs and pays for design and permitting, schedule for completion, status of using union labor, etc. More specifically, define responsibility and clarify at least the following:

   - Permits
   - As-builts
   - Surfacing
   - Landscaping
   - Shape of the land
   - Environmental analysis
   - Who drives the schedule
   - Providing other dry utilities
   - Storm detention requirements
   - Erosion control responsibility
   - Rough-gross and finished-net square area

- Responsibility for fencing and site security
- Who pays for engineering and survey support
- Architectural style of any buildings or structures
- Responsibility for cost overruns as a consequence of working out of season

Figure 3-1 graphically shows that an owner's ability to influence a project diminishes over time and nothing quite illustrates this more than finding optimal locations for vaults and appurtenances early during planning.

Owner obligations

Here's a list of possible areas in which an owner may be obligated to pay for dry-utility system installation.

- Obtain permits.
- Provide easements and land.
- Commission geotechnical work.
- Move, extend, or add power poles.
- Upgrade utilities to the edge of the project.
- Handle all design, engineering, and coordination of meetings.
- Provide temporary power, phone, or wet-utility service connections.
- Pay for materials for and installation of custom festoon street lighting.
- Pay for a neutral third-party consultant to provide a cost estimate or construction schedule.
- Provide all materials including, but not limited to, vaults, structures, conduit, sleeves, and other appurtenances.
- Coordinate, manage, and pay for joint-trench work and installation of vaults and appurtenances including survey and control, potholing, sawcutting, excavation, bedding, backfill, and paving.
- Pay for and provide other special services that any one utility provider may require, such as permanent phone for the power provider, power for the cable provider, or storm stubs to underground electrical or phone vaults.
- Clear, strip, and grade vault-farm areas. Provide all-weather truck access, precast, cast-in-place, or paved pads for surface-mount appurtenances, cut slope treatment, erosion control, and retaining walls for structures and vault farms. Possible costs for fencing, landscaping, and irrigation.

Utility obligations

Depending upon the service provider's policy, the local custom, how they view your project, or as to how the project was negotiated, a utility-service provider may provide any of the following services:

- Obtain permits.
- Upgrade existing utilities already present on-site.
- Coordinate and draft joint-trench plans and details.
- Handle design and engineering of their own distribution systems.
- Provide inspection and quality-control services for new construction.
- Be responsible for final sign-off and acceptance by the permit agency.
- Provide complete turnkey installation as negotiated with the project owner.
- Provide, pull, and connect any and all wire, fiber, or cable required for service.
- Provide all materials including, but not limited to, vaults, structures, and other appurtenances, conduit and sleeves, hardware, gas pipeline and gas-line equipment.

Private and municipal ducts and fiber systems

Owners may install a private conduit system to provide exclusive community-wide intranet services for an alternative voice-, data-, and video-service provider or, to sell or lease excess conduit capacity. As opposed to installing wet utilities or service-provider-owned dry utilities, planning and installing private conduit is straightforward and easier. Why? Because owners control their destiny. It's within their power to design, schedule, and install ducts in any shape, form, or configuration desired. Compared to other dry-utility systems, private ducts tend to be small and supple, and that's good because they're typically the last system installed. They have to go wherever they can fit.

Many municipalities are installing public fiber backbone networks. Owners who plan to link to a municipal fiber network may be required to connect their private systems to predetermined fiber nodes. So, it pays to research how and where a municipal fiber backbone network will be installed as it can impact on-site fiber design and connectivity. Unlike private conduit or service-provider-owned fiber systems, strict standards and specifications apply to the design and installation of public fiber systems. The schematic diagram depicted in *Figure 14-4* shows the general order of how fiber finds it's way from broad-based regional networks into a project and, ultimately, to private, retail, and corporate users.

In-ground private conduit systems follow the same procedures as other dry-utility systems when it comes to layout, design, and installation. Conduits in a private system can be of any number or size, however, for most planned communities and mixed-use projects, a *quad* bundle of four one-inch (25 mm) pipes strapped together, or one four-inch (10 cm) conduit with four one-inch (25 mm) interducts will usually do the trick. Anything beyond residential and light retail, such as commercial, industrial, schools and high technology, likely requires greater conduit capacity, pipe redundancy, and, perhaps, multiple connections into and out of the project.

When it comes to occupying space in public right-of-ways, private conduit systems are low priority. As opposed to publicly owned municipal utility systems, private systems represent a non-essential service and are not required for occupancy. As the last system installed in a joint trench, private conduit systems have to go wherever they can. This may entail installing them under existing conduit or weaving them around shallower pipe at the top of the trench. As with any utility system, modifying or expanding private conduit systems cannot happen later without incurring significant cost, causing cosmetic damage, impacting schedule, or disrupting street activity.

Because private systems are a low priority to everyone else doesn't diminish the burden for project owners. For owners, planning and establishing a private conduit system should remain equal with, and as important as, installing any other wet or dry utility system. No matter how supple, private ducts cannot be tossed into a trench with the expectation that they'll automatically service every lot or possess enough capacity to carry sufficient fiber or cable to key project locations. Plan and set aside vault and junction box locations early to serve as destinations for incoming stubs and connection points for structures or neighborhoods. Lack of planning can lead to areas missed, dead ends, stubs at wrong locations, and compromised user capacity. Without plans to work from, contractors are apt to install private junction boxes and vaults wherever possible, which usually means shoehorning them into landscape planters or clustering them in someone's front yard.

Getting ducts to users

The project owner or seller's role is to stub utilities to lots or parcels so that buyers and builders can make connections. Once a vault is set at the edge of right-of-way or a stub is laid across a property line, it's the buyer or builder's responsibility to trench to it and run conduit to their structure.

While it's possible for sellers to run utilities to structures, it's more the exception than the rule.

Installing redundant conduit is wise, but it's especially important when running utilities across private land where pipe can be susceptible to damage.

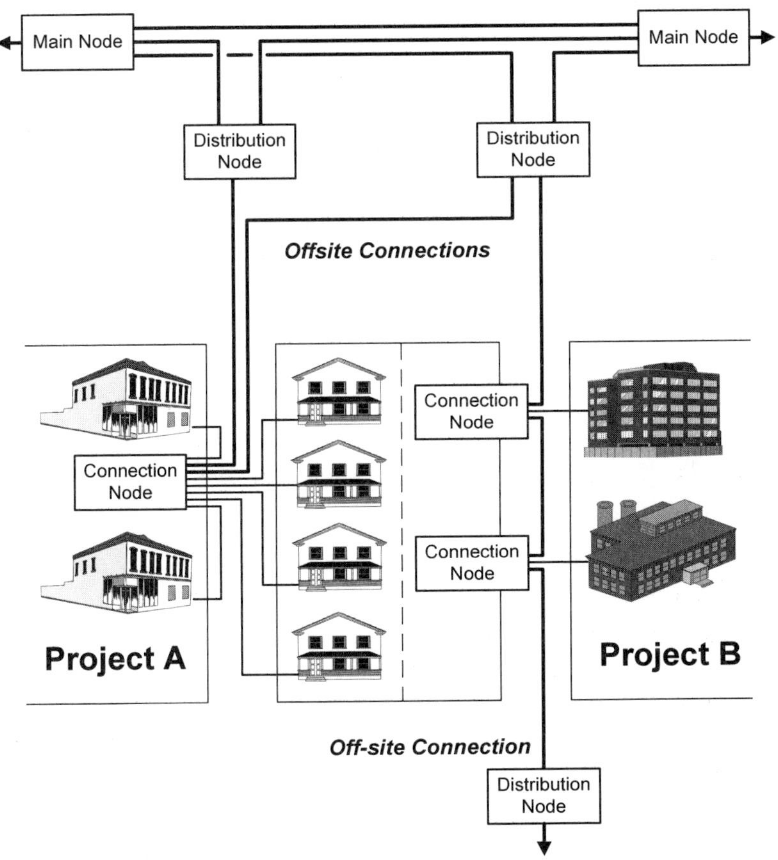

Figure 14-4

Why? One reason involves documentation. As opposed to public agencies, which are compelled to maintain underground utility records in perpetuity, private landowners are more likely to lose underground utility records, or not include them when the property changes hands. Another reason is because contractors may carelessly backfill uninspected private utility trenches which can result in subsidence, and a trench unable to support future loads. In either case, pipes can be crushed. When a private landowner drives a stake, fence post, or irrigation line through conduit damaging the fiber or cable inside, redundant conduit provides a possible route for re-establishing connections. Moreover, redundant conduit allows sellers and buyers to better meet higher and more diverse service demands common in dense urban developments. More conduit means that competing utility companies can potentially gain access for delivering service. A final reason entails "mandatory access" laws and regulations. Government-mandated access for telecommunications service providers (TSPs) can force private managers and building owners to allow new and competing telecom providers access to their structure. A smart strategy is to install redundant conduit to structures during the site work phase of construction thereby saving a TSP from later having to tear up mature landscaping and cut hardscape in order to gain access. When room doesn't exist for running multiple conduits into small vaults or junction boxes at or near the edge of right-of-way, individual pipe can be dead-ended, and capped, short of a vault and stubbed in later. Retail, commercial, industrial, and office parks may already require the installation of multiple conduits for redundancy purposes. Installing extra PVC is cheap insurance.

If a buyer or a builder requests upgraded or additional utility leads from locations other than those provided per agreement with the seller, that buyer or builder should be responsible for coordinating and paying for any additional services requested. Utility companies can then pull wire or fiber and fulfill the request.

Who handles what?

When bringing privately owned utilities into a project, the owner usually contracts with third-party service providers who supply equipment such as surface mount pedestals and belowground vaults. Providers also pull wire or fiber, make connections, inspect work, and provide in-house hardware and electronics required to make the system work. Once activated, the provider may also maintain the system.

With private systems, paying for design and engineering, supplies, and appurtenances such as boxes and conduit is often negotiable. However, the project owner usually manages on-site construction and pays for surveying, layout, trenching, installing conduit and boxes, and backfilling. If a service provider finds a certain locality, project, or market particularly attractive, they may handle everything turnkey without charging the project owner a cent. Once they occupy private conduit, third-party providers gain exclusive rights to sell ongoing community services and owners enjoy a state of the art project. Most projects must meet a minimal threshold of occupancy in order for private utility systems to succeed financially.

Direct-bury versus conduit

Installing conduit in open trenches is the safest and most versatile way to carry fiber, wire, or cable. An alternate method of installing lines is to direct bury them by use of static or vibratory plow. Direct-burial techniques are common when installing long lengths of shallow underground lines in areas slated to remain natural, not at risk of future development or vehicular loading, or in hard-to-access, unstructural areas. Direct-bury is also a quick and cost-effective method for installing temporary lines and solitary services across private land to homes.

Possible problems when plowing in direct-bury lines include pulling rocks to the surface, creating underground voids, kinking lines, backfilling unevenly, and possibly damaging already existing utilities. Compacting around direct-bury cable isn't easy and, many times, it can prove impossible.

Conversely, conduit works as a sheath protecting cable, wire or fiber from loading, water, construction backfill, future intrusion, and rocks. Conduit also acts as a sleeve allowing lines to be changed and upgraded without excavation. Protected utility lines backfill easier and accept structural compaction without failing.

Utility coordination meetings

Regular coordination meetings with representatives from all utilities should begin once an owner contacts them to initiate design and planning. These meetings must be productive sessions where the project owner disseminates information, schedules, and plans to everyone. The key is keeping utility people engaged. If they believe that an owner is wasting their time, they may lose all motivation and either not take your job seriously or divert their resources to better managed projects.

Coordination meeting minutes should list each utility representative's name along with mailing address, telephone and fax numbers, and e-mail addresses. A condensed contact page is invaluable as both a record and a reference sheet for anyone needing to reach somebody. Other team members instrumental in planning, layout, engineering, and construction should also attend these meetings and be included in the contact listing. Distribute meeting minutes to builders and people in other departments who are interested in the schedule or curious about the locations of utility stubs and boxes.

Construction issues related to dry-utility installation

Dry-utility contractors can damage other utilities and surface infrastructure, so do whatever is necessary to minimize impact. Also, keep in mind that stark differences exist between installing wet and dry utilities. Here are some ways to prepare for dry-utility construction.

1. If lot-line stubs are to cross under sidewalks and hardscape, install sleeves.

2. Lay sleeves under intersections *in lead with* lines coming in and leaving the crossing.
3. Clearly mark stubs with painted stakes labeled to identify each utility at that location.
4. Keep utility company construction crews from dumping stockpiles of imported backfill, precast structures, and bundles of conduit wherever they please.
5. If requested to move underground lines by a third-party service provider, first check to see if that service provider has a legal right or franchise to even be there.
6. Carefully monitor the installation of sleeves and crossings. Check for adequate pipe, end caps, locator tape, tracer wire, and proper trench backfill and compaction.
7. Sequence piping and structures correctly. Dry utilities install easiest when subsurface vault or structure installation precedes piping. It's easier to make pipe fit a block out than it is to set a structure both horizontally and vertically perfect to meet an existing stub.
8. Maintain dry-utility work schedules. Keep all parties apprised of upcoming work, provide an ongoing warning system that ties design and permit performance to deadlines, and continually remind everyone about the overall construction schedule. Schedules and updates are documents of record that will be useful later should an owner want to sort out why milestones were missed or to determine responsibility for a bungled dry-utility construction schedule.
9. Protect all pavement during dry-utility work. Work activity that can result in damaging pavement includes:
    - Grinding in mud and dirt
    - Dropping pallets of materials
    - Walking or spinning tracked vehicles over pavement
    - Dumping, scooping, and scraping piles of material over pavement
10. The next item flows from the last in that any work occurring near or above curbs can result in damage to planters, irrigation, shallow wet-utility stubs, hydrants, and sidewalks. Dry-utility work should precede landscaping and sidewalks but only rarely should it precede wet-utility installation. Don't allow contractors working out of sequence to store materials, backfill, trench waste, or to sit equipment on curbs, sidewalks or on any other finished portion of work.

11. Budget for and be prepared to survey stake underground utility vaults that fall within hardscape. Refer again to *Figure 14-3*. Here, a utility vault installed prior to sidewalks doesn't match final sidewalk grade. Since road, curb, or finished lot grades drive sidewalk elevations, the likelihood of sidewalks changing grade is remote. By the time a project is ready for dry utilities, you can gamble that infrastructure will remain static so install vaults to match future grades.

12. Refer to *Figure 14-2.* Utility easements in dense communities may only be as wide as five feet (150 cm). Narrower easements leave little margin of error when trying to keep dry utilities within the right-of-way. Tighter control in keeping utilities in narrower easements may mandate increased survey staking and layout. Also, note that narrower easements mean less working room for equipment, workers, and materials. This can result in private property damage, unanticipated change orders, and construction cost overruns.

13. When installing dry utilities in roads, take whatever means is necessary to protect prepared subgrade. If structural compaction is required, don't allow backfilling to proceed without adequate geotechnical inspections. This means budgeting ample funds to geotechnically inspect dry-utility installation. Regardless of where they exist in the right-of-way, bed all pipe to specification. In structural areas, dry utilities should receive the same treatment as wet utilities when it comes to backfill and proper lifts. Insist on as-builts for everything.

14. Refer to *Figure 14-5,* which ties the construction schedule to individual contractors on-site. The scheduled item numbers and construction activities cited in the first two columns match the same activities and numbers noted on the project master construction schedule. (Naturally, this assumes that you have and are working from a well-documented construction schedule.) It's a good idea to attach to *Figure 14-5* a letter-sized plan of the entire project that illustrates where each activity is to transpire. Update documents like *Figure 14-5* as necessary, but certainly no less than weekly, and integrate them into the construction and project team meetings.

15. At times, a project owner may be required to supply materials for installation. When this happens, owners can minimize mistakes by proceeding as follows:
    - Only use a supplier approved by the utility company.
    - As a reference and for recordkeeping, always include with your order a complete set of utility plans, shop or structural drawings, and specifications.
    - Schedule material deliveries when crews are on the construction

Dry-utility Update that Dovetails with the Construction Schedule

Sched Item No.	Individual Project Contracts	Civil Engineer	Dry-utility Lead Designer	Date of First Submittal	Date of Permit Approval	Scheduled Dry Utility *Start*	Scheduled Dry Utility *Finish*	Comments
	Black Mountain Contractors							
6	Maple Drive, Sta.10+90 - Sta. 22+70	R3MH	Valley Electric	7-Jun	20-Jun	In Process	10-Aug	With weather, should complete work early. Pole bases going in.
8	Division 12 through to the Park	R3MH	Valley Electric	7-Jun	26-Jun	In Process	28-Jul	Need staking and both apron and stub locations for north retail.
5	Lower Willis Avenue	R3MH	Valley Electric	7-Jun	26-Jun	10-Aug	1-Sep	Contractor to begin trenching once a crew breaks free off Maple.
	Summit World Construction							
15	Complete Brush Hollow Estates	Bayside	Valley Electric	21-Jun	Pending	Pending	15-Sep	Crossings are in. Builder to transmit final lot stub locations.
19	Gas Main Extension along Green Belt	R3MH	Consolidated	3-Aug	7-Aug	In Process	21-Aug	Easement has been logged, some grubbing and chipping remains.
22	Off-site Fiber Connection - Lewis Hill	Metrolite	Metrolite	Pending	Pending	Pending	15-Sep	Final design hung up due to county re-design of 7th Avenue.

Figure 14-5

site. Materials left too long in one place can become damaged or mysteriously disappear.

- Order early and follow up regularly to check that you're getting what you want when you want it. Forward copies of orders and paperwork onto the utility company.
- Send the supplier a site plan clearly illustrating where to deliver materials. It's crucial that owners give delivery notice in a timely fashion or else they become liable to pay for reloading, restocking, or moving materials on-site.
- If something arrives defective, incomplete, or wrong, send it back immediately. Coordinate deliveries so that the owner or the owner's representative can inspect and prevent anything that is damaged or undesirable from being loaded off the truck.

16. Perform complete as-built drawings for all dry utilities installed. As-builts should encompass vertical elevations and horizontal alignment for all pipe and structures. As-builts also verify that conduit has been installed per plan and specification. This magnifies the importance of scheduling as-builts concurrent with layering and backfill and not after the fact, when only the top layer of conduit is visible. Conflicts and lost time between surveying and construction can get expensive. Owners may find it cheaper to keep construction crews moving and to pay survey crews to sit and intermittently measure progress rather than vice-versa. Or, employ a competent member of the construction crew to record installation. Keeping work moving and utility trenches backfilled serves another purpose in that filled trenches are less likely to have vehicles driven into them, to cave in, or to fill with water. Concerns over having enough surveying resources to keep work moving and to perform as-builts are a good reason to closely schedule and coordinate survey and utility crews.
17. Protect curbs. Cast-in-place curbs usually follow in-street wet- and dry-utility installation but precede subgrade preparation and paving. When dry utilities are to be installed in planters or along the edge of right-of-way rather than in streets, then curbs can be installed anytime independent of them. However, curbs that precede dry-utility installation remain most vulnerable to construction damage when contractors have to trench and install conduit into a planter that lies between the pavement and curb, and a sidewalk. In this scenario, contractors will sit heavy equipment in the street or straddle curbs rather than risk damaging the sidewalk. Equipment that straddles the curb can bounce, turn, and advance, resulting in striking, chipping,

fracturing, and marring the curb. Equipment can also fracture and push extruded and precast curbs out of place. Although tracked vehicles possess the greatest potential for damaging curbs, low riding rubber-tired vehicles can bottom out and angle into portions of curb thereby shaving and chipping away without the operator noticing. Rubber tires can turn an otherwise clean curb into a black-stained nightmare. One way to protect curbs is to lay boards or dunnage over the portions of the gutter or pavement that are exposed to track or tire damage. Dunnage elevates equipment and insulates hardscape from damage. Be careful though because dunnage laid over minute ridges, joint lines, and concrete irregularities can effectively concentrate equipment weight and dynamic force resulting in cracks and fracturing.

Booms, busts, and changes in technology can result in a proliferation of utility contractors who may not be dedicated or qualified to install dry utilities. Some contractors follow trends in development masquerading as needed in order to find work or just pay the bills, which can create fierce competition and abnormally low pricing. It's a mistake to judge contractors only on the basis of a low bid. Consider using contractors who your service providers trust and with whom they have working relationships. Dry-utility contractors must be experienced, available, and possess sufficient resources. Coordinating and installing dry utilities is tough enough without dealing with a contractor who can't get it done.

General dry-utility planning and design tips

1. Hold regular planning and design meetings.
2. Consider including conduit for site security.
3. Develop vault-farm options as early as possible.
4. Where applicable, budget for multiple connections and looped systems.
5. Coordinate with one primary point of contact at each dry-utility company.
6. Notify the franchised service providers in your area before designing or planning.
7. Leave utility right-of-ways and easements in as good or better condition than found.
8. Design runoff to leave and not to channel into utility right-of-ways or easement areas.

9. Dry-utility systems are planned from the top down but they are built from the bottom up.
10. Integrate the excavation of vault farms and utility pads on side slopes with grading, drainage, and retaining-wall design.
11. Keep builders and buyers apprised of vault and stub locations or where they are planned to go relative to lots or parcels.
12. Service providers should be on-board as early as possible to learn about the project, review plats and road plans, and provide feedback.
13. Develop and circulate street lighting, signalization, metering, and other electrical power designs to all design consultants as soon as possible.
14. When planning utility layout, consider that most utility companies interested in wireless meter reading may want to install conduit for data and communication.
15. Keep utility lines and crossings between curb and lot lines out from under, and away from, retaining walls, tree wells, streetlight boxes, and signalization poles.
16. Don't design metal culverts over cathodically protected pipelines. Metal may interfere with the cathodic protection system possibly resulting in premature culvert failure.
17. Power outlets installed on streetlight poles for festoon lighting typically require a second power-line design, installation, and additional crossings. Budget accordingly.
18. Try to substitute subsurface power transformers for standard surface-mount minipad units. The cost differential necessary to hide these "chestnuts" is typically money well-invested.
19. For purposes of telemetry and communications, plan to install power and phone conduit to public works facilities such as water reservoirs, pump stations, lift stations, and storm vaults.
20. If they exist across a project, consider using high-voltage power-transmission and gas corridors for trails and recreational access. Where possible, integrate these into trail and access plans.
21. Plan in advance for the installation of dry utilities to community and park amenities such as:
    - phone and power service to parks;
    - cable and fiber for recreational facilities; and

- lighting for walking trails, playgrounds, sport courts, and project entries.

22. Include tracer wire over non-metallic pipe and locater tape over *all* pipe. Tracer wire must be accessible at multiple points above ground so that a locating instrument can energize the loop necessary for detecting non-metallic pipe.
23. Where tolerances are tight or for cheaper construction, install street lighting and irrigation conduit in the same trench. This value-engineering concept works well in parking lots where planter islands require both landscaping and lighting.
24. Contact service providers at least a year in advance if you want to move or heighten existing power poles, to install temporary power and phone, or if off-site power service needs to be upgraded to meet expected power demands of a facility such as a pump or lift station.
25. When submitting plans to utility companies for review, include at least, plans and profiles for wet utilities and roads, expected scope of work, details, specifications, legal descriptions, lot layout, expected loadings, square footage to be built per product type, and a list of project team members including their phone numbers and e-mail addresses.
26. Small-diameter, temporary, and individual direct-bury lines get no respect. Contractors plow through them if they're unaware of their existence or if lines are unmarked and appear out of service. Use of conduit and colored marking tape helps to warn contractors of existing lines. Conduit says, "I care" and tape says, "don't dare." Regardless of diameter or use, document and as-built all lines installed.
27. Instead of leaving the decision up to an equipment operator, plan the location for all utility stubs beyond right-of-way. When stubs are incorrectly located, expensive and time-consuming options to properly place them can include rerouting the lines into newly created private easements, following or doubling back along the outer edge of public right-of-way, or looping them under and across aprons and driveways.
28. When a service provider sends an owner as-built information, dimensions, or technical data to be used by the owner's design consultant or engineer, always respond with a letter of confirmation describing in detail what the service provider has sent and why it was sent. Ask that the service provider review the content of your letter

and provide prompt notice if your design team is proceeding erroneously or with wrong information.

29. The chances that a utility provider's design effort and customer service will satisfy a project owner may depend on the owner's style and actions. When choosing to allow a service provider to represent your interests and oversee dry-utility planning and permitting, remain in constant contact with both the participating local permit authority *and* the service provider in order to ensure that communication and coordination are proceeding as expected.

30. Unless your service providers use colored conduit, have someone on-site spray paint conduit as it's delivered on-site. Paint the bundles different colors to correspond to each provider who owns pipe. Also, count and measure what each provider has delivered. Some dry-utility companies will hold project owners accountable for pipe that is seemingly lost or used by another utility company. Intermittently spray-painted pipe is easy to spot when transported or if used by someone other than the party that owns it.

31. Fortunate owners have dry utilities ready and waiting for connection at edge of their property. When they don't, bringing dry utilities to a project may entail extending them through a third-party jurisdiction. This frequently results in having to procure additional permits and may introduce the owner and service provider to a new satchel of regulations. When faced with this situation, request, in writing, a copy of the provider's schedule and target dates for providing service to the edge of your property. Request dates for finalizing design, obtaining permits, beginning construction, and completing work. Document everything.

32. Locating dry utilities through and under retaining walls may require special design, expensive wall penetrations, concrete backfill, and protective sleeves. Combining dry utilities and walls can also create logistical, coordination, and liability issues for owners, service providers, and contractors. Sequencing becomes an issue if upfront utility work results in accelerating other site-work on yet-to-be-developed ground. Coordinating wall and utility contractors is also critical. Impacting either of the contractors can disrupt work, delay backfilling, and expose trenches or excavation to deterioration if exposed to moisture. Most service providers prohibit installing their utilities and easements through or under walls, so check before designing.

33. Strike quickly on deals. Dry utility installation isn't and never has been a set and predictable procedure when it comes to defining responsibility, terms for payment, reimbursement for materials, inspection services, and construction coordination. And, it changes from provider to provider. The situation compounds when a utility company is interested in acquiring land within your project or when they want an owner to provide an easement or right-of-way for future use. When hammering out time-sensitive agreements with service providers, complete the deal as quickly as possible. Then sign it. It's never known when the winds of change will introduce new people or circumstances into the process, resulting in carrying you back to square one.

34. Dry-utility procedures and standards vary widely. Many national retailers and builders use consultants spread throughout the globe who may not understand the codes, procedures, and dry-utility regulations particular to your area. As an owner or seller providing dry-utility stubs, you want the buyer or the builder's architect to place utility vaults and connection points where they will work. This includes: designing areas large enough and in the right place to accommodate vaults and aboveground transformers and boxes; placing wet utilities where they are unlikely to conflict with incoming service lines; ensuring that incoming conduit radii will work; specifying acceptable materials; and, meeting surface drainage requirements. Do the setback distances work? Has access been provided for utility trucks and servicing?

35. When placing vaults (and utility structures) in the right-of-ways and easements, service providers prefer to stunt (splay) their appurtenances at varying distances according to loading and use requirements, and to accommodate connections. Each provider has different requirements for splay distance so find out what these dimensions are before laying out aprons, planters, trees, and roadside features and of course, vault farms. Splay distances are control points. When locating vaults, begin by finding places for those that have inflexible splay distances beginning with the largest units. Next, locate places for vaults more flexible in splay distance. Last, fill in with the most flexible vaults and consolidate them. It's easier to control, screen, and locate places for vaults when they're in groups. Minimize stringing them out.

36. Project owners often have to grant service providers authority to construct, operate, maintain, protect, remove, and replace any part of

their system as a right of access. Before signing anything with a utility company, clearly understand how easements differ from fee, right-of-way, and "set aside" land. Also, evaluate the long-term consequences and benefits of issuing non-exclusive easements. For instance, a builder or buyer may encourage a new and possibly *undesirable* utility provider to enter a community or development through a non-exclusive easement. Another issue can involve seniority and rights concerning the design and location of road right-of-ways that cross existing easements or other right-of-ways. Seek the advice of a qualified land-use attorney on matters pertaining to easements, right-of-ways, and set-asides.

37. When bringing data, communication, or video conduit to a central "switchboard" location for distribution, remember that the number of conduits leaving the data center will usually be greater than the number of conduits entering it. Anticipate that some utility systems will require running dedicated home runs from the data center directly to customers. Also, simultaneously install in the same trench, incoming mainline conduit to vaults or to a central hub with outgoing smaller distribution conduit leading out of the same vault or hub. If lot stubs, home runs, or secondary mains can't be installed because lot lines aren't known or established, still try to stub distribution lines as far down the same trench as possible before backfilling. The objective is to consolidate trench work, to excavate subgrade as few times as possible, and to accelerate work that precedes curbs, pavement, and sidewalks.

38. Plan private utilities as though they were public. Apply specifications such as:

    - minimal depth of trench;
    - capping redundant conduit;
    - minimal distance from wet utilities;
    - type of bedding and backfill required;
    - minimal distances between layers of conduit;
    - minimal distance between gas lines and all other utilities; and,
    - sequence for placing conduit in trenches in bottom-up fashion.

39. Craft a construction-procedures document for dry-utility contractors. Highlight the project owner's expectations for items such as:

    - On-site behavior
    - Handling field changes

- Rules for leaving trenches open overnight
- Locations for employee parking and for stockpiling material
- Special requests from utility companies or permitting agencies
- Agreed upon methods for construction and measurement of work
- Specifications for content, field measurements, and presentation of as-built data
- On-site management. Have the contractor list all individuals who will be part of the owner's on-site management team.

40. When choosing streetlight poles, prepare to evaluate the following criteria:

    - Festoon lighting
    - Type of luminaire
    - Type of arm structure
    - Type of pole material and color
    - Pole height
      Pole height can be a factor if permit conditions require low-impact lighting. Requirements may restrict or prohibit streetlight wash on projects seen from a distance. Shorter and more numerous poles may be a solution to countering restrictions.
    - Minimal levels of lumination
      There exist at least three ways to attain minimal levels of lumination. The first is to develop without overhead lights and, instead, use entryway lanterns and in-pavement lighting. The two other solutions include erecting fewer poles of greater height, or erecting more poles of shorter height. Other considerations affecting the relationship between height and number of poles include road width, relationships of poles to structures and setbacks, security and shadow factors, expected heights of adjacent trees at maturity, available locations for pole bases, distances between intersections, and aesthetics.

41. When a utility service provider won't respond, here are a few tips to help expedite the process:

    - Don't ride people—give them time to react.
    - Try to make your requests short and to the point.
    - Fax or e-mail your request to multiple people within the targeted organization.

- Always respond with a clearly written memo encapsulating the results of design sessions or meetings and noting who's responsible for what.
- Carbon-copy records of communication to someone else on the owner's team and to at least two other people involved with dry-utility planning and design.

Chapter 15

COPING WITH HAZMAT AND CONTAMINATED SOIL

Hazardous materials, or *hazmat*, and contaminated soils find their way into projects by one of two routes: as a pre-existing condition, or as a newly created problem resulting from a fuel spill or other on-site calamity.

Pre-existing contamination

People who insure, finance, and permit land development must know the potential environmental risk and liability before supporting a project. As a result, project owners perform environmental studies and audits to document and to better understand a property in the following ways:

- Historical past and land use record
- History relative to storage, handling, and use of hazardous chemicals and pesticides
- Presence of equipment containing polychlorinated biphenyls (PCBs) or other hazards
- Presence of soil and groundwater contamination including natural hazards such as radon
- Features such as storage tanks, septic fields, landfills, asbestos, pipelines, and industrial or other manufacturing facilities

Phase 1 and 2 environmental site assessments

If a Phase 1 Environmental Site Assessment (ESA) is available for a property, verify, or have your environmental and geotechnical consultant check, that the report meets specification ASTM D1527. If the report does not meet this standard, it may be missing important information. The Phase 1 report usually includes a history of previous development known to have occurred on a property. This information is crucial to planners since it allows them to miss, or prepare for encountering, old and buried features such as foundations, tanks, and septic fields. Estimators also need this type of information for quantifying costs and scope of work. Other information contained in the Phase 1 report may include verification of obvious surface contaminants or the potential risk of soil and groundwater contamination. Sampling isn't normally conducted during a Phase 1 environmental assessment.

Soil and groundwater sampling, and testing for contamination happen during a "Phase II" ESA. Describing the multitude of different tests pertinent to, and implications of, a Phase II ESA are beyond the scope and intent of this book, however, if a Phase II ESA has been conducted, owners will be better positioned to understand:

- if an environmental consultant will be required on-site during site construction;
- whether health and safety regulations will have to be followed during site work;
- other geotechnical information if boring and well log data from the Phase II ESA provides bonus information concerning non-contaminated native soils and deeper subsurface conditions;
- their options. Knowledge-in-hand affords owners time to prepare for potentially long-lead approvals such as being able to leave and cover contaminated soil, or to pump contaminated groundwater into the nearest sanitary sewer manhole.

Environmental and geotechnical sampling

Similar techniques can be used for environmental and geotechnical sampling. The primary goal of environmental sampling is to access the presence of contaminants or hazardous materials. The goal of geotechnical sampling is to develop a subsurface soils map delineating rock, water, and soil variability. Both exercises determine an owner's parameters for developing a property and typically include design, engineering, and remedial recommendations. Besides possibly discovering that a property is going to be far more expensive to remediate than though, the act of exploring, performing environmental studies, and writing reports is expensive and time-consuming. The good news is that it may be possible to glean geotechnical and environmental data from the same borings or test pits provided that samples taken are deep and frequent enough to serve both purposes. Ease of coordination, billing, and responsibility would suggest using the same firm for both geotechnical and environmental work, however using separate firms that specialize in one discipline or the other may result in higher quality work and more creative construction solutions. Whether an owner chooses to use one or two firms, it can pay to consolidate exploration and gather as much data as possible.

In both environmental and geotechnical exploration, there can be uncertainty regarding the number of samples needed to conduct a statistically sound analysis. In general, the number of samples taken depends on the degree of variability present and whether geological or hazardous in nature. In general, the level of confidence increases and management risk decreases with increased sample-size. For example, the presence of clay may limit or absorb contaminates whereas sandy areas may allow contamination to freely migrate. The contrast in soil types helps determine how and where ground exploration is to take place. Whereas free draining and variable soils may call out for more frequent and deeper exploration work, under similar circumstances, fine-grained and homogeneous soils on the same site may require less effort in exploration.

One way to avoid the cost and time of oversampling is to proceed on a phased basis. That is, cast a wide net and take samples intermittently in the search for evidence of contamination. Map the site according to the degree of pollution detected and focus additional sampling on those areas that exhibit contamination. A qualified environmental consultant should be able to iterate this process and determine the optimal degree of sampling required to satisfy statistical analysis, if required. At the same time, this exercise can produce information useful for site planning and for budgeting construction or mitigation work.

Options for owners

Brownfield, in-fill, and redevelopment are all prone to encounter contaminated soil, contaminated groundwater, and hazardous materials. How these materials are remediated depends upon the project and the development objectives. Cost and risk of performing environmental remediation can be extremely high. Here is a list of options available to owners faced with on-site contaminated soil:

1. Investigate the use of chemical or biological on-site soil treatment alternatives.
2. Excavate and export all contaminated from the site and import replacement material.
3. Cement-treat the soil to seal and stabilize it in-place, or to provide structural support.
4. Contain the contaminants in-place and cap them with clean soil or an impervious material.
5. High enough quantities of contaminated material may make it economical to "cook" the contaminants out and reuse the soil on-site.
6. If contaminated soils are granular and compactable, utilize them as-is, or mix them with ground asphalt, crushed concrete, or clean structural material for use as structural base course under pavement and hardscape.
7. Deal with contaminated groundwater as soon as possible because groundwater typically, is more expensive to remediate than contami-

nated soil. And, it's more difficult to gain approval to remediate contaminated groundwater than it is with contaminated soil.

Figure 15-1 illustrates a decision tree for evaluating practical and economical ways to handle pre-existing on-site contaminated soils. As with most dirt-related issues, the most expensive alternative is usually wholesale excavation and export. Owners may find it economical to plan development around contaminated areas and elect not to disturb them, or to cover them with pavement and hardscape. Methods for dealing with contamination typically depend upon the nature of the contaminants, its location and depth, quantity, availability of dumpsites, costs for export and import replacement material, project objectives, governmental regulations and restrictions, a desire to reduce or eliminate a hazardous condition, recommendations from consultants, temperature, moisture, and weather, and—as always—the owner's overall budget.

On projects where contaminants vary in substance, toxicity, and form, define and catalog known on-site contaminants in material safety data sheets (MSDS). Post the MSDS sheets on the jobsite at kiosks, bulletin boards, inside the cabs of trucks and equipment, and at hazmat stations. Distribute them during preconstruction, kick-off, toolbox safety, and weekly construction meetings.

Owners may find municipalities eager to develop industrial areas and brownfield sites. Ease and expediency in permitting, an array of tax credits and incentives, and insurance products tailored to reducing an owner's exposure make developing high-risk properties an attractive option. Benefits aside, owners still must concentrate on smart site planning and economical construction.

Contamination happens

No process is trouble-free and that certainly applies to land development. The question with site construction isn't so much whether fuel spills and contamination will happen. Rather, it's how can fuel spills and contamination be *prevented* from happening. Not that hazardous spills are inevitable, but during the course of land development, it's highly likely that some form, however small, of soil contamination may occur from one or more of the following situations:

1. Vandalism
2. Leaking vehicles
3. Broken hydraulic lines
4. On-site traffic accidents
5. Toppling of temporary toilets
6. Spills while delivering or fueling equipment

Dealing with Contaminated Soil

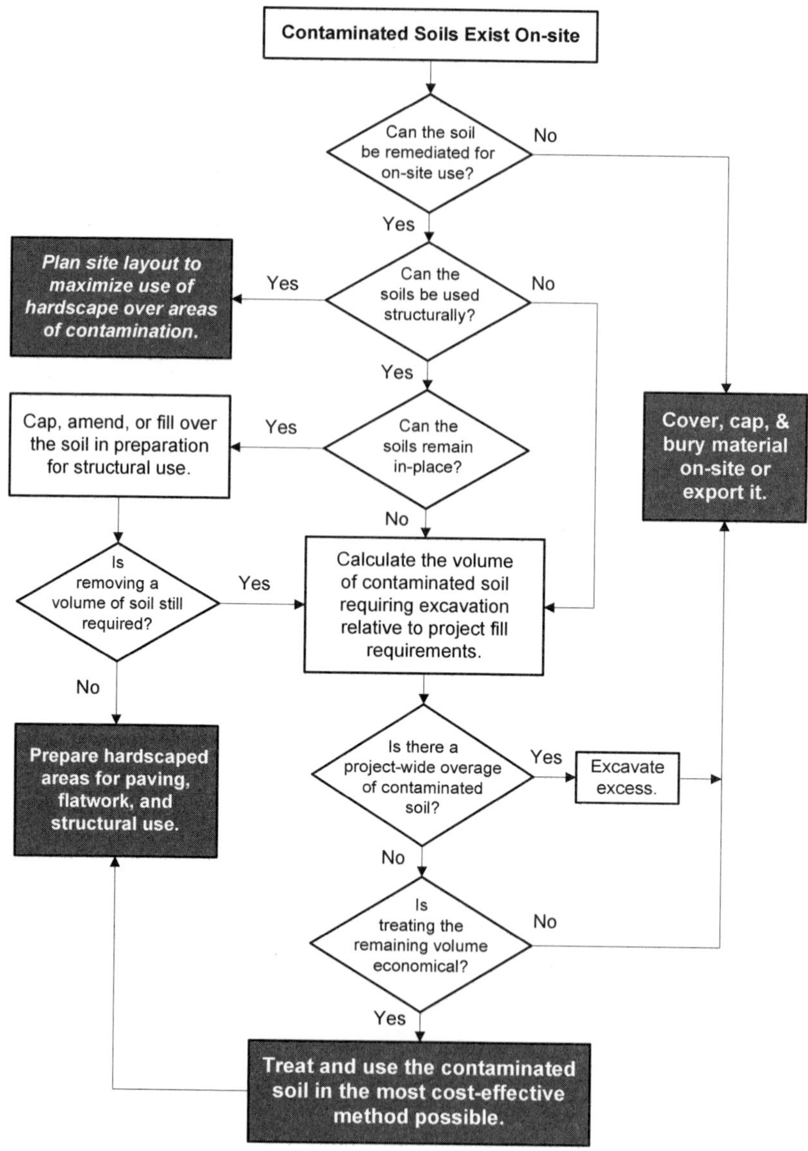

Figure 15-1

7. Rollovers, fires, and accidents with machinery
8. Hitting and rupturing live sanitary or gas lines
9. Spilling preservatives and other petroleum-based products
10. Loss of fluids and poor ground protection while servicing equipment
11. Careless activity causing on-site stored hazardous materials to discharge

Rather than expect contamination to occur, field personnel must take steps to prevent and control spills or discharge of hazardous substances. For owners, the task in accomplishing this is two-fold. First, spell out procedures aimed at preventing accidental discharge. Second, specify how to handle and report an incident should contamination or a spill occur. And, owners and their contractors must be ahead of the pack by never positioning themselves to allow a third-party or government entity to discover or report an on-site environmental problem without themselves first knowing about it. Carelessness and contamination can bring heavy consequences. Surprises are worse.

Preventing hazardous contamination

Here are rules to consider imposing in order to prevent spills and enhance site-work job safety:

1. Secure exposed contamination.
2. Prohibit the cross fueling of tankers anywhere on-site.
3. Work near existing gas pipelines and facilities as directed by qualified gas personnel.
4. Don't allow vehicles or construction equipment on-site that is leaking fluids or materials.
5. Post material safety data sheets indicating the presence of hazardous substances stored or present on-site.
6. Allow fueling only at approved areas and require the use of pads and containment devices under the point of transfer.
7. Require that contractors regularly and properly maintain on-site vehicles, construction equipment, and rented gear.
8. Set safe on-site speed limits and require per contract that all contractors and their subcontractors observe and drive as posted.
9. Check that truck drivers hired to haul hazardous material possess a HAZMAT endorsement on their Commercial Driver's License (CDL) license.
10. As part of their supervisory and safety duties, all contractors and suppliers should routinely instruct personnel about the hazardous

material response plan.

11. Utilize standard and proven utility construction practices. Require that contractors employ the correct type and size of machinery with qualified operators when performing work. This reduces the risk of equipment failure and breakdown.
12. Prohibit on-site vehicles and construction equipment from storing and transporting surplus fuel or other hazardous materials within project boundaries. Only allow fuel trucks operated by certified or approved operators to roam and fuel equipment and vehicles.
13. Post maps showing all areas of intermittent and permanent on-site water including conveyance routes and ditching. Note on each map, the reader's location relative to the site. This allows personnel to understand how spilled fluids can flow and migrate to open water.
14. Prohibit the on-site storage of hazardous substances including, but not limited to, fuels, hydraulic fluids, petroleum products, industrial oils and fluid compounds, oil refuse, batteries, and sludge. When hazardous materials aren't stored on-site, sources of contamination diminish leaving vehicles and construction equipment as the primary risks for spills.

Additional measures that should be taken by contractors include routine monitoring of:

1. Excess grease around fittings and the condition of hoses and gaskets
2. Drainage ditches, traps, and sumps that can accumulate oil and other hazardous substances
3. Wood chips, soil, and asphalt for evidence of oils, petroleum products, or other hazardous materials
4. Piles of straw bales regardless of who owns them
   The reason? Dry available straw is useful in corralling hazardous spills. When an emergency happens, anybody's straw is everybody's straw. Keeping blocks of straw neat, elevated off the ground, and protected from moisture is cheap insurance.

Contractor spill-response materials and on-site kits

Depending upon the type of hazardous materials on-site and the regulations imposed by governing permitting agencies and other environmental authorities, require contractors to distribute, protect, and clearly mark spill-response kits that are available 24-hours-a-day to all on-site-workers. Materials and methods may differ from project to project, however, the task of controlling site construction hazardous-substance spills and contamination is straightforward.

Require that site contractors keep a deep supply of loose sorbent material spreadable over fuel spills, pillows or pads that can be used to sop concentrated fluids, and booms to contain larger spills. Like first aid kits, prepackaged spill-response kits are available for purchase.

Spill response and on-site reporting

When contaminants are unexpectedly encountered, or are discharged on-site, at least the following general steps should be immediately taken:

1. Protect personnel. Check to ensure that all personnel are safe and accounted for. Call for help immediately if anyone is injured, trapped in a roll over, or requires emergency treatment.
2. If fire or the potential for fire exists, get help.
3. Evaluate, contain, and attempt to stop all spills at their *source*. Examples of containment techniques may include building berms or dikes, or quickly constructing a spill diversion area. Use special hazard abatement materials as approved for your project.
4. Evaluate the magnitude of the spill and call for additional workers if needed.
5. Secure the area; erect warning signs' and, post MSDS information.
6. Report the spill to the governing and permitting authorities and the project owner.

In most cases, equipment operators and laborers will be a contractor's first line of defense to control an accidental spill and prevent growing contamination. Once an "event' transpires, contractors must provide leadership until a qualified environmental consultant, inspector, or supervisor arrives to take charge. When reporting an incident, an on-site contractor should be prepared to transmit at least the following information:

1. Action taken
2. Nature of accident or spill
3. Time, duration, and exact location of accident or spill
4. An accounting of others already notified about the spill
5. The names of anyone injured including the extent of their injuries
6. Description and quantity of substance discharged
   If the contaminant is unknown, identify the color, smell, viscosity (thickness), and temperature of the substance.

One possible scenario for abatement and remediation in site-work

Here is an example of a hypothetical and generalized course of action that could follow a site-work incident provided that nobody is in a life-threatening position and that the hazard poses no risk to natural resources, wildlife, or health.

1. Stop the flow or plug the spill at its source.
2. Ensure that the contamination is contained and secured.
3. Notify all government agencies, an environmental consultant, and the owner.
4. Wait for an environmental cleanup or consulting firm to respond and take the lead.
5. Upon cleanup and remediation, the environmental consultant will review the situation, analyze the result of cleanup, and compose a report documenting the entire incident.
6. A follow-up meeting attended by all responsible parties may shortly follow.

Contractor written notification to the owner

Supplementary to the environmental consultant's report, request that the responsible contractor forward copies of the daily reports and compose a separate report explaining, from the contractor's point of view, what happened, why, and how similar incidents can be avoided. As regrettable as they are, accidents and spills can be powerful tools for education. The contractor's report may contain:

- A listing of those notified
- Action taken for containment
- Time, weather, and duration of spill
- Approximate known quantity of substance discharged
- Nature of work being performed when the incident occurred
- Planned corrective measures to prevent another like occurrence
- List of those who were involved in spill containment and cleanup
- Notes regarding the exact location and reason for the accident or spill
- The names of any injured people including how and where they were taken for medical attention
- Upon being notified, the time it took consultants, hazardous-material contractors, and government regulators to reach the site

Hazmat spill-response plans

In the context of site work, hazmat plans outline how to prevent, contain, remediate, and report spills or conditions of contamination that happen during the course of construction. Many project owners have no hard and fast rules for composing hazmat plans and jurisdictions can differ widely as to what they require in a hazmat plan. This means that contractors typically get to formulate their own hazmat plans in whatever style, format, and content they see fit.

Still, require all contractors to submit a hazmat plan. Besides outlining procedures and policy relevant to hazardous spills, a contractor's hazmat plan serves a secondary purpose for owners. It can reveal a great deal about that contractor's operation, priorities, and preparedness. Because hazmat plans can be project-specific, owners may consider generating one in-house plan for consistent for use by all on-site contractors. Regardless of who develops the hazmat plan, attach it to the contract documents.

One alternative for owners is to create a standard one-size-fits-all project hazmat-plan set up in template format with blanks or empty lines for contractors to fill in their company information. Then, the contractor can distribute copies to company personnel. Regardless of the dollar amounts of their contracts or their purchase orders, all subcontractors should also be required to sign, submit, and follow the same hazmat plan as the prime contractors. No exceptions, no mistakes.

The elements and direction spelled out in a hazmat plan can vary greatly depending upon any the following conditions:

- Project type
- Soils present
- Land-use strategy
- Expected weather
- Regulatory requirements and criteria
- The goals and objectives for cleanup
- Type of contractor and machinery employed
- Local options for disposing hazardous material
- Nature of the substance, toxicity, and risk potential
- Proximity to water, sensitive areas, wildlife, and people
- Distance from health, fire, safety personnel, and on-call resources
- The developer or contractor's track record and reputation for adequately handling or managing prior hazardous spills, toxic materials, and contamination

Hazmat plans can take many forms. Here is a list of items that owners could require in a contractor-generated hazmat plan:

1. A block to sign and date the plan
2. A list of unknowns or contingencies
3. Rules and policy for vehicle fueling and maintenance
4. Phone numbers of all on-call parties that could respond to an incident
5. A plan sheet depicting on-site water, drainage, and conveyance locations
6. A list of heavy equipment, machinery, and personnel assigned to the project
7. A list of the contractor's preferred geotechnical or hazardous-material consultants
8. An overview of the site's soils as interpreted by the contractor before beginning work
9. Site information including project data, owner information, and contractor information
10. An outline describing how the contractor will notify and report spills and contamination
11. A list of materials, fluids, and demolition debris that may be stored or encountered on-site
12. The phone numbers and e-mail addresses of the contractor's on-site hazmat-trained employees
13. A statement describing how the contractor will provide security, site inspections, and self-audits of the jobsite and their equipment
14. On small projects or in confined areas, a description detailing how and where the contractor would store or treat contaminated material
15. The contractor's plan for developing staging areas, truck washes, and fueling pads including construction details and where the contractor would prefer to locate said facilities
16. A reference to regulations governing hazmat plans in your geographic area or jurisdiction including a list of federal, state, and local governing authorities who control, permit, or influence the environmental aspects of the project
17. Through the contractor's eyes, a list of pre-existing site conditions that could influence a discharge or spill including a summary of on-site features deemed by the contractor to be at risk should a discharge occur (This gets contractors thinking and generally results in productive conversation.)

18. A detailed list describing the quantity and description of spill-response materials that the contractor plans to store in each spill-response kit (There should also be a plan depicting where each spill-response kit is to be initially placed on-site and the type of spill-response materials to be stored on heavy equipment, trucks, and in vehicles.)

Communication is important to controlling incidents involving spills and discharges of hazardous materials. Hazmat plans should, therefore, list the phone numbers and e-mail addresses of all appropriate parities who could respond to or who must be notified in the event of a hazmat incident or medical emergency including:

1. The project owner
2. Related permitting agencies
3. Ground and airlift ambulances
4. One-Call utility system number
5. Numbers to call in case of a forest fire
6. Direct number for the local fire department
7. The hazardous-material consultant on retainer
8. The local public works or highway department
9. Federal, state, and local environmental regulators
10. Direct number for the local police or sheriff department
11. Numbers for dry-utility companies including gas, power, and fiber
12. One-Call emergency numbers for personal injury, fire, and the police
13. Heavy tow truck or mobile crane numbers for lifting equipment and materials

Suggestions for avoiding problems with contaminated soils

1. Attach a signed hazmat plan to every site-work contract.
2. Contaminated soils are expensive to handle and dump. Estimate accordingly.
3. Be ahead of incidents. Locate potential sources of contamination and neutralize them.
4. Thoroughly bore and test in-fill and redevelopment property before committing to a project.
5. Don't allow piles of contaminated soils to loiter on-site—get rid of them as soon as possible.

6. Supply MSDS information to contractors and truckers handling contaminated soil or material.
7. Seek permission to use contaminated granular soils for structural purposes *well in advance* of construction.
8. Require on-site contractors or builders working for buyers or other property owners to submit hazmat plans.
9. Require subcontractors to complete separate hazmat plans identical in format to those submitted by prime contractors.
10. Long-term monitoring may be a required in order to receive a "clean bill of health" and obtain a *document of closure*. Budget accordingly.
11. The discovery of existing surface contamination should automatically trigger an immediate response for subsurface exploration or borings.
12. Don't mix water and contaminated soils. Protect and cover piles of contaminated material from moisture, disturbance, and from being mistaken as available fill material.
13. Know whom to call. Set up channels beforehand with firms that specialize in burning, remediating, and treating, or in loading, hauling, and dumping contaminated soils.
14. Since owners can be held responsible for the illegal disposal of contaminated soils originating from their site, require that contractors *promptly* submit truck and dump certificates for all hazardous export.
15. When hauling contaminated soil, motion and moisture may turn marginal material into liquid swill. To safely transport hazardous material, specify the use of specially designed, tight and flow-proof trucks.
16. On larger projects or on ground with high contaminant variability, remediate in phases in line with development. This approach saves on carrying costs and gives owners time to value engineer alternative ways of handling material.
17. Unless planned, don't work ground that is expected to hold contaminated soil. Disturbing contamination begets more contamination, testing and the accuracy of soil readings can suffer, and pollution limits can become less defined. Disturbing contaminated soils often leads to exporting more volume of contaminated material and importing more replacement material than planned. The result is schedule impact, increased exposure to poor weather, and, always, greater expense.

Hazmat inspections

Government officials and local activists may not give advance notice before visiting a jobsite to inspect hazmat work, to test soil, water, and air, or to view ongoing construction activity. They may choose to come early and stay late to observe fueling and maintenance work or to check parked equipment for fuel and hydraulic leaks. Inspectors may also request to review the owner or contractor's on-site office records and field reports.

Owners can request that inspectors produce a written scope of work before examining a site. Also, inquire beforehand what office documentation an inspector may want to review or whom the inspector plans to interview. Ask inspectors to outline their test sampling methods and describe their criteria or thresholds for gauging contamination and thoroughness of work. Request copies of the inspector's field reports. Note the testing laboratory that the inspector uses and request like samples of any material, water, or air taken by the inspector for evaluation. These actions better position an owner to later verify or dispute an inspector's lab test results or an inspector's conclusions.

Always provide an individual experienced with environment inspections and versed in the hazmat plan to escort inspectors. This escort should videotape the inspection tour, recording scenes and circumstances in which the inspector takes notes, pictures, or shoots video. It's easier for inspectors to trust and develop relationships with one in-house person than with a group of people. Using one individual also breeds consistency. Owners of numerous projects may find it advantageous to hire a dedicated, full-time environmental specialist who can concentrate on environmental audits, archeological finds, erosion control, and the growing weight of regulations and restrictions.

Penalties

Those who ignore the presence of known or newly detected contaminated soil or that practice negligence when encountering such material could face civil and criminal charges and even imprisonment. Owners may be liable if a contractor dumps or spreads hazardous soil or material over uncontaminated portions of a project. The same holds true if a contractor mishandles or disposes of an owner's contaminants and hazardous materials at unlicensed (unpermitted) off-site dumps or other inappropriate locations. Tough penalties can apply when hazardous materials are illegally stored without the appropriate permits. As opposed to private projects, many public owners attempt to protect themselves by requiring that contractors obtain permits for environmental work. In any and all cases related to understanding culpability, grounds for violations, and how an owner or contractor should proceed in a safe and controlled manner, always seek the counsel of both an environmental consultant and an attorney well-trained in environmental law and procedures.

Chapter 16

RECYCLING, TOPSOIL SCREENING AND PIT DEVELOPMENT

There's absolutely no question that under the right conditions *and* with the right people pulling levers, a bonanza awaits owners who can process and recycle raw material produced during site construction. Recycling can lessen environmental impact, reduce development costs, and improve profitability.

Now, this may sound anti-climatic, but recycling is nothing new. It's been a part of site-work for a long time. Losing asphalt and concrete demolition waste in fills, using logs and wood chips to stabilize ground, and finding holes and places for raw topsoil are standard techniques employed by successful and thrifty site contractors. And, recycling has always made sense to those working in remote locations without the benefit of a hardware store or supply house. I am referring to loggers, miners, heavy-construction contractors, and the military. Many of the old timers have a sleeve full of tricks when it comes to using what doesn't seem reusable and to turning a profit out of seemingly, nothing. Fortunately, the secret is getting out.

Projects rich in raw resources including quality native topsoil, free-draining inorganic soil, pit-quality sand and rock, excess storm water, and thick timber can yield a mother lode of useful products and net revenue if handled properly. As we'll cover in this chapter, processing and recycling can:

1. enhance project image;
2. reduce dependency on landfills;
3. reduce the need for import material;
4. improve the odds of meeting schedule;
5. reduce liability of mishandling waste material;
6. provide non-potable water for construction use;
7. cut costs, enhance cash flow, and in general—save money;
8. give owners more control over supply and widen uses of on-site materials;
9. provide an avenue to meeting Best Management Practice (BMP) requirements;
10. provide cheap alternative construction material for site contractors and give on-site builders a place to dump woody debris, topsoil, and

waste construction materials; and,

11. reduce the volume of trucking into and out of a site resulting in less traffic on adjacent roadways and spending less to maintain and upgrade on-site haul routes.

When most people in the industry think of recycling, they picture building demolition—contractors wading through mounds of structural debris sorting out valuable items useful elsewhere. While a demolition contractor might grab wooden or steel beams for use on-site, with the exception of concrete and asphalt, few materials from building demolition are reusable in site construction.

So—putting building demolition and urban re-development off to the side—owners interested in processing and recycling during site construction must evaluate their projects in terms of raw materials. A minimum critical mass of raw material must exist in order to first, cover fixed costs, and second, make on-site recycling worthwhile. If there's enough demand locally and if the price per unit can cover costs for processing, stockpile maintenance, loading, hauling, and management costs, then off-site sales of on-site recycled product can develop into a profit center. The real moneymaker however lies in using on-site recycled materials for *on-site* construction.

Making recycling work for on-site products

It takes four ingredients to make on-site recycling work in site construction. Two ingredients are sufficient quantity and sufficient quality of raw recyclable material to produce a desirable end product. The third ingredient is an on-site use or market for recycled product. The fourth is economics. Refer to *Figure 16-1.* This logic diagram traces economic, use, quality, and quantity issues in deciding whether to recycle or burn (woody debris), lose on-site, or export raw site materials. It also brings into focus the need for ample working room. Depending upon desired end product, *Figure 16-1* links with *Figures 16-3* and *16-4.*

For it to be profitable, the benefits of recycling have to exceed the combined cost to export waste product and purchase new product. Once the monetary value of recycling breaks even with processing costs, it doesn't take long for savings to mount. Here's a list of items to consider when evaluating if recycling is feasible:

1. What can be recycled and used on-site?
2. If required, is tap water and power available?
3. Are there suitable recycling contractors available?
4. Are on-site builders interested in purchasing end product?
5. Will a recycling operation create noise, dust, or odor problems?

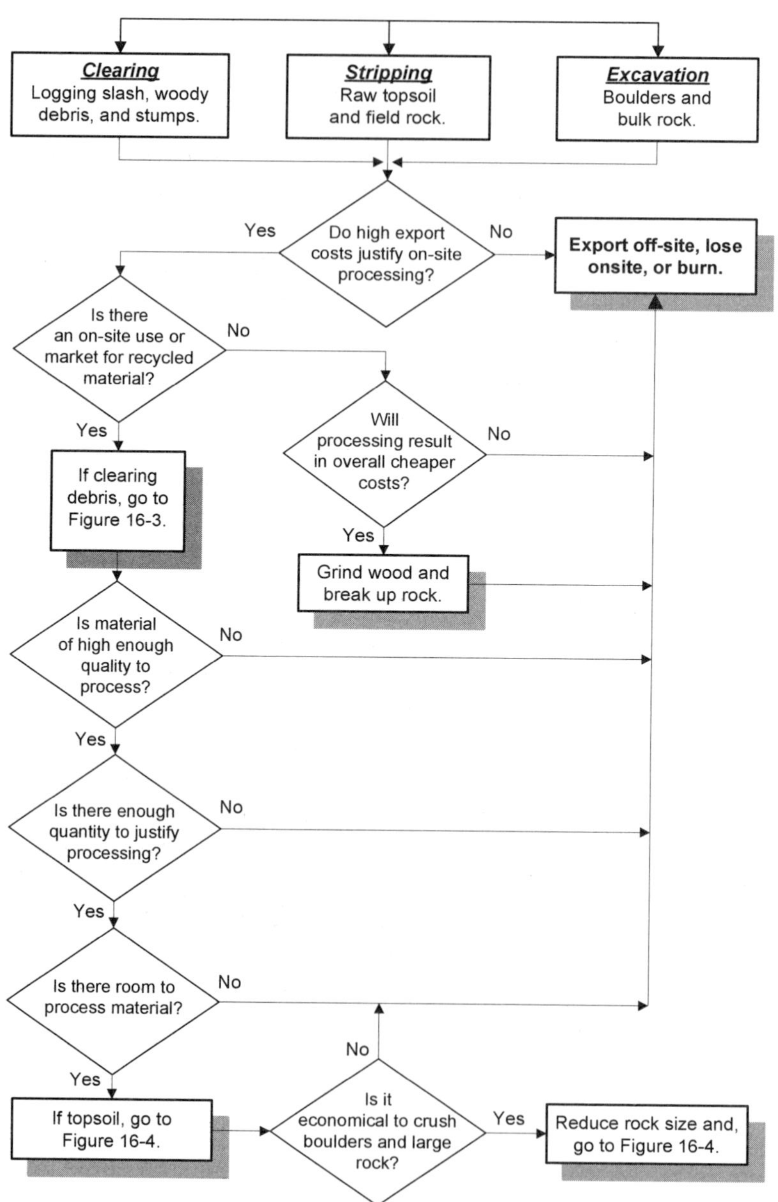
Topsoil and Rock Recycling Logic
Clearing
Logging slash, woody debris, and stumps.
Stripping
Raw topsoil and field rock.
Excavation
Boulders and bulk rock.
Do high export costs justify on-site processing?
Yes
No
Export off-site, lose onsite, or burn.
Is there an on-site use or market for recycled material?
No
Yes
If clearing debris, go to Figure 16-3.
Will processing result in overall cheaper costs?
No
Yes
Grind wood and break up rock.
Is material of high enough quality to process?
No
Yes
Is there enough quantity to justify processing?
No
Yes
Is there room to process material?
No
Yes
If topsoil, go to Figure 16-4.
No
Is it economical to crush boulders and large rock?
Yes
Reduce rock size and, go to Figure 16-4.

Figure 16-1

6. What quantity of end product can a recycling operation produce?
7. Can recycling eliminate or curtail off-site truck traffic or haul distance?
8. Is there enough supply of raw material to support a processing operation?
9. Do local or on-site competitors produce a cheaper or higher-quality recycled product?
10. Would your lender or insurance company have reason to discourage on-site recycling?
11. Without recycling, where are dumpsites located and how long will they be operational?
12. Do the load, haul, and disposal fees for exporting material make recycling cost-effective?
13. Is adequate or surplus land available on-site for supporting a recycling and stockpile operation including a detention pond?
14. Will local authorities and permit agencies encourage recycling by promoting flexible and alternative specifications and quicker permitting?
15. Does the danger or possibility of contamination, environmental degradation, or human endangerment exist by recycling a particular resource?
16. How scarce is your commodity? In other words, who outside of your project wants what you have? Where are they located and how much are they willing to pay?
17. What permits will be required to run a recycling operation? Will these permits impose onerous restoration and reclamation constraints making recycling uneconomical or even a liability?
18. Will recycling be viewed as a detriment to marketing and sales, or can it be used to demonstrate that the owner is performing a service that is helping to maintain or lower lot prices while benefiting both the land and community?

Recycling *only* works with clean materials so it's imperative to keep both raw and processed materials separate and uncontaminated. For example, logging slash that is ready for grinding shouldn't contain topsoil, rocks, or chunks of concrete. From a safe distance and with the approval and recommendation of your grinding contractor, watch a wood grinding operation. Notice how the equipment operator grabs piles of stumps and clearing debris and bangs it to shake out dirt and rocks before feeding it into the mouth of the grinder. This serves two purposes. It knocks out material that may damage the machines' knives and it results in a cleaner end product.

The same applies when gravel and sand are mined for use in producing concrete and asphalt. To be suitable for producing concrete and asphalt, sand and gravel cannot contain woody debris or fine-grained soils such as clay. Similarly, recycled asphalt and crushed concrete suitable for subbase material shouldn't contain topsoil, vegetation, and other deleterious matter. Raw topsoil destined for screening should be purely organic with little or no inorganic soil or woody debris mixed in. Raw organic material mixed with clay or silt resulting from overstripping or careless earthwork produces a screened topsoil product that is low in nutrients, short on growing medium, doesn't drain, and isn't easily compactable.

Contamination also penalizes production. When topsoil containing fine-grained inorganics, such as silt and clay, gets wet, it clods and gums up the processor's screens producing less screened topsoil and more unusable material per equipment hour. At a point, it's becomes too expensive to work and rework topsoil that is bound by wet clay so greater volumes of waste material get produced and production suffers. Unless it's dry and freshly screened, landscapers will find silt- and clay-laden screened topsoil to be heavy and difficult to rake and level.

On-site vegetation

Recycling includes transplanting native trees or re-using pre-existing landscape stock. This subject is covered in the next chapter, "A Word About Native Trees".

Demolition and dumps

As discussed earlier, the economics of recycling and hauling off demolition debris are apparent. For the cost of sorting and stacking, owners can usually locate entities that are willing to purchase and cart away unprocessed demolition debris such as concrete, bricks, wooden beams, and structural steel. In this way, owners create a revenue stream while saving on hauling-and-dump fees. Using demolition debris for on-site construction saves owners money since they avoid having to import an equivalent amount, or volume of new materials. Sorting and dismantling materials also works to resize debris resulting in efficient hauling. Sorted materials stack more uniformly than if loaded as pulled from the wreckage. Smaller material and garbage that remain after sorting also loads easily and packs well resulting in trucks being able to haul more volume and weight per load to the dump. The result is less truckloads and ultimately, less construction cost.

Figure 16-2 shows a process for sorting demolition debris. This process is only one of many equipment configurations that are possible for sorting and refining demolition debris. Choice of machinery and its arrangement depends on size of raw material and desired end product.

Depending upon magnitude and content, an unexpected garbage dump discovered during site construction is bound to impact budget and schedule. Mitigation can entail excavating and backfilling a hole larger than required in order to chase and remove buried debris that extends beyond project boundaries. It can also entail encountering contamination and hazardous materials or penetrating the water table. With landfills, one thing is for certain—

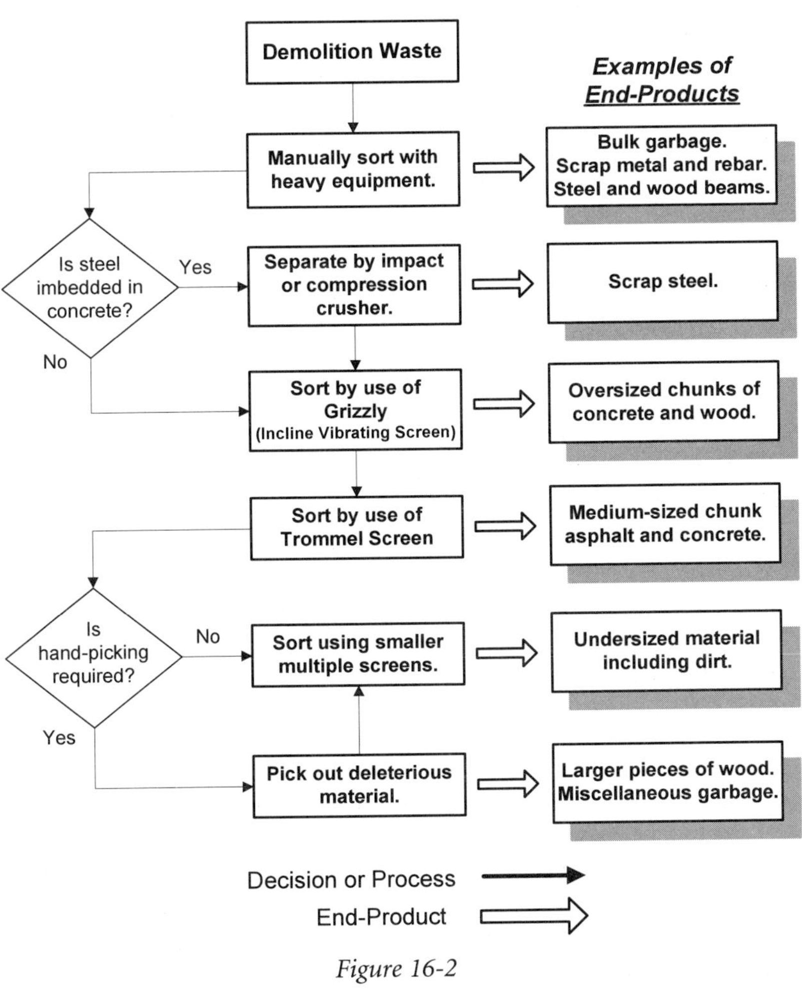

Figure 16-2

whatever is unearthed will be dirt impregnated.

There are few options available when dealing with piles of garbage intermingled with dirt. Cost as always, is the issue. The most obvious reaction is to excavate the mass and export it as-is to another dumpsite. Still, this approach does nothing to minimize cost, which would require keeping as much material on-site as possible. The solution is to separate the debris from the soil manually or by using a vibratory shaker screen. Although the quality and type of soil resulting from dump screenings may be questionable, at least the trash component can be isolated, segregated for value, and shipped to a dumpsite as a smaller volume minus the dirt component. Recovered dump or demolition dirt of structural quantity may be reusable as on-site fill. Non-structural areas can accept both topsoil and mixed unclassified materials. Rich, fermented organic topsoil may work well in planters. Use wood grindings for plating construction areas and haul roads, for erosion control, or export it in tightly packed and lighter, soil-free loads. Screened soil suitable for fill and placed back into the hole will reduce import requirements needed to complete backfilling.

Screening and handling operations can take as many forms as necessary to gradate material. Single-pass systems, including a vibratory screen, a grizzly, or trommel, typically handle one phase of material separation sending larger products over conveyors to piles. Multiple deck screens further refine material. Individual pieces of equipment can be daisy-chained together or used separately with other specialized machinery, such as a wood grinder, to produce any level of separation desired.

To limit the affect of moisture, immediately process dry dump material or cover and protect it in tight conically shaped piles. Lower soils value and the nature of buried debris make drying the dumpsite material by disking it unlikely. Other limiting factors that affect screening include type and size of debris, soil gradation, land constraints on stockpile and working area, and room to accommodate trucks and heavy equipment. Dust, noise, and the ability to work at night may affect working hours and impact production. On most projects, owners are going to want to quickly backfill an unexpected dumpsite to minimize schedule disruption and prevent the hole from collecting water.

Asphalt

According to the National Asphalt Pavement Association, asphalt is the most recycled product in America. It's recycled more than paper, aluminum, plastic, or glass. Asphalt and concrete can be broken into random-sized chunks of material by use of heavy equipment. When schedules are tough to predict and completing work has priority over cost, charging into asphalt or concrete with brute force and hauling away the debris is quick and easy. However, exporting chunk asphalt foregoes the opportunity to resize it into base material suitable for applications in site-work.

Asphalt can be resized in-place by grinding or cold planing. Either process produces mounds of gravel-sized material useful for base course or for recycling into new asphalt. Grinding and cold planing differ and are application-specific. Grinding completely removes entire portions of asphalt through to subgrade or subbase material. When asphalt grindings are meant to be left in-place as base material, it's wise to grind a few inches below the bottom of asphalt so that the grindings can knit themselves into the subgrade. Dipping hot grinding teeth into the earth also works to cool them promoting machine longevity. Grinding machines can be used to blend stabilizing agents, such as portland cement, fly ash, or dust inhibitors, into native subgrade.

Cold planing removes asphalt to a uniform depth leaving a trough ready for fresh asphalt or concrete fill. A belt system carries the ground material up and into waiting trucks. Grinding and cold planing are different applications giving the same result—they reform asphalt while producing mounds of reconstituted asphalt product ready for re-use elsewhere.

Given the right equipment and warm weather, ground asphalt can be laid, graded, rolled, and compacted for use in temporary haul roads, access aprons, fueling and maintenance pads, and job shack parking. Because reconstituted asphalt doesn't knit or compact as well as hot-mix asphalt, it's much less structurally sound, prone to leak, and can be difficult to place and shape on compound slopes and grades. Soil contaminants in ground asphalt will further compromise its integrity.

Other purposes for ground asphalt lie in blending it with clean granular material for use in fills and embankments. Inquire first as to whether your jurisdiction will allow the re-use of asphalt grindings on-site. When agencies view grindings as a contaminated product, a solution may be to use them as subgrade plating or base material in roads that are destined to be covered with hot-mix asphalt pavement. The savings that result from not hauling off old asphalt and importing an equal volume of base material can be significant.

Portable pugmills can blend crushed concrete or ground asphalt with granular material. Pugmills produce a uniform and well-blended mix of ingredients based upon volume or weight. Being able to produce an exact mix is important when meeting specifications for engineered and blended material. Pugmills are also handy for blending soil with cement treated base (CTB), or for mixing soil-bentonite mixes to meet specification for sealing ponds. Pugmills come in a variety of sizes and can usually be moved anywhere on-site.

Clean and large quantities of ground asphalt are also desirable for re-use in producing new hot-mix asphalt. Asphalt plant operators will usually pay owners to remove and haul off their asphalt whether it's in-place, in chunk form, or in piles of grindings. Deals typically depend on quantity of material and haul distance to the plant.

Concrete and brick

Recycling concrete is very popular. More tons of concrete are recycled annually than any other material and for good reason; concrete is abundant and useful for many purposes. In site-work, sources of raw concrete chunk material typically include sidewalks, curbs, obsolete utility vaults, catch basins and manholes, septic tanks, cement blocks, slabs, pavement, concrete structures, and building foundations. Concrete can be recycled in many ways. As a function of demolition, broken down and pre-sized chunks of concrete can be fed into portable crushers resulting in a finer crushed product. Two primary types of crushers used for on-site-work include impact crushers and compression (jaw) crushers.

Impact crushers are able to fracture larger material and separate concrete from imbedded rebar and wire. The speed and force exerted by impact crushers results in an end product high in fines. Alternatively, compression crushers accept smaller material and don't process concrete imbedded with metal as well. As opposed to impact crushers, compression crushers manufacture a more uniform rock-sized crushed concrete product with fewer fines. Together, both types of crushers compliment each other.

Fracturing slabs in-place by means of *rubblization* can also produce chunk concrete material. Rubblization is a construction practice that utilizes weight or a striking force to turn concrete into rubble. Depending upon the thickness of the slab, rubblization fractures concrete in pieces ranging from fine dust to approximately six inches (15 cm) in size. Upon fracture, rubblized material can be left in-place and re-used as base course, can be collected and used for fill purposes elsewhere, or can be fed into an on-site crusher for re-sizing. As with any crushed concrete product, rubblized material is a marketable commodity and may be desirable for other off-site purposes. Rubblization is a quick and cost-effective means of recycling slabs of concrete.

Like asphalt, crushed concrete is commonly mixed with free-draining structural material for use in fills or for base-course material. It may also have limited use in pavement foundations and as road surfacing. Unlike asphalt, concrete imbedded with rebar, mesh, and plastic is tough to handle and can be expensive to dump. If a recycling center will accept steel-laden concrete, it may be cheaper to dump it there and pay a *tipping fee* than to haul it to a landfill or municipal dump. A *tipping fee* is the term that recycling centers use when levying charges for accepting materials. Unfortunately, recycling centers aren't located everywhere as they're most commonly found in urban areas. The result is that dump fees for imbedded concrete can escalate.

Disposal costs aside, when comparing the expense of importing crushed pit aggregate to the expense of using on-site recycled concrete, weight and haul distance can play important roles. Crushed concrete material is lighter per loose cubic yard (LCY) than most crushed aggregates making it an economical product to haul. Lighter weight equates to being able to legally haul more volume of crushed concrete per truck than certain types of pit material or crushed rock. The result is fewer loads of material to do the same job. Also, consider the haul distance for importing pit material. The longer the distance between the project and the source of crushed aggregate or pit, the more economical it may be to recycle and substitute crushed concrete for import material.

As stated earlier, two factors exist that can justify crushing and using recycled concrete. First, there has to be a use or an off-site market for the end product, and, second, there must be sufficient raw quantity to make crushing economical to pursue. The rest is basic arithmetic. Compare the cost to haul and dump concrete and to import pit material against the cost to crush and use recycled material on-site. The analysis will prove one way or another whether recycling concrete is the owner's smartest option.

Depending upon it's content and quality, crushed brick may present more problems than it's worth. Freshly crushed brick, when exposed to weather, moisture, and the weight of construction equipment, may continue to fracture and break down into powder. Although it's colorful, powdery and soft chunks of brick or clay can work their way into the subgrade and weaken it or cause otherwise free-draining material to hold water. In site-work, unless they're used whole as bricks in non-structural ways, consider their reuse with the utmost caution.

Pipe

Pipe isn't easy to recycle. With the exception of ductile iron, steel, welded high-density polyethylene (HDPE), and newer reinforced concrete pipe, most pipe breaks up or disfigures when pulled out of the ground. As with any recycled product, there has to be enough available pipe and the demand for its reuse to make recovery worthwhile. Concrete pipe aside; pipe cannot be recycled into another usable product through processing. It's valuable only as pipe. Re-usability depends on length, diameter, material, prior use, and salvage condition.

Salvaged pipe is rarely re-used for its designed purpose. Typical applications include non-pressurized and less demanding uses such as temporary culverts, pond piping, tight-lining surface runoff, utility sleeves, or as a substitute for swales and ditches. The strength of concrete-lined, cast-iron waterline makes it superior pipe for utility crossings or for shallow temporary culverts in roads under construction or in haul roads. Cast-iron pipe can usually handle loaded scrapers and off-highway haulers without crushing. It comes out of the ground in long lengths, it's sanitary, and doesn't require much headroom when buried. Depending upon re-use, depth of bury, and loading, the higher the classification of cast iron, the better.

HDPE is versatile and reusable for many purposes in site construction. It's resistant to corrosion and salt water. Welded HDPE pipe can be cut to length for use as one-piece culverts or for tight-lining runoff over steep and erosive slopes. Long lengths of recycled HDPE pipe are useful for erosion control and handy for transporting materials. On one occasion, we used HDPE pipe to slide rock down a long slope hundreds of feet long to an inaccessible area suffering from erosion and water damage. Dragging long lengths of HDPE pipe are a challenge and can disrupt work. Therefore, it may make more sense to schedule "night drags" after working hours. Non-sanitary sewer, heavy-gauge bell-and-spigot PVC pipe also has use in erosion control or as temporary culverts. PVC pipe, however, can be fragile and easily fractures when pulled out of the ground. Given a choice, it's cheaper to remove, cut, drag, and place HDPE pipe than it is to carefully excavate and try to reuse segmented bell-and-spigot pipe.

Conversely, concrete pipe is short, clumsy, and expensive to lay. The fact that it's heavy and can chip or easily crack nullifies its use as anything beyond crushed base course provided that it's free of steel reinforcement. Ribbed plastic pipe, thin-walled PVC, and galvanized culvert are apt to bend, split, or crush when removed and should not be considered for reuse. Copper has off-site salvage value.

Besides being useful for on-site construction, it must make economic sense to recycle pipe. Long-length and large-enough-diameter pipe that has a high probability of coming out of the ground in good condition is typically a candidate for recycling. Short-length and smaller-diameter or heavy pipe that easily crushes, fractures, or disfigures isn't worth pursuing. Recycling pipe has to be worth the cost to carefully remove, haul, and stockpile it. Discard any pipe contaminated from passing unsanitary, flammable, or hazardous fluids or gases.

Hog fuel and wood chips

Logging, clearing, and grubbing are main components of site-work. From these activities flow a steady stream of woody debris, stumps, logging slash, whole trees, and brush. Volumes vary depending upon timber type, soils, moisture regime, and prior site history. The latter can add mounds of demolition and woody garbage to the mix. Clearing-debris can also include rock and topsoil mixed with wood. Refer to *Figure 16-3.* This flowchart illustrates alternatives from grinding logging and clearing material into *hog fuel* through processing

Processing Woody Debris Mixed with Dirt

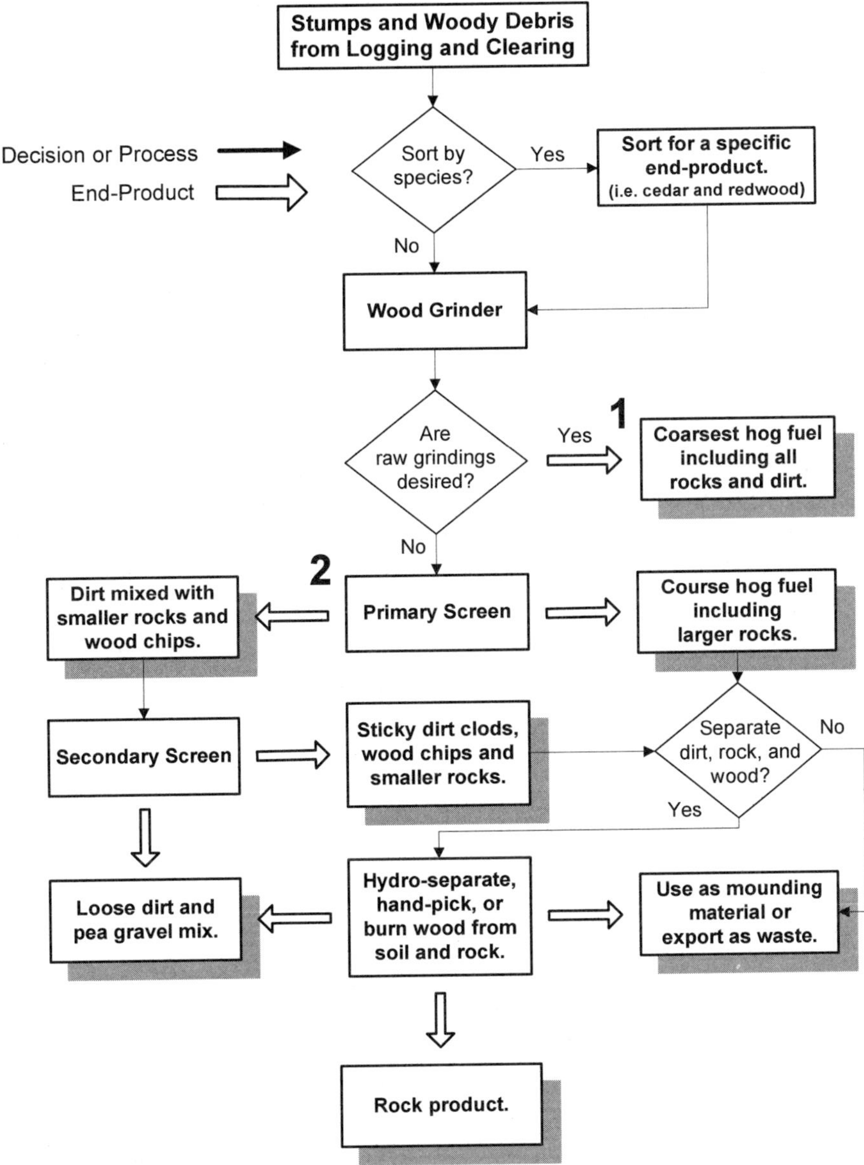

Figure 16-3

mixed wood and topsoil into more refined end products. *Hog fuel* is a term used to describe the larger wood shreddings produced by a wood grinder.

Clearing-debris and logging slash that is raked and put through a grinder will invariably contain a mixture of woody material, rock, and topsoil. The higher the degree of soil placed through the grinder, the dirtier the end product or hog fuel. As opposed to clean, woody hog fuel, dirty hog fuel is much less effective for construction purposes. Box number one on *Figure 16-3* illustrates the primary result of grinding clearing-debris—and that is hog fuel.

Sorting out rocks and dirt mixed with hog fuel or producing more refined wood products requires additional screens, chippers, or specialized equipment. For instance, landscape bark consists of smaller and cleaner wood chips, not hog fuel produced from a grinder. When aromatic, colorful, and weather-resilient wood species, such as redwood and cedar, aren't available, light-colored chips can be dyed a mulch color and used for landscaping and ground cover. Color can be added to wood during or after processing. Box two on *Figure 16-3* signifies the beginning of secondary processing that transforms rough hog fuel into other end products.

Besides grinding, options for dealing with woody debris include burning it, hauling it off-site, burying it, and mounding it for decomposition. Only the first three options work to maximize land use and minimize land degradation, and, of the three, only grinding can turn an otherwise negative cash flow activity into a profitable one by minimizing export costs and by producing useful recycled products.

On any particular project, until the aggregate value of on-site wood processing rises enough to offset grinding costs, short of shoving it into a preexisting hole and walking away, burning is typically the cheapest option for getting rid of woody debris. Burning is more weather-dependent than grinding and requires round-the-clock supervision, heavy equipment suitable for maintaining a fire—possibly a fan unit—and a source of fire control water. Both grinding and burning can negatively affect peripheral neighborhoods. As more jurisdictions prohibit burning, grinding and chipping will become more attractive. And—burning debris isn't the only issue on the docket at town hall. A movement is growing to limit hauling and dumping of demolition thereby leaving some owners with no option but to recycle what they can. In fact, some jurisdictions may *base* permits and approvals on mandatory recycling.

Anything woody can be ground or chipped including vegetation, stumps, logs and slash, pallets, crates, dunnage, scrap lumber, and carpentry waste. Unlike demolition and clearing work that produces enough woody debris to justify an on-site grinding and processing operation, wood waste that is sole-sourced only from new-structures construction rarely justifies mobilizing a chipper. In addition to volume, raw-material size dictates the type of wood-processing machinery required. Tub and horizontal grinders are production machines. They are well-suited for larger limbs, stumps, long butts, coniferous tops, heavy foliage, and dirtier material. Chippers handle smaller and cleaner material. Whereas grinders produce large shreds of hog fuel, chippers quickly produce dimensionally smaller chips that are more predictable in size and typically cleaner than grindings. Two types of chippers are available. Drum chippers can handle larger material and produce more variable and stringy wood whereas disc chippers produce cleaner and more-uniform chips suitable for sale as fuel. Brush that is too small or uneconomical to be processed through a grinder or chipper can be fed into a mulcher or mixed with larger material and ground. Desired end product affects the type of machinery used.

Hog fuel has many temporary uses in site construction. Primary uses include controlling erosion, plating soil, and armoring haul roads. Hog fuel is light in weight and, when supported by fabric, works to keep pumping wet soil from rising and contaminating a road's running surface. When laid at adequate depth, it can support the hauling of hundreds of thousands of yards of dirt in all types of weather. It also floats making it useful during wet conditions or when bridging wet areas without use of fabric. By keeping vehicles out of the ground, hog fuel helps to limit dust problems. Hog fuel saves owners time and money because it provides contractors with a volume of useful material and because it is an effective substitute for imported pit material or crushed rock normally purchased for building temporary haul roads. It decomposes and rots over time and that's okay since hog fuel is a temporary construction material. As opposed to raw material, decomposed hog fuel is significantly smaller in volume making it more efficient to haul off later in fewer loads. Depending upon locality, hog fuel or clean chips may also be marketable as fuel for cogeneration (the production of electrical power and thermal energy), for livestock bedding, for producing fuel pellets, and for wood products such as medium density fiberboard and particleboard.

It may not be required to remove woody debris from remote portions of, or long utility easements within, a project. When plans include processing and scattering wood chips in-place, a chipper with tracks may be the best choice for competing work. However, it's more common to haul woody debris to one or more central locations for grinding or chipping by a unit, towed into place. Processing operations require adequate area for: a grinder, a loader or excavator, piles of raw material, piles of processed material, trucks, and even more perimeter distance for safety. You cannot be too safe around a wood-processing operation as tub grinders can throw a chunk of wood or rock hundreds of feet in the air. Here's a recap highlighting ways that wood grinding can benefit project owners:

1. Hog fuel can be sold as fuel and is a useful on-site construction material.
2. An element of brush, weeds, and useless vegetation can be lost through grinding.
3. Wood grindings are a unique solution for plating wet areas during site construction.
4. Cost of producing hog fuel for temporary roads and erosion control can be offset by not having to purchase mulch or import rock and material.
5. Grindings are versatile. They're useful for surfacing temporary roads, controlling erosion, stabilizing construction areas, and, in deep landscape areas, providing subbase below more expensive imported landscape bark.
6. Government agencies may label stagnant piles of raw, woody material as solid waste. Eliminate this problem by promptly chipping woody material. A site free of woody debris is a site ready for permits, stripping, and earthwork.
7. Owners faced with hauling off woody debris will experience cheaper export trucking costs when grinding first reduces green material. Grinding or chipping consolidates material, reduces mass, and lowers haul weight by releasing water content. The result is more efficient loading, fewer well-packed truckloads, and lower dump costs.

It's not uncommon to view smoke rising from piles of hog fuel particularly on cool days. Due to internal combustion, large piles of dry and compressed hog fuel are a risk of catching on fire. Watch stationary piles of grindings or chips daily. If they appear warm or begin to smoke, piles should be broken down, wet, exported, or put to use. It's also prudent to keep woody debris clear from the sides of a chipper in-use. A hot machine can ignite woody debris piled against it.

Logs

Unmerchantable logs have use as firewood, grindings, or decorations. Species, diameter, taper, and length determine log value but they also govern construction use. Logs long enough and uniform in size are handy for lining trails, accenting parks, cribbing topsoil, or as *corduroy* in temporary haul roads. *Corduroy* is a term used for burying logs side-by-side and perpendicular to centerline in a roadway. As part of subgrade, corduroy can stabilize a temporary road and support loads over wet and sloppy soil conditions. Stumps are also useful for bridging chronic wet spots in temporary roads. Here, stumps placed upside down in a wet hole can provide a quick solution for keeping trucks moving. Stumps and whole logs are popular materials for creek, habitat, and wetland restoration.

When owner's posses enough quality logs and an on-site need for rough-sawn lumber and beams, they may consider milling their own wood. Mobilizing a portable on-site mill is an option when:

- Insurance isn't an issue.
- The price of alternative lumber is high.
- Noise and wood storage aren't problems.
- Log-grading services are readily available.
- Drip-treated and preserved wood isn't required.
- On-site lumber needs are high and log prices are down.
- There are enough quality logs to make the venture worthwhile.
- Large-dimension or custom-cut rough-sawn lumber and beams are desired.
- There's ample room to set up a temporary mill, run equipment, and sticker-dry wood.
- The project is located on a remote site or is a significant distance away from a commercial off-site mill.

Carefully handle the selling of logs to vendors other than off-site commercial lumber mills. For example, selling logs for firewood is fine provided that the wood is loaded and hauled to an approved *off-site* location for manufacture and resale. Firewood cutters can be an unpredictable lot showing up at all hours in unattractive rigs and sometimes with no regard for safety. For the sake of liability, project owners and their contractors should not allow firewood cutting anywhere on the project. Insist that only on-site licensed and insured contractors handle loading and hauling of firewood.

Owners may also find that there's risk in being benevolent. For example, free loads of logs delivered to a municipality for stream restoration purposes or for charity can end up in the wrong place at the wrong time. Unfortunately, if the person accepting the wood is irresponsible, working with on dysfunctional schedule, or clueless as to what to do with the wood, project owners may be asked to reload and rehaul the wood elsewhere at their own expense. Owners can protect themselves by writing up a short agreement with whoever is receiving the logs. Set the price or terms for delivery, or either require that the receiving party hire its own self-loading log truck to pick up the logs or arrange to have your on-site contractor load their full-sized logging truck for them when it shows up.

Pond water

Potable water has traditionally been the answer for controlling dust and moistening soil to make it compactable. Things however are changing. Water is becoming a guarded commodity. Increased potable water costs combined with decreased availability is forcing owners to search for other sources of water. With enough planning and foresight, one solution lies in manipulating surplus on-site storm water.

Projects large enough to support storm-detention ponds, vaults, or infiltration galleries may hold an expendable oversupply of non-potable water useful in construction. Using excess storm water for construction purposes saves money and is convenient. Transferring water is quick and easy. Pump the surplus water directly into tankers from storm ponds, vaults, or infiltration galleries, or erect an elevated stationary tank and pump water up and into it. Elevated tanks allow drivers to fill at their leisure by pulling under the tank and loading their trucks with water by gravity. Savings result from not having to purchase potable water, rent meters, or haul water in from off-site. A potential side benefit of this activity is that dead and sometimes, live water storage capacity in facilities can be increased. Stormwater facilities must be built prior to construction and may require regulatory approval for diverting surface water. Stable and vegetated facilities are preferred since they will yield cleaner water than will freshly cut ponds. In order to make this strategy worthwhile, facilities must receive adequate runoff and storm flow. Install valves and flow-control devices where balancing storm-water levels and monitoring of facilities is required.

Depending upon where the storm ponds are located, pond water may be used for irrigating on-site nursery stock and peripheral pond plantings. Provided that reliable and redundant portable pump facilities are available, pond water may also be useful as a backup source of water for supporting combustible building construction and wildfire control.

Hydric soil

Hydric soils are recyclable and can be quite valuable. Refer to the chapter titled "Wetlands and Other Sensitive Areas".

Screening topsoil and rock

Stripping topsoil and organics from inorganic material is a prerequisite for foundations and structural earthwork. Estimating the stripping volume can be difficult and getting rid of strippings in a cost-efficient manner isn't easy. As a result, stripping can easily overrun in cost.

Exporting raw topsoil strippings and using the same trucks to back-haul import material is expensive. However, one-way export hauling is even *more* expensive. Owners need options when faced with the costs for exporting raw topsoil and importing large quantities of blended topsoil. While most owners can't escape importing blended topsoil altogether, techniques such as topsoil processing can dramatically reduce the costs of exporting raw topsoil and reduce, or at least minimize, the volume of blended topsoil imported.

When deciding whether to process native on-site topsoil, begin by finding out if the topsoil needs to be screened or if it's usable in its current native raw state. Topsoil rich in nutrients, loamy in nature, and low in roots, woody debris, and rocks, may be suitable for limited on-site landscaping as it sits. However, it's doubtful.

Unscreened raw topsoil containing roots, organic matter, sticks, and rocks is difficult to work and doesn't provide suitable growing medium. The greater the proportion of "foreign" materials in raw topsoil, the worse it is for landscaping and growing stock. Topsoil with limited pea gavel and sand *may* work for landscaping if it is free of sticks, larger rock, and other deleterious material. But, it also must drain properly and possess enough organic material and nutrients to support plantings. The chances of an unscreened sample of topsoil meeting the above criteria are slim.

On-site topsoil processing entails using a portable screening plant to separate raw topsoil from woody debris, rocks, roots, coal, and clods. Screen size determines end product. Optional multiple deck screens can further segregate material into various sized products. *Figure 16-4* shows a configuration that was used on-site to process more over three million tons of raw topsoil strippings and mixed unsuitable material. The project was a master-planned, urban-type development built partially, on steep virgin land and portions of a retired open-faced rock and gravel pit. The mine tailings processed were primarily topsoil, but they contained a rich proportion of quality rock—perfect for screening and crushing.

Refer again to *Figure 16-4.* The process begins by first separating larger material from raw topsoil and, then, progressing to finer and finer mesh screens until the desired product is produced, or, until it's no longer possible to separate the remaining tailings of wet material, clods, wood, coal, and rock. The wood is further separated by burning, handpicking, or by being floated apart in water. In some areas, burning off wood may be prohibited, may require special permits, and may be useful only during certain portions of the year. Tediously handpicking out large woody material is expensive and inefficient but it works. Soaking or working the mixture mechanically in water can be a cost-effective solution for separating wood from rock. *Figure 16-4* also illustrates

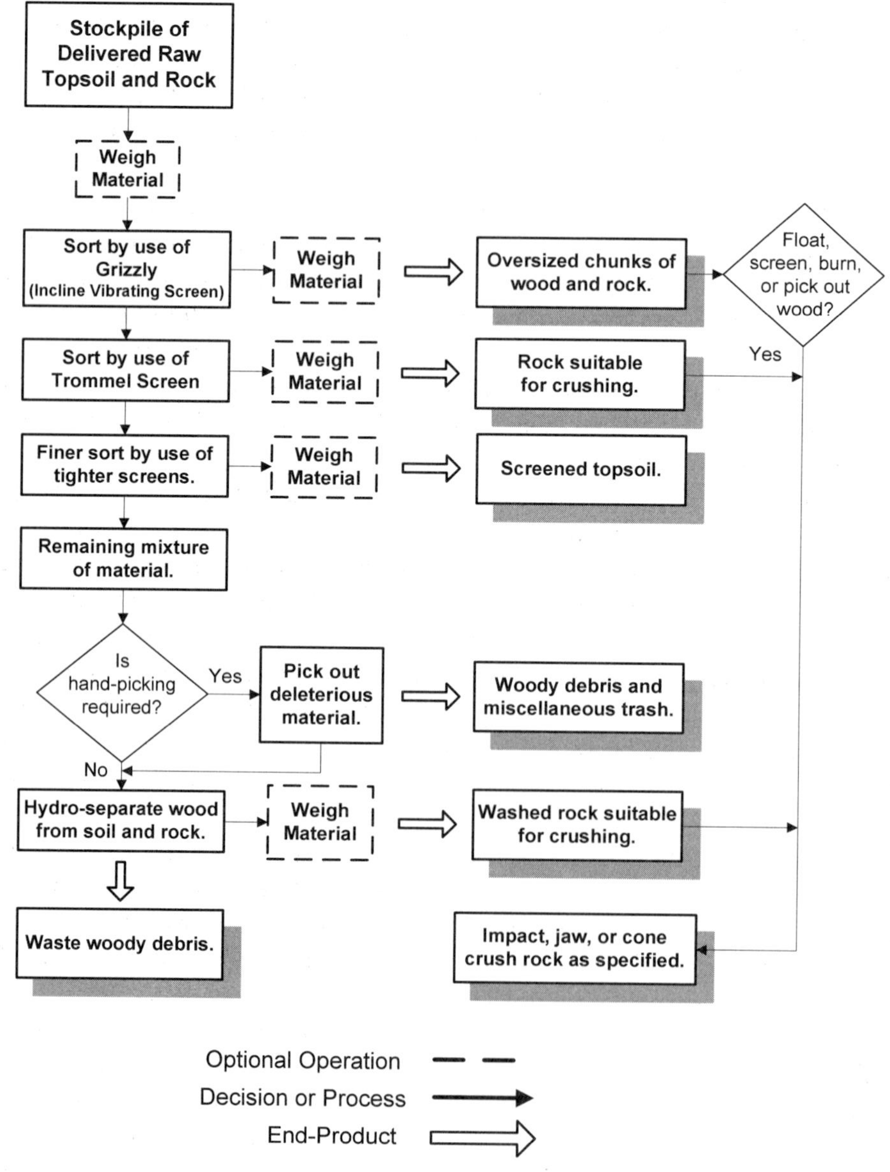
One System of Processing
Raw Topsoil and Rock
Stockpile of Delivered Raw Topsoil and Rock
Weigh Material
Sort by use of Grizzly (Incline Vibrating Screen)
Weigh Material
Oversized chunks of wood and rock.
Float, screen, burn, or pick out wood?
Yes
Sort by use of Trommel Screen
Weigh Material
Rock suitable for crushing.
Finer sort by use of tighter screens.
Weigh Material
Screened topsoil.
Remaining mixture of material.
Is hand-picking required?
Yes
Pick out deleterious material.
Woody debris and miscellaneous trash.
No
Hydro-separate wood from soil and rock.
Weigh Material
Washed rock suitable for crushing.
Waste woody debris.
Impact, jaw, or cone crush rock as specified.
Optional Operation
Decision or Process
End-Product

Figure 16-4

belt scales for each end product. This is important for a couple of reasons. First, contractors are paid by tons processed. Scales verify weight. Second, weighing material daily provides owners with a running account, which allows the team to anticipate quantities and plan how and where each end product can best serve the project at that time.

Although not always possible, raw topsoil should be left in place until it can be processed. Moving topsoil prematurely to stockpiles can expose it to moisture and other wet soils. Topsoil high in clay and silt is tough to screen especially if the material is wet. Wet clay balls up and sticks to screens limiting an operator's ability to process topsoil on the first pass. Second and third passes may be necessary before material will screen and, even then, some waste is inevitable. Costs increase with increased passes.

Material too wet or sticky to screen or that isn't worth additional processing time may be marketable as *mounding material.* A by-product of screening, mounding material is typically a mixture of wet topsoil, clay or silt, small woody debris, and fine-sized rock. A "cocktail" that is too wet to screen is going to be too heavy to rake making mounding material a prime candidate for deep nonstructural fills such as subbase planter material. Mounding material is still largely organic and, certainly, a better choice for deep landscape fills than inorganic clays and silts.

Crushing may not be the best option for screened rock. Larger rock often requires the use of multiple crushers to break it down into a smaller crushed product increasing the processing price per ton. Another option for using oversized rock includes hand-placing in streams as riprap, by armoring ditches, building check dams, and for general erosion control use. Larger rock placed with finer binding material can work for supporting deeper sections of a road fill. The savings in using oversized screened rock in lieu of imported quarry spalls or other larger-sized crushed material can be startling. Uses for smaller screened rock include pipe bedding and drainage material. By using screened rock straight off the conveyor, owners avoid greater processing fees and avert the time, schedule, and working area needed for mobilizing additional crushing equipment. Presorting eliminates larger rock and debris from going through and possibly damaging the plant. Handle boulders as described in other sections of this book. Burn, chip, or grind large woody debris.

Use of screened products offsets the need to purchase similar quantities of alternative imported materials to do the same job. Here's a summary of end products and uses for uncrushed materials common when processing raw topsoil:

- Ground woody debris → hog fuel, woody mat material, erosion control, and landscape chips

- Topsoil → planter fills, wetland enhancement, park and ballfield fill, and finish landscaping
- Washed pea gravel → pipe bedding and additives for mixing small batches of concrete
- Washed round rock → pipe bedding and drain rock for walls and French drains
- Oversized rock → erosion control, check dams, and deep subgrade ballast
- Mounding material → deep planter fill and non-structural fills

Topsoil screening and on-site rock crushing requires operators who know what they are doing. I have had the pleasure of presiding over a couple of portable topsoil-screening and rock-crushing operations as well as being part of a team that removed, barged, unloaded, and erected a rock crusher, a concrete batch plant, and an asphalt plant in a remote corner of Alaska. All of this has taught me that having people with enough experience to overcome day-to-day operational problems is mandatory. So is having the right equipment for the job. Bill, a recycling contractor of ours is considered by many on the west coast to be the best anywhere. He built custom hydro-machinery (out of older used equipment) to separate rock from woody debris that performed so well that heavy equipment manufacturers climbed all over the project to see how he did it. Bill and his crew handled millions of tons of material from screened topsoil and oversized rock to woody debris, which he removed, and rock, which he crushed to the Washington State specification. Having abundant crushed rock allowed us to proceed with a thicker lift of base course and a thinner lift of asphalt than specified, yet still provide an equivalent, but much cheaper, road section. But best of all, the rock stayed on-site resulting in less off-site trucking with no export-and-dump fees. Savings compound with smart recycling.

Besides employing good people and having the right equipment, topsoil and rock screening operations require certain conditions to succeed. Here are a few:

- Have an erosion-control plan and enforce it.
- Preplumb power, phone, and water to the site.
- Measure performance by installing belt scales.
- Plan to build a settling pond to detain muddy runoff and to treat it with an outlet system designed to handle overflows.
- Don't allow on-site contractors to "beg, borrow, or steal" washed product for petty uses and wasteful applications. When used on-site, processed materials are only valuable if they replace something of more value.
- If the site-work contractor has windrowed or piled dry raw topsoil but cannot deliver an even supply to the processing operation, permit the

screening contractor to load and haul whatever volume of material he needs to maintain production.

- The sticks and woody debris resulting from screening can be used as hog fuel for building dependable temporary roads. Because mats of intertwined sticks and wood are buoyant and elastic, they work well in bridging wet areas and plating access roads over raw topsoil.
- Provide ample room for processing, for stockpiling raw and finished topsoil, and for handling incoming and outgoing truck and trailers. As a rough rule-of-thumb, small processing operations require at least three to five acres (1.5 hectares) of *flat* land plus room for a temporary settling pond.
- Maintain a consistent supply of dry raw material. This may mean balancing stripping with topsoil screening or delaying stripping until the advent of dry weather. The key is to deliver raw topsoil at a rate commensurate with its processing, leaving minimal amounts in stockpiles to consolidate or spoil with moisture. Stockpiled wet material mixed with freshly stripped dry topsoil will screen and process better.
- Maintain a well-trenched and sloped work area that sheds water away from all activity and stockpiled material. Standing water wicks very easily into topsoil making it imperative for operators to maintain proper drainage around stockpiles. Don't be fooled if a topsoil processing operation is set on free-draining native material. It doesn't take long before free-draining native soils clog and evolve into a wet sloppy mess.
- Continually work to reduce the water-holding capability of stockpiled topsoil. "Track-walking" the slopes of stockpiles with a wide-track bulldozer will consolidate material and promote positive drainage. Build piles in a conical, geometrically efficient manner and, where possible, roll and compact them. This also means placing piles so that they can be safely manipulated, are accessible to loaders, and do not channel runoff back into the floor of the processing operation.
- Monitor the material brought into the screening operation. Strippings must be clean with woody debris preferably absent or at least no larger than one-inch (25 mm) in diameter. Stripping operations that result in overexcavating and mixing topsoil with inorganic material produce a lower grade of screened topsoil. Silts and clays can ruin topsoil by choking it of nutrients and robbing it of its ability to drain. They also gum up the screens, belts, and everything else that they come in contact with thus impacting production.

Quality imported topsoil is a blend of topsoil, mulch, and nutrients. By using a pugmill, the same quality product can be produced on-site by blending processed raw topsoil with additives, such as imported sand, mulch, or compost. However, issues need to be addressed before deciding whether to amend on-site, screened topsoil. Here are a few:

- How will blending occur?
- What's the plan for enforcing quality control?
- Who will specify amendment percentages or weights?
- How will the operator prevent amendments from being contaminated?
- Who will order, deliver, and handle billing for imported amendments?
- Who will test blended topsoil for quality and draw up MSDS information?
- Storage of more valuable end products such as blended topsoil can mean a higher risk of theft. How will the site be secured?

Screened and unblended indigenous topsoil may be preferred when native stock or wetland species are being transplanted on-site. Check with your landscape architect, wetland specialist, or landscaper before committing all raw screened material to blending. You may want to save some.

Crushed rock

Rock size determines the speed and the cost of crushing. Large chunks of blown rock and boulders must first be reduced in size before they can fit into a portable impact crusher. Compression crushers require even smaller-sized rock than do impact crushers further increasing costs to reduce large rock. Rock produced by an impact crusher may not be uniformly small enough resulting in the need for more screens and a compression-type jaw or cone crusher. In all, taking large rock down to a small crushed product may require at least three separate operations. All of this costs money and takes time. Although the return on crushed on-site rock can be substantial, here are items to consider before proceeding:

1. Is there an off-site market for crushed rock?
2. What quantities of crushed and washed rock do you need?
3. Can you run a crusher without exceeding allowable noise levels?
4. Will the bridges to your project support trucking in crushing equipment?
5. Does the cost to purchase and import crushed rock justify on-site crushing?
6. Does it make sense to import additional off-site rock or chunk material for crushing?
7. Will the on-site rock fracture without degrading into a pile of fines or non-angular material?

8. Is native rock too porous for use as base course, or, if sold, for producing asphalt or concrete?
9. Can your contractor meet the specifications for crushed rock? If it will cost more to crush to specification, is it worth the expense?
10. Do other on-site materials such as concrete or asphalt require recycling? The more volume put through a crusher, the lower the cost per unit.
11. Do you have enough washed rock to make crushing economical? What is the break-even volume to justify mobilizing a crushing operation?
12. Pea-gravel sized material may be too small or may not fracture adequately to meet crushing specifications. Larger material works better for producing smaller crushed rock. Evaluate the value in mobilizing additional machinery capable of crushing smaller rock.
13. Do you expect to crush ripped or blown bedrock or boulders? Extra care in blowing or ripping rock to attain fracture that is suitable and sized correctly for hauling and crushing can be expensive. What is the bottom-line aggregate cost-to-benefit ratio when dealing with bulk rock?
14. Evaluate timing. First, will a capable mobile crushing contractor be available when there's sufficient washed rock to proceed. Next, will the contractor be able to profitably sustain operations given the amount of washed rock available or being produced by ongoing topsoil screening? And last, can the contractor produce crushed product fast enough to handle incoming flow and still satisfy contractor needs?

Pit development

Developing an on-site pit can be a very smart strategy provided it makes economical sense to overexcavate a hole and then backfill it. Pit development requires having ample material of high-enough quality to justify overexcavation. The real leverage, though, for land developers comes in creating a place to dump unsuitable material or other less desirable on-site fine-grained soils. Pit development can be a mistake if the market value, quantity, or quality of on-site material is overestimated. But, higher penalties exist if the hole is backfilled with the wrong material or in a manner inconsistent with future site requirements. While it can make sense to overexcavate projects with marketable and useful sand and gravel, projects with abundant clay, excess fine-grained soils, and topsoil can usually use a place on-site to lose material. When both desirable and awful soils exist on the same project, conditions are ripe for developing a pit. Large multiyear projects having the time and area available to accept a flow of clean and suitable material from off-site locations can produce even higher financial returns.

When it makes sense, pit development is a brilliant move that, on one hand, creates cash flow and, on the other, saves both on export-and-dump fees. Consider at least the following constraints before developing an on-site borrow pit or a temporary commercial dump:

- Will the presence of a pit penalize sales and leasing?
- Can the local roads handle increased two-way truck traffic?
- Can you get permits for a borrow pit that will double as a dump?
- What about the water table? Will an excavation intercept subsurface water?
- Will safely laying back the slopes still allow for sufficient material recovery?
- Can you isolate the land long enough to make the operation financially successful?
- What dumps are you competing with? How close are they and what do they charge?
- Can cheap dump fees and quality pit material be used to entice buyers and builders?
- Will the operation profit enough to cover final land-reclamation costs such as landscaping?
- Is there enough off-site demand for a commercial dump or would it only serve your project?
- Is the on-site material consistent enough in quantity and quality to justify wholesale removal?
- Will the excavation drain? Are there places to pump water in an environmentally acceptable manner?
- Can a pit be developed large enough and remain in operation long enough to handle the project to build-out?
- Are non-structural areas available on-site for overexcavation such as a park, playground, golf course, or ball field? If so, does the shape of these areas support economical mining? For instance, long and narrow or triangular areas don't work as well as round or square areas.

Digging a hole with the intent of filling it comes with risk. Once a hole is created, an owner is halfway to completing that site. If infrastructure shifts or land use changes, the cost of backfilling and bringing the hole up with granular material in compacted lifts can be tremendously expensive. Examples of other risks include topographical error or survey mistakes that allow overexcavation to bleed laterally into adjoining roads and buildable area; incomplete fills that impact project schedule; or the accidental dumping of contaminated soils or hazardous materials.

There's also the upfront geotechnical cost and expense of boring to verify subsurface conditions and the cost to monitor fill and take compaction tests.

Not all sites possess homogeneous soils. Many projects are simply a mixed bag of materials ranging from expansive clays to pure rock and gravel. On projects with variable soils, where possible, plan non-structural areas, such as parks and ballfields, above veins of quality material suitable for mass borrow. Non-structural areas are cheaper to fill since they can accept any type of material, require little or no compaction, and serve as places to dump unsuitable on-site material year-round. Non-structural areas typically are also non-revenue-producing parcels making them likely targets for development later in a project—all the better for a pit location! Locate pits early during project planning, as they will become control points.

General practices for on-site processing and recycling operations:

Regardless of the type of processing operation considered, here are some tips to help ensure a successful recycling operation:

- Keep any on-site operation simple.
- Maintain a clean working environment.
- Maintain piles of material in the smallest footprint possible.
- Have a plan for cleanup and closeout when processing ends.
- Budget costs for mucking out and backfilling settling ponds and ditches.
- All operations must be held to a hazmat plan and maintain spill-response kits.
- Consider hydroseeding topsoil piles that lay dormant for extended periods of time.
- Institute a program *early* in the project to ensure that site contractors take recycling seriously.
- Appropriate hard hats, ear protection, and safety gear should be readily available for visitors.
- Profitable recycling only works if there is ample volume of high-enough quality material to justify it.
- Set up easy-to-access, project-wide, and builder-friendly collection bins for woody recyclable materials.
- Obligate and provide incentives to on-site buyers and their contractors to purchase your recycled materials.
- Keep raw and processed topsoil in a dry and usable condition. Contamination nullifies the use of recycled products.
- Keep raw material considered for recycling pure, clean, and uncontaminated during extraction, stockpiling, and hauling.
- If setting up an on-site portable concrete or asphalt plant is an option, be prepared to handle issues pertaining to water and air quality, site

security, noise, smell, traffic impact, and hours of operation.

- Something's wrong if the processing contractor is constantly rescreening topsoil. Raw material is being mismanaged, or it's coming out of the ground too wet, or there's too much silt or clay in the topsoil, or the contractor has inadequate equipment to do the job. Or—all of the above.
- Large utility jobs that produce a high percentage of granular material mixed with fractured rock may justify mobilizing a bedding machine. Following trench excavation, padding and bedding machines can screen trench material to produce volumes of finer native material suitable for immediate re-use as pipe bedding and trench backfill.
- Stockpiles of fine material require more protection from water than do piles of courser material. Make sure that piles of fine material are adequately drained to shed water. Also, slope all operational areas to ensure positive drainage. Hard surfaces are superior for protecting stockpiles from ground moisture and contamination, however pavement can also quickly channel water into places where it can do damage.
- Errors can proliferate when measuring material by the truck cubic yard. Errors are a certainty when measuring material across inaccurate or miscalibrated weigh scales or by visual observation. As a check, employ a third-party surveyor to cross section stockpiles on tight intervals. Compare the gross processed tonnage or the cross-sectioned loose cubic yards in stockpiles to the raw bank cubic yard strippings as surveyed in the field. Then develop conversion factors.
- Hiring a processing contractor goes beyond bid. The comparison spreadsheet *(Figure 16-5)* is an easy way to consolidate and compare contractor conditions for performing work. Recycling contractors differ in their style, their approach to processing, and their utilization of resources. Besides price, examples of criteria to consider include each contractor's experience, erosion control record, quality and diversity of processing equipment, number of acres required to work and stockpile material, and guaranteed time commitment. From a contract and qualification standpoint, choosing a recycling contractor is no different from choosing a site contractor with one exception; the entrepreneurial nature of recycling makes everything negotiable. An operation that creates products of inexact value from approximate quantities of various raw materials can be contracted with, and proceed on variable terms. Flexibility can be the key to making recycling and processing work.

Comparing Processing Contractors

	Screen King	Bobby Dee Trucking	Recycling Bros.
1. Minimal number of acres required?	10 acres	15 acres	4 acres
2. Minimum time commitment required for a processing contract?	4 years	5 years	3 years
3. Will contractor share in cost of access road maintenance?	None	Some	Some
4. What type of material would contractor import for use?	Compost	Saw Dust	Compost & Manure
5. What hours of operation would contractor expect to keep?	14 hours	11 hours	Open to negotiate
6. What days will contractor want to work?	7 days	6 days	6 days
7. Will contractor purchase screened topsoil from owner?	No	Possible	No
8. Contractor's price for delivering processed topsoil on-site for owner:	$5.50/ tcy	$7.50/ tcy	$6.50/ tcy
9. Price for processing topsoil per pile measured loose cubic yard.			
3-way topsoil.	$3.90	$4.10	$4.25
2-way topsoil.	$4.00	$4.25	$4.25
Screened unamended topsoil.	$2.75	$3.00	$3.10
Winter mix topsoil w/ heavy sand.	$3.50	$3.75	Varies
Planting topsoil soil mix.	$4.50	$5.25	$4.90
10. Price for processing and crushing rock per weighed ton.			
5/8 inch minus crushed.	$4.75	$5.00	$5.50
1 1/4 inch minus crushed.	$4.75	$5.00	$5.25
Washed pea gravel.	$3.60	$4.00	$4.25
1 1/2 washed rock.	$3.60	$4.15	Varies
11. Will contractor supply a cost free loader to load raw material on-site?	Possibly	Yes	Yes
12. Will contractor supply a cost free truck to haul raw material on-site?	Possibly	Yes	Yes
13. Will contractor handle their own permits and approvals?	No	Yes	Yes
14. Will contractor request purchasing processed material at cost?	Yes	No	Yes - pea gravel
15. Will contractor load and haul tailings on-site as directed by owner?	Yes	Possibly	within 100'
16. Will contractor clean up and grade area as directed when leaving?	Yes	Yes	Yes
17. Will contractor share profit with owner if tailings are processed?	Yes	Yes	Yes
Owner conditions for all contractors:			
a) Owner will deliver raw material to stockpile as directed.			
b) Any raw topsoil on-site will be considered suitable for processing.			
c) Contractors are permitted to import sawdust, woody material & compost.			

Figure 16-5

The mix of recycling equipment needed for processing recyclables depends on the type and volume of material being handled and the desired end product. Here are some common processing operation configurations:

For wood processing:

- A grinder
- A chipper
- A track excavator or loader
- A water truck (optional)

For screening raw topsoil from wood and rock:

- Float tank or hydro screw

- Conveyor belt systems in and out
- Bulldozer (preferably wide track)
- Track excavator or front-end loader
- Primary grizzly feeder/ trommel screen
- Secondary deck screens for finer grading
- Optional–scales, wash plant, wood grinder, power plant, lights, and job shack

***Add* for amending topsoil or for blending materials:**

- Pugmill or portable mixer
- Disk or harrower attachment

***Add* for rock or concrete crushing:**

- Wash plant
- Conveyor belt system
- Jaw, impact, and/or cone crusher
- A second front-end loader, usually rubber-tired
- Elevated hopper (optional)
- Cone crusher for oversized (optional)

Off-site sales

There's much more to handling off-site sales than simply loading trucks and collecting checks. Set up a sales procedure and an accounting system before putting end product or recycled materials on the market. Items sold individually, such as pipe, steel beams, or decorative boulders, typically aren't a problem. They're a known quantity, their price is set, and transactions are quick and easy. Weighed and bulk items, such as crushed rock and screened topsoil, can pose a different set of issues. Here are tips to consider when selling bulk-recycled material:

1. Develop a sales schedule for quantity and wholesale discounts.
2. Require all material purchasers to complete a credit report and application.
3. Off-site sales may require providing a full-time loading and trucking component.
4. If volume and sales justify the cost, consider renting on-site weigh scales for trucks.
5. Define early how agreements will be structured and which party will draw up the contract.

6. Set up a system that continually tracks sales progress, delinquent payments, and accounts payable.
7. Protecting valuable end products may require investing in security fencing, night lights, and locked gates.
8. Leave nothing to chance. Erect appropriate signage that dictates policy and driving directions for all truckers and visitors.
9. Determine policies for measuring bulk sales and amendments received. In the absence of truck or weigh scales, options include measuring by truck cubic yard or by the loader bucket.
10. Provide and post lab results for all materials that require testing. Periodically test end products to ensure that items, such as wood chips, topsoil, and crushed rock, are per specification and what purchasers expect to receive.
11. Develop a plan for advertising, product sales, and cash flow. Perform continual topsoil mailings such as flyers, to contractors, landscapers, schools, government agencies, landscape architects and recycling agencies. Tell them what you have.
12. Provide a weatherproof register book for all truckers to sign. Have them include the date, time-in, time-out, driver and company name, phone number, rated truck volume, truck number, and account status. Locate this book in a kiosk with other pertinent information.
13. Heavily discount stockpiles of burdensome material. It's better to receive below market fees for material than to give it away, and it's always better to give material away than to pay to get rid of it. Don't play games with the market. The time and cost required to dispose of unwanted surplus processed material is burdensome and can discount upfront savings.

Chapter 17

A WORD ABOUT NATIVE TREES

Sales and marketing people know that people will pay more for property that possesses mature, healthy, and attractive trees. Depending upon the species, season, and latitude, trees can increase energy savings in buildings, improve air quality, attenuate rainfall, improve site security, provide screening, and filter noise. In land development, leaving native trees can result in lower costs for clearing, grubbing, chipping, grading, importing topsoil, and landscaping. Many project owners consider native-tree retention a key objective provided that it doesn't impact lot value or penalize net developable area.

Some jurisdictions impose strict ordinances and regulations concerning tree retention. Inflexible laws that govern the preservation of native trees without considering the economic impact to owners or whether such vegetation is even salvageable can be problematic. As an example, consider a group of native trees requiring preservation as a condition of approval. If grading lowers the area around them, trees left on higher ground have a tendency to blow over or suffer from lack of moisture and exposure. Trees left in a depression due to grades rising around them are likely to collect too much moisture, drown, rot, or have their outer roots crushed by fill. In either case, if the trees perish, the intent of the permit will falter and the permit agency may hold the owner or contractor accountable even though the same agency reviewed and approved the plans that depicted grade changes around those trees. In reality, the trees tabbed for retention may have been the wrong species to save or they may have pre-existed in a condition that rendered them unable to adapt to change, regardless of grading. The grading design may have also been in error or the base topography may have been off enough to result in deeper cuts or higher fills than anticipated. Or, a change in planning and layout subsequent to the original permit may have occurred resulting in wiping out the stand—and nobody noticed until the damage was done. Something changed and now, the owner is in violation of the development agreement or the grading permit.

The fact is, trees aren't like boulders, artifacts, and water. They live, die, and reproduce according to the dynamics of their environment—and land development changes environments. This chapter addresses trees as they relate to land development by framing the principles required for better ensuring native tree survival.

Want trees? Then plan now.

As prevalent and popular as native trees can be, they usually don't receive the level of attention required during initial planning. Nevertheless, it's during the early phases of site layout that action to save native trees is most effective. Unless a site is to be completely cleared and graded, it's critical to integrate strategies for native tree preservation with overall planning. Tree retention plans are important base templates because save trees become control points that affect layout, grading, and road location. All consultants and, thus, contractors must understand where trees are to be preserved. It only takes one errant design, one innocent slip in something remote and unsuspected—such as in the erosion control, dry-utility vault, or retaining wall plans—for a contractor to mistakenly eviscerate trees and shrubs slated for retention.

Tree retention and location survey plot plans accurately describe the exact location, species, and approximate size of trees desired for keeping. This work includes evaluating the health of all "significant" trees on the site to determine if those trees are worth retaining. Significant trees can mean many things to many people, but they're typically trees worth saving because they meet criteria established for political, environmental, historical, biological, aesthetic, practical, or personal reasons.

Many tree retention plans only consider plan view. This practice is fine on flat terrain, however, as ground steepens, not considering slope can result in undercutting or overcutting dripline and impacting root systems. Test the topography plan by taking ground elevations around save trees. How the ground cuts-and-fills relative to save tree roots should influence grading design. The value created and the landscaping cost saved by not having to replace native trees and vegetation should justify the added survey expense. As topography increases, an owner's chances for saving native trees can also diminish since earthwork, road cuts, and underground utility trenching performed on steeper ground can alter surface runoff patterns and intercept subsurface water flow. Grading that works to release sources of water, or to impound it, can dry-out or oversaturate tree root systems. In addition, opening a forest on sloped ground may unnaturally expose the wrong species to increased sunlight and wind resulting in sunscald or mechanical damage to limbs and tops. Any of these factors lessen the likelihood of native tree survival.

The flatter the ground, the easier it is to save native vegetation and trees, and possibly forego some of the expense of importing topsoil, irrigating, and replanting areas cleared. Road networks on mild terrain usually can lead anywhere which allows development to contour around individual or groups of save trees (or leave trees) with minimal grading and disturbance. Under these conditions, some species may not only survive—they may flourish. Attributes that should be considered when evaluating save trees include:

- Soils
- Health
- Species
- Appearance
- Hazard potential
- Damage or defect
- View corridor impact
- Risk of combustibility
- Root system condition
- Balance or percent of canopy
- Ecological position for long-term survival (An arborist or forester can address this issue.)
- Probability for being able to adjust or thrive in response to sunlight, wind, and moisture change.

In general, the smaller the tree—the easier it is to preserve.

Differences in stands of trees

The forest itself can dictate what trees should stay or go. How? First, either trees grow homogeneously or they grow as a mixed stand. Second, trees have a hierarchy for succession. The first tree species that grows on a site may not ultimately be the same species growing there when a full mature forest is finally established.

Trees growing in a homogeneous stand of timber don't possess as much individual character as trees growing in a mixed stand. Therefore, in a homogeneous stand, the choices narrow quickly when choosing individuals as save trees. Trees that comprise homogeneous timber stands typically are the same age, or *even-aged,* so they usually vary little in diameter, height, color, canopy, or spacing. As a result, homogeneous stands of trees are more prone to catastrophic disease and insect attack than are stands of mixed, or heterogeneous, species of trees. Variability in itself offers biological and physical protection.

So, with homogeneous stands of timber, differentiating which trees to preserve on the merit of individuality can be a meaningless exercise. When any tree will do, altering grading and roads to accommodate individual specimens or groups of trees stands loses importance. Instead, cutting pattern gains significance. Homogenous trees depend upon the stand for physical protection. This mutual protection vanishes with logging, making it critical to cut homogeneous stands in a fashion that minimizes exposure to micro-climatic changes in wind patterns, rain, temperature, and sunlight while ensuring that groups of saved homogeneous trees be as large as possible.

Individual trees in heterogeneous or mixed stands of timber possess more character than those in homogeneous stands. Trees in mixed stands tend to vary widely in species, spacing, diameter, height, color, and size or shape of canopy. This is one reason why it makes sense to select save single trees from stands of heterogeneous timber. Here, softwoods (trees with needles) and hardwoods (trees with leaves) mingle according to their place on the successional (ecological) ladder. For owners looking to save residual specimens, character trees are abundant in multiaged mixed stands of timber. Trees that vary in age also exist at different heights making them more wind firm and predisposed to the elements.

Reading timber

To this point, we've explored two ways to read timber: by the stand or by the individual tree. Besides judging trees based on appearance, are trees of a certain species, type, or size more desirable than others? For instance, are conifers preferable to deciduous trees, or is a mixture of trees important? How will nursery stock look against native residuals? Are there disease and pathogenic issues to consider? Here are examples of physical characteristics to consider when evaluating individual trees and their potential to survive on a site being developed:

Coniferous trees

1. Is the tree showing inner wood and skinned bark? Is it broken or missing its top? Any of these symptoms indicate an unhealthy tree or one that may soon be a hazard.
2. Are there multiple stems rising from the main truck? This condition may indicate a missing top with lateral branches competing to replace the top that's missing. Trees possessing this characteristic aren't attractive and may be rotting from the top down.
3. Does the tree appear to be sitting higher than the forest floor? If so, the tree may be growing on organic debris, an old log, or a stump resulting in a shallow root system. Trees in this condition are prone to wind throw and may have their root systems damaged easily.

Deciduous trees

1. Are there any open cavities? Large open cavities can lead to an array of unhealthy conditions including rot and insect damage.

2. Are large sections of the tree missing? Sheared-off limbs and trunks lead to dead wood, frost damage, and other health problems.
3. Are the trees bending or crooked in the direction of open sunlight? Trees bending indefinitely towards sunlight eventually fracture or topple over. Until they fall, they're a safety hazard.

All trees

1. Dead trees also called "snags" are an obvious liability.
2. Do any trees have exposed roots? If so, they may not be prime candidates for retention.
3. Evidence of sawdust triggers an investigation for insects or disease; neither of which indicates a healthy tree.
4. Broken trees can indicate heavy wind or the presence of rot that may also be weakening other trees in the stand.
5. Ghostly stands of gray dead trees sitting upright may have died due to rising water, poor drainage, or an evolving wetland. Forget the trees, tackle the water problem first.
6. Does the bole (stem) of the tree exhibit cracks, lumps, bunched limbs, a dead top, or missing bark? The tree may be suffering from disease, rot, insects, an old wound, or fungal attack.
7. Does the tree appear hollow? Banging it with a soft mallet or coring it with an increment borer can reveal the tree's interior condition. Hollow trees are a hazard and at risk for failure.
8. Premature defoliation. Is the foliage weak, off-color, dropping, or sparse? This condition reveals biological or physical stress. Trees in this condition are weak and typically not worth retaining.
9. Sharp changes in timber type may indicate past logging or agricultural practices. More importantly, it may indicate a change in soil or subsurface moisture conditions that may negatively affect future grading and construction.
10. Does the stand exhibit an abundance of downed trees? This is usually a sign of high winds sometimes combined with overly moist or thin topsoil, or it can indicate a shallow-rooted species. Groups of trees leaning in unison can signal uniform and strong wind from one direction or unstable ground.
11. Old road cuts and trails through a forest may indicate that adjacent trees may have already had their root systems cut or filled-over in the

past. Dangling roots and stubs are a sure sign of prior root damage. Full-canopied trees with damaged or missing roots may lean or topple when exposed to increased wind. While some species are more resilient to root damage than others are, there's no way to accurately predict how an individual tree will react to root damage over time. Generally, trees that survive root damage usually don't thrive.

12. On sloped ground, trees leaning at an angle with the trunk remaining relatively straight or in a long bow can indicate soil movement. It can also suggest that a one-time, recent catastrophic slide event has transpired. Another condition common on sloped ground involves J-shaped trunks where the lower part of the trunk is sharply curved upward. J-shaped trees can indicate the presence of a slow and deep-seated moving hillside where the tree, in a constant battle to remain plumb, slowly bends at the base forming what's called a "pistol-butted stem."

(The above guidelines are offered as information only as trees should always be evaluated with the assistance and oversight of a qualified forester or arborist.)

Owners can also tell a great deal about trees by studying the soil supporting them. Thin soils above bedrock or hardpan will force root systems to grow shallow and usually indicate the presence of moisture. Deep, well-drained gravels encourage roots to go straight down. A greater abundance of vertical roots indicate a higher probability for tree stability. Petition your geotechnical consultant and arborist for more information.

Hazard trees

Hazard trees can plague projects. Trees remaining along the periphery of open-space areas, greenbelts, parks, native protection areas, wetlands, and other sensitive areas can die, weaken, or threaten to fall into structures and other adjacent facilities. Evaluating hazard trees is far from being an exact science. More than one person has witnessed the crash of a healthy-looking tree when the half-dead tree next to it was the one considered hazardous. Nonetheless, there are ways to evaluate which trees should be removed in order to reduce potential hazards.

Certain tree species don't react well to changes that result from clearing and grading. Minimal changes in moisture or wind can send them crashing to the ground. Some fast-growing deciduous trees will bend towards a newly created opening in the canopy in order to gain sunlight until wind or their own weight causes them to topple into the opening. Other trees that appear stable may be slowly dying from the inside out due to stress or conditions naturally occurring in the forest. In all of these cases, the safest policy is to examine danger trees and remove those that pose an immediate, or soon-to-be, hazard, before they fall into the work zone.

Before logging and clearing, and with your permit agency, establish a policy for handling hazard trees. More specifically, draw up and agree upon a plan to reduce or eliminate hazard-tree liability as construction unfolds. Begin by understanding which native tree species are most prone to being unstable. A qualified tree professional can target and evaluate which trees residing along the cutting boundary are at risk for becoming hazard trees. Remove trees that are suspect, dead, dying, already leaning the wrong way, appear weak, or that remain potential candidates for falling into the project regardless of health condition. Given the latitude, don't take chances. The most efficient and cost-effective time to remove hazard trees is during, or right after logging because:

- overall construction activity is low;
- the slash will be handled during subsequent clearing operations;
- little, if any, additional damage will occur to the site by felling and skidding;
- and there will be equipment and qualified loggers available to perform the work making it more likely that trees will be cut and removed properly, and that merchantable logs will be manufactured in a manner that will yield maximum value.

Here are some reasons to pursue prompt removal of hazard trees:

- Dead trees in the air or on the ground are fire hazards.
- Builders or buyers may later require hazard trees removed as a condition of sale.
- Before they can become "politicized" and possibly affect a project, remove hazard trees.
- Hazard trees can come down anytime causing physical damage to structures and land improvements.
- Anyone walking or occupying a structure in close proximity to hazard trees is at risk of personal harm. Risk increases with wind, increased weight due to ice and snow, and with extreme temperature changes.
- After land improvements have been completed, hazard-tree removal and cleanup costs increase dramatically. Taking trees down piece-by-piece within striking distance of landscaping, hardscape, or existing structures is expensive work.
- Dead, dying, and leaning trees are a constant danger to workers and equipment. Vibrations caused by logging, clearing, grading, and utility operations can accelerate the degradation of weak or dead trees causing

them to fail when there are higher concentrations of people on-site.

- Hazard trees in the vicinity of site construction may slow or curtail work. Using the two-tree-length rule, a 60-foot-high (18 meters) hazard tree left uncut can literally freeze site-work in an area 240 feet (36 meters) in diameter, or approximately an acre (.42 hectare) around that hazard tree. Owners don't need the schedule impact, inconvenience, or liability that results from working under and within two tree lengths of hazard trees.

Opponents seeking to prevent the removal of hazard trees from greenbelts and native growth areas may argue that dead and decaying trees have wildlife value. While this may be true, the objective isn't to deprive creatures of a place to live. We remove hazard trees to ensure the safety of workers and property. Counter the arguments against removing hazard trees in a couple of ways. For one, a falling hazard tree that shatters on impact does nothing good for its occupants be they bugs or animals. So, wildlife and human habitat both are at risk of damage or destruction when hazard trees crash to the ground. Another solution may be to offer double the number of hazard trees slated for removal by strategically placing logs on the ground within buffered areas. Or, create twice as many new wildlife trees by removing the tops and limbs off live trees effectively transforming them into soon-to-be, dead stems suitable for bugs and wildlife. With this alternative, the key is to top them at a height shorter than their stump-distance to the edge of buffer. Also consider a species ability to resprout when cutting live trees for wildlife habitat. Trees likely to resprout limbs and produce foliage after cutting and limbing don't die and therefore make poor choices for wildlife trees. Topping and limbing trees can be accomplished by use of an excavator and chain saw or by employing specialized logging equipment such as a feller buncher. *Unmerchantable* hazard trees can be removed by use of standard logging practices or by simply pushing them back into the velvet green with a canopied bulldozer, skidder, or excavator. Above all, the key to removing unstable hazard trees is doing it safely.

When building a case for removing hazard trees, owners must make it clear that they only want to remove trees posing a current or most-likely risk to the project and field personnel. These include trees that can reach beyond the limits of clearing and into the development or work zone. Any hazard tree that exists horizontally at a distance deeper into the forest than the tree is high can stay provided that it doesn't pose a danger to existing or future trails.

A combination of extreme winds and saturated forest floors makes large trees a high-risk for toppling over in any direction. Some larger trees may have already had their root systems loosened or damaged by past winds. Regardless of how owners try to mitigate the loss of such trees, it's a matter of time before they come down or die due to natural reasons. If these trees can fall and reach into the project or beyond the buffer's edge, remove them. Pruning and controlled thinning is an effective technique used to neutralize menacingly large hazard trees. One objective of pruning and thinning is remove as much "sail" and wind resistance from the canopy as possible. However, be careful not to allow overcutting. A general rule-of-thumb is not to remove more than 20 percent of live foliage per year.

Preparing save trees for logging and clearing

All trees aren't created equal. Some species are more resilient to bark damage, root impact, and wind. It's counterproductive when somebody, such as a planner, buyer, or permit authority, requires saving the wrong native tree species. When this happens, the story typically plays out by having the owner mark individual or groups of trees as required. Logging proceeds, land is developed, and structures construction begins. After a year or so, exposure to sunlight, wind, and a changed moisture regime results in trees dying off, blowing over, or having their limbs sheared and their tops knocked out. From here, the tragedy unfolds in one of two ways. The save-tree area is either left in an unaesthetic state piled deep in litter and debris, or it's replanted at great expense. Whose expense? Who's responsible for choosing the wrong trees in the wrong location for retention? How can owners be better prepared to intercept this chain of events and, instead, with knowledge, retain the best and most resilient species of trees?

One way is to rely on the opinion of a tree specialist who can match resilient tree species with the owner's vision. Tree specialists can choose native species that possess the highest chance of survival after logging and that also enhance the project's landscape. Another way is to survey existing trees on-site that have been weather-tested and that remain hardy. Weather-hardened and full-canopied trees already occupying exposed ground are typically the right species and are worthy of retention. An option also exists to choose "candidate" trees per species, from within a stand, and prepare them for future disturbance. Another reliable source of hazard tree information is the local power company. Ask them what species in your area commonly takes out their power lines.

Go to *Figure 17-1* and follow the logic flow depicting one process for saving native trees. The logic begins with delineating save tree areas and works its way down through the process by evaluating the consequences of grading and construction, exposure to the elements, views, and finally, pruning and thinning.

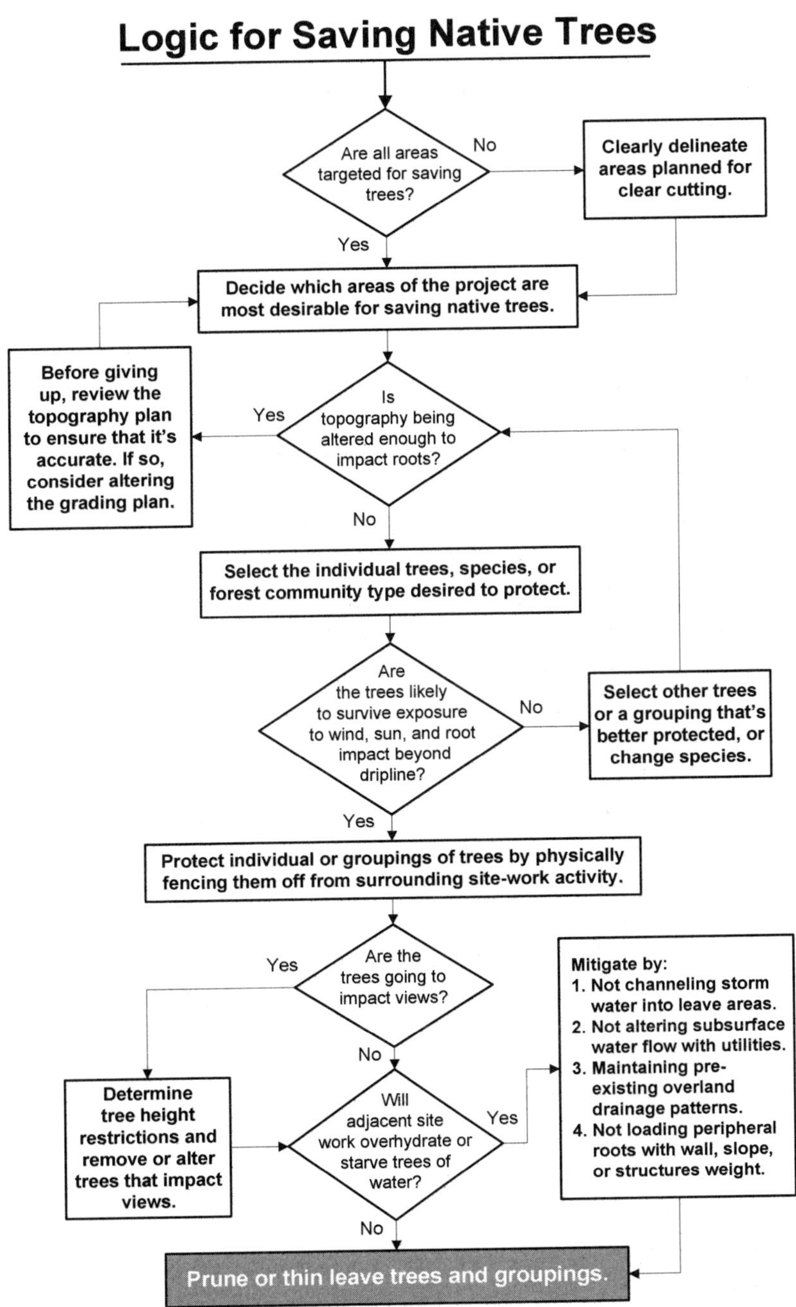
Logic for Saving Native Trees
Are all areas targeted for saving trees?
No
Clearly delineate areas planned for clear cutting.
Yes
Decide which areas of the project are most desirable for saving native trees.
Before giving up, review the topography plan to ensure that it's accurate. If so, consider altering the grading plan.
Yes
Is topography being altered enough to impact roots?
No
Select the individual trees, species, or forest community type desired to protect.
Are the trees likely to survive exposure to wind, sun, and root impact beyond dripline?
No
Select other trees or a grouping that's better protected, or change species.
Yes
Protect individual or groupings of trees by physically fencing them off from surrounding site-work activity.
Yes
Are the trees going to impact views?
No
Mitigate by:
1. Not channeling storm water into leave areas.
2. Not altering subsurface water flow with utilities.
3. Maintaining pre-existing overland drainage patterns.
4. Not loading peripheral roots with wall, slope, or structures weight.
Determine tree height restrictions and remove or alter trees that impact views.
Will adjacent site work overhydrate or starve trees of water?
Yes
No
Prune or thin leave trees and groupings.

Figure 17-1

To slowly prepare save or "candidate" trees for survival, preempt logging by thinning and pruning the stand. Prior pruning and thinning "preshocks" save trees. Thinning the stand allows save trees to acclimate to smaller doses of wind and exposure. Thinning and pruning invigorates trees, reduces stress, cleans out dead material, and lowers demands of the root system. The natural response to pruning and thinning is growth, and that's good because save trees in peak condition are more likely to survive disturbance. The catch though is to perform thinning and pruning well enough in advance to allow save trees the time to overcome the initial shock of cutting and react with vigor when confronted with site logging. Remove prunings and thinnings to reduce fire hazard, improve aesthetics, and allow native shrubs to flourish. It may not be necessary to prune all save trees in order to meet development goals. Seek the counsel of an arborist or forester when scheduling thinning and pruning.

Thinning and pruning also enhances views by removing entire trees or by altering canopies to improve the passage of sunlight and increase sight distance. The benefits of individual tree pruning are limited to clearing the ground-to-bottom of canopy field-of-view, or gaining line-of-sight above the tops of shortened trees. Shortening or topping trees is a common method for removing part of a tree's upper canopy. The intent of topping is often to improve views or control growth, however, the opposite can occur since topped trees usually respond by growing a dense collection of smaller replacement limbs. There's less risk of permanently damaging a deciduous tree through pruning and thinning than there is with a coniferous tree. As a rule—don't top conifers. Topping conifers severs the trunk leaving the cut exposed to mechanical damage from water, sunlight, temperature, and insect invasion. It also interrupts the internal nutrient and water transfer system by which coniferous trees sustain themselves. In short, they weaken, become stressed, and can die. Here are some benefits of thinning and pruning trees:

Thinning:

- Can produce log revenue
- Can arrest fungus and disease
- Improves views and sight distance
- Improves tree spacing and appearance
- Prepares save trees for logging and clearing
- Removes ugly and overgrown dominant trees
- Creates open areas and "edge effect" for wildlife
- Creates park-like open space for nature or hiking trails
- Enhances airflow and sunlight between lots and parcels

- Removes dead snags, hazard trees, undesirable, and unhealthy or disfigured trees
- Creates open space for new growth including transplants and imported landscape stock

Pruning:

- Reduces windsail area
- Improves individual tree health
- Balances a tree canopy's weight
- Prepares save trees for logging and clearing
- Creates park-like open space for nature or hiking trails
- Enhances airflow and sunlight between lots and parcels
- Reduces fire hazard by eliminating fuel ladders up into the crown

Depending upon stand condition, pruning accompanied by thinning yields the greatest benefits and value. There is a difference between pruning trees for commercial value and pruning them to enhance land development. Use consultants and contractors who understand the difference.

Getting help

Who chooses the save trees? The answer depends on whether or not there's enough merchantable timber to warrant commercial logging. Municipal preferences, owner objectives—such as maximizing view potential, overall stand condition, and the planner's vision for trees—also come into play. Selecting retention trees can fall on any, or a combination of the following individuals:

1. Forester
2. Arborist
3. Landscaper
4. Project owner
5. Landscape architect
6. Permitting authorities
7. Loggers or timber buyers
8. Tree grower or nursery operator
9. Neighboring landowners, builders, or buyers
10. Local municipal professionals and agricultural extension offices

Most of the above entities possess a different perspective on tree retention.

The best plan for owners is to accept varied opinions and distill all input into a plan that makes sense—but more importantly—works. Professionals qualified to assist in choosing save trees offer other related services including:

- Landscape architecture
- Urban forestry consulting
- Schemes for thinning stands
- Logging and timber plan layout
- Procuring logging and clearing permits
- Hazard tree inspection and removal strategies
- Logging management and mill receipt verification
- Management of pests, weeds, and tree-fertilization programs
- Developing tree pruning, cabling, and canopy-thinning criteria
- Managing the planting of commercial tree seedlings over areas conducive to reforestation
- Consulting services for selecting nursery or on-site stock suitable for replacing native trees

Locating protective fencing

Temporary fencing is normally specified to protect individual or groups of save trees. When protective fence is missing, trees are at risk of being damaged or destroyed. The outermost dripline or tree canopy edge is legendary for being the best-guess place to locate protective fencing. The problem with this practice is that some trees send out roots three times or more the distance of their dripline. And, species isn't the only factor that governs root length. Root systems vary according to location, tree size, stand density, soil type and depth, moisture regime, and an individual tree's reaction to wind or moving ground.

Arborists and foresters can suggest locations for protective fencing relative to individual trees. Also, consult with your permit agency. They may have minimum specifications for saving trees during construction. There's also a generic formula to locate protective fencing. The formula stipulates locating fencing out from the tree at whichever distance is greater: dripline or 12 inches (30 cm) from the main trunk for every inch (25 mm) of tree diameter at breast height. For example, an 18-inch (46 cm) diameter breast-height tree would have protective fencing located 18 feet (548 cm) out from the tree's trunk. Breast height is defined as a point measured 4.5 feet (137 cm) up a tree trunk from the ground. Ultimately, the limit of roots depends upon species, tree size and age, and ground conditions. Seek an arborist's assistance when locating fencing around older and large trophy specimens.

Transplanting native stock

Transplanting native trees is growing in popularity. It is, however, an expensive proposition and one that some landscapers resist. Transplanting native stock is expensive and makes choosing the right species and, moreover, the right individual trees, critical. Proceed only under the guidance of a qualified tree specialist.

The primary challenge to transplanting on-site trees is having enough quality native stock to choose from. The second challenge is size. As tree size increases, costs mount exponentially and survivability decreases. Successful transplanting requires cutting and removing enough ground volume around the tree to ensure maximum root retention. Even after removing a large root ball, the majority of a tree's root system remains. This can leave the tree in shock for a number of years after transplanting. In most areas, transplanting works best in spring and fall making ground conditions a critical issue when using wheeled or track-mounted transplant equipment. Hillside transplants are tough and can result in much higher costs and significant access grading. Getting a transplant vehicle in and out with stock may demand custom grading and it may also require building throwaway access trails constructed with fabric and crushed rock. When budgeting for transplanting stock, include access costs.

Meet early with a transplant contractor to discuss strategy, choose viable trees for moving, stake planting locations, and review access. Tree spaders may bid by the job, but they typically charge extra work by the hour so preconstruct adequate access to and from spade locations prior to work. Discuss beforehand how to prune each tree to reduce weight, provide spading room, minimize root system, and better balance the tree. Also, consider water. Transplanted trees may not be located near hydrants or irrigation lines so plan and budget for delivering water to transplants. Where the spade has taken trees, promptly backfill the holes to prevent them from filling with water, vehicles, animals, and people. If power or phone lines within a right-of-way have to be removed to allow transplants to pass, budget accordingly and contact service providers early enough to be ready when the spade arrives.

Under the best conditions, moving larger native trees on-site and expecting them to survive can be a risky proposition. Depending upon species and spade size, tree survivability can range from 10 to 60 percent. Statistically and financially, if transplanting is worth the cost, plan well in advance.

Construction ramifications

My dad would often say, "Things worth having are worth taking care of." As the years would testify, he was right. The same logic applies to site construction. When you've paid to survey, plot, plan, design, fence, thin, prune, transplant, grade around, and pamper saved native trees; why in blazes would you allow someone pushing levers behind 200 horsepower of churning iron to destroy it? In short—you wouldn't. Here's a list of potential problems and avoidance strategies aimed at protecting individual and islands of leave trees when performing peripheral site-work:

1. Never prune with the bucket of an excavator.
2. Keep leave-tree areas free of garbage and debris.
3. Identify the maximum dripline of a tree *before* pruning it—not after.
4. Don't allow even a thin skiff of topsoil or organic matter to be lost around leave trees.
5. Regulate the volume and timing when channeling runoff into leave-tree islands and plan for drainage.
6. Protect the immediate zone beyond save-tree fencing from traffic, equipment, material storage, concentrations of water, and debris.
7. Warn crane, concrete-pump-truck, lift-truck, and boom-truck operators not to swing equipment into and across zones of protected trees.
8. Control incidental compaction by keeping compaction equipment as far away as possible from leave trees and, certainly, nowhere near dripline.
9. Ensure that there's a separate base plan for leave trees. If not, show them on a topography plan with proposed infrastructure shaded in the background.
10. When options exist, prohibit landscapers from installing underground irrigation lines through leave tree zones. Surface lay or hand-dig shallow lines that have to pass through protected leave areas.
11. Seek the opinion of an arborist before irrigating save trees before, and during site construction. During droughts, seek advice as to how and when transplanted trees and shrubs should be watered.
12. Install and maintain bright colorful protective fencing around save trees and islands before, during, and after site construction concludes. Repeated encroachments may justify driving posts and hanging chain link fence. Always include signage advising people to keep out.
13. Before flagging clearing limits, check survey the existing ground elevations around leave trees. Using accurate topography is a step in the right direction to ensure that clearing and grading as staked, doesn't conflict with leave-tree areas. This further reinforces the case

for accurately plotting leave trees and islands onto the grading, road, and utility plans.

14. Making a tree well by surrounding a tree with walls and fill material can kill or weaken leave trees. Not only is root damage a possibility, concentrated ponding can also drown a tree. Solutions may involve mobilizing an auger to drill vertical weep holes to drain water. Tree wells can also starve a tree of surface runoff by disrupting historic flow patterns.
15. Lookout for concrete truck drivers who may view a patch of leave-trees as an excellent spot to clean out or dump remaining mix. Drivers accustomed to shade may find leave-tree islands as pleasant places to sit or park idle trucks, which can result in potentially more compressive impact to root systems. Plus, the heat from a truck's exhaust system can be hot enough to ignite dry needles, leaves, and branches scattered around leave tree islands.
16. Tunnel under large exposed roots connected to leave trees when utility trenching near protective fencing and dripline. When tunneling isn't possible, require that the roots be hand-cut instead of yanked through with a bucket. Pulling roots to break them results in unseen stretching and fiber damage further up the root system in the direction of the tree. Utility trenching around root systems requires digging and pruning by hand or using air excavation methods.
17. Fills or walls placed immediately outside of protective fencing may crush, suffocate, or channel excess water into root systems. Though expensive, walls adjacent to leave trees impose less damage when sitting on grade beams supported by piers. Wall-footing excavations can also sever root systems or leave them exposed, drying them out. An economical solution may be to place minimal setback limits on walls and toes of fill. The cheapest alternative, though, is *solution through design* starting with a smart grading plan.
18. It's better to curve around protective fencing than it is to run tangents under and through dripline. Designing work away from protective areas or getting on-site contractors to follow the rules is easy. There can be problems though with public service, municipal, or buyer and builder utility crews who don't understand *your* protective fencing. Wildcat crews and subcontractors showing up unannounced sometimes remove fencing and cut across root zones without posting notice. This can be resolved by maintaining constant communication with buyers, service providers, and municipalities, and by posting

plenty of warning signs including a copy of the landscape appraisal value for each save tree assessed. Also, promptly investigate *any* unknown and new staking or freshly painted utility markings. Follow up when trailers loaded with utility pipe and materials suddenly show up on-site.

Trees and water

Removing certain tree species can greatly affect soil moisture. Fast-growing deciduous tree species that rely on abundant moisture move a great deal of water out of the ground and up into their canopies by a process known as transpiration. Though such trees may not be wetland indicators, it's wise for owners to consult with an arborist, ecologist, or forester to find out what local tree species indicate subsurface moisture. Removing these tree species can result in a rising shallow water table causing moisture to accumulate and impact earthwork, grading, and utility work. When they are cut, water-sucking trees exhibit much wider light-colored wood rings (early wood) than dark-colored rings (late wood). Early wood or *springwood* is light-colored wood developed annually during the growing season and primarily made up of water vessels. In some species, water-sucking trees will squirt water out of these vessels when cut.

Here are some general maintenance and native-tree replacement guidelines to consider:

1. Don't mow seedlings.
2. Remove vines from trees.
3. Leave tree work around power and utility lines to service providers.
4. Topping trees can result in rot, disease, weakened limbs, and an ugly tree. Avoid it.
5. Don't allow turf or grass to grow where mowing could damage shallow roots and trees.
6. When replacing native stock, remember that fast-growing species tend to die out quickly.
7. Smaller growing varieties of trees are less prone to reach out with destructive root systems.
8. Native trees that are retained after opening up a dense forest may require temporary irrigation to survive.
9. Too much overland flow or storm runoff into a stand may kill the trees, develop a wetland, or promote root rot.
10. Understand the risk in agreeing to transplant native tree species that

are tough to relocate and that possess a high mortality rate once disturbed.

11. Don't agree to leave individual trees based solely on tree size. Reserve judgment and draw conclusions according to species, condition, location, and adaptability to change.
12. Clear-cutting homogeneous sites of water-loving trees may have minimal impact to overall site construction if mass grading later will cut deep enough to intercept, and remove, perched or other sources of groundwater.

Remember—there's a difference between protecting trees and saving them.

Chapter 18

LEVERAGING PRE-EXISTING CONDITIONS

Profitability in land development begins by preserving the value of raw land as it exists in its primitive state. This may mean taking precursory action to prevent trespassing, encumbrances, and degradation. Or, it may also denote preserving land value by thinning trees, erecting access controls, or developing a commercial pit well before submitting plans for construction permits. This chapter highlights simple prepermiting techniques aimed at improving and protecting a property's condition, increasing or maintaining its value, and better preparing it for design, permitting, and construction.

Introducing neighbors to someone who knows site-work

While project owners need to consider politics, civic responsibility, and timing when interacting with neighbors, they must also demonstrate, at the most elementary level, that they're open, honest, and knowledgeable about what they're doing. The importance of gaining allies amongst neighbors who either share access to or surround a project cannot be overly emphasized. Gaining neighborhood or local approval is a double-edged sword. When meeting in public forums either to display or to introduce projects, too often owners focus on the top edge—the often politicized issue of selling the broad benefits of a project. In doing so, they neglect the hidden edge of questions and personal concerns that neighbors harbor such as ongoing access, site-construction methods, and boundary impact. These issues should be nipped-in-the-bud as soon as they surface. Project owners and their team usually only get *one* chance to develop certain neighbors as allies. It's crucial that they capitalize on and nurture these delicate encounters, and that they handle each issue with the highest regard. In large part, this means being prepared and having answers.

When attending public meetings, feature a friendly, plain-talking, nuts-and-bolts site-work professional as part of the team to interact one-on-one with neighbors. Although some of them may fit the bill, I'm not referring to planners, architects, or civil engineers. Instead, consider featuring a seasoned site-construction authority that is knowledgeable about the project. Without saying a word, the presence of a construction individual bolsters the owner's credibility, demonstrates preparedness and concern, and retains someone in case construction issues arise. Besides, neighbors may gravitate to and trust someone who is thought of as being lower on the ladder—a sort of casual, salt-of-the-earth construction type—capable of handling everyday concerns and grievances. This tactic is even more effective if the person attending the meeting will also oversee or manage actual on-site-work.

A site-work professional can serve as a buffer and prevent an owner from responding with wrong or misleading information to questions concerning technical field issues. Remember that neighbors and project opponents have long memories. While project owners cannot and should not absolve themselves from communicating with neighbors, a site-construction-oriented representative can tackle specific issues, bridge problems with troublesome neighbors, address scheduling, detail and carry neighbors' concerns to other consultants, and insulate the owner if issues surface concerning unfulfilled promises or capital improvements. A good site-construction representative can resolve a multitude of personal and technical issues with neighbors and government officials, and make the owner and project team look awfully good while doing it.

And—don't forget—municipal authorities attending pre-entitlement town meetings also want confirmation that the owner has capable people ready to carry out whatever they permit. Many government officials look past the upfront people and the owners and, instead, base their decisions for approval on the strength of the owner's supporting cast. They focus on the individuals or firms that are going to make the project happen on a day-to-day basis in the field. Can they pull the project off or is issuing permits a liability for the agency at hand? How neighbors and civic officials perceive the site-work specialist goes a long way to building confidence that leads to trust and, ultimately, permits.

Prior to construction, owners can maintain and improve neighbor relations by following a few quick and practical steps. Here are some to consider:

- Keep your property clean of garbage.
- Secure property boundaries and prevent trespassing.
- Don't store materials on neighboring land without permission.
- Offer to remove hazard trees that threaten neighbors or their property.
- Keep meticulous and accurate records of all conversations and promises made.
- When accessing a project from private roads, offer to grade them, fix potholes, and generally improve the road for all parties.
- Without neighborhood consent, don't allow contractors to access the land in off-hours. Keep all ingress and egress to normal working hours.
- Don't allow contractors to park equipment, personal cars, and materials in front of neighbors' property without the neighbor's permission.
- Alert neighbors of work that may alter, or appear to impose on, their

property such as boundary survey work, property-line fence installation, or logging.

- Unless the neighbor has use for the water, rechannel drainage that is flowing unnaturally onto adjacent property. If a neighboring property is unnaturally draining onto your property, casually check to see if the neighbor possesses (or wants) a drainage easement.
- Agree with neighbors and commit to writing exactly what and when you'll deliver whatever it is you've agreed to provide or deliver. Don't let it sit. Time can allow a neighbor to think and later demand more than agreed upon. When completing work for a neighbor, send them a nicely written letter of completion including pictures portraying before and after conditions. For your own records, videotape everything. Offer to inspect any work performed for a neighbor long after your warranty or grace period expires. Do quality work.

In like fashion, neighbors can provide owners with many benefits including potable water, alternative access routes, areas for staging equipment and materials, utility and retaining structure easements that can increase net developable area, and a willingness to allow off-hour and extended working conditions without complaint.

In as much as it pays to work with neighbors, project owners may have to challenge them or find common ground concerning long-time infringements, trespassing, and other property issues that neighbors may have, historically, taken for granted. In fact, some neighbors may believe that they own portions of the project owner's property. For project owners, it boils down to asking, "What can we do to better position our project for development?" and then—follow through.

Securing vacant property

On-site security begins with gating entrances and blocking trails into the project. Protect gate hardware, such as locks and hinges, from vandalism or being shot off. Also, common-key all locks so that *one key* works universally throughout the project, and weld lock boxes to all gates blocking vehicular access. Lock boxes are a useful place to store a copy of the project owner's gate key. Police and fire personnel that have copies of the same lock-box key can, at any time, open the lock box and use the owner's key to open the gate. Purchase lock boxes through your police or fire department so that they will be the correct boxes and so that they will work with keys already in use by both departments.

Using bare cable or chain as a gate or an access impediment is dangerous and not recommended. An alternative is to attach cables or chain to concrete blocks or trees and string the chain or cable through brightly painted (fluorescent), easily seen, larger diameter, schedule 40 or thicker PVC that runs from one point of attachment all the way to the other point of attachment. Extend the PVC sleeve so that the chain or cable remains totally hidden. A more expensive solution may include blocking access with concrete blocks, Jersey barriers, rocks, or bollards. Unfortunately, these measures are only effective at stopping four-wheel vehicles. Stopping horses, bicycles, and motorized recreational vehicles may require a number of trial-and-error attempts until a final solution is found. Post no-trespassing, danger, and trail-blocked signage.

Trails through an owner's property usually lead everywhere including to some neighbor's backyard. In these instances, it's good policy to speak with the neighbor about the issue, post no-trespassing signs and remove or plug the point of access. One technique that works well is dumping hog fuel and screened sticks onto trails. Don't grade the trails—just place the material in the trail rough and uneven. Hog fuel and mats of sticks are highly unpopular with errant travelers. Another alternative is to dump stumps in narrow portions of the trail. Unsightly, but effective.

Videotape

Videotape all pre-existing trails and points of access into your property from off-site neighboring land. Also videotape all pre-existing adjacent infrastructure including at minimum, roads, pavement, curbs, sidewalks, landscaping, streetlight poles, storm grates, sanitary lids and water valves lids, and fencing. In short—video everything within the right-of-way at least a quarter mile (400 meters) or more beyond the limit of the project.

The best time to video is right after the surface of wet pavement has dried but before the cracks dry. Dark wet cracks and defects contrast well against lighter and drier flat surfaces. The idea is to verify pre-existing hardscape damage before the advent of hauling. Before shooting, make sure that the video camera is displaying the correct time and date. Also, photograph areas of significant damage from every angle. Duplicate and safely store the videos and photographs in an off-site vault. There are a multitude of reasons to take video but the primary motive is to prepare for possible attempts by a municipality to unfairly get an owner to pay for and restore pre-existing infrastructure damage. I know of at least one instance where this strategy saved a project owner from rebuilding a mile or so of paved highway.

Revegetation

Since it takes time to establish thriving vegetation, hydroseed bare spots and establish plantings in areas that are ready and slated for future landscaping. This prepares the site for erosion control, provides aesthetics, and suggests quality ownership. Hydroseeding may require a minor degree of ground preparation and there must be sufficient rain or moisture to germinate and support seeding. Early landscaping works if pests like deer aren't an issue, theft is unlikely, impending construction won't damage plantings, and irrigation is available. The key to maximizing returns in site-work is to be ahead of the curve, and nothing demonstrates this more than revegetating land and performing erosion control before issuance of construction permits.

A cheap solution for owners planning to keep broad open areas ungraded and wild is to plant commercial tree seedlings. Commercial seedlings are cheap and they go in quickly. When matched to your projects microclimate, they perform well and can help control erosion. A forester or arborist can recommend reputable contract tree planters, species, and spacing suited for your project.

Another problem facing owners is damage to boundary trees caused by a neighbor's logging, construction, or general negligence. An owner may have to invest in planting replacement stock when trees formerly expected to act as a buffer or screening, are unexpectedly damaged or destroyed. Without planting, an owner may have to forfeit additional green space or bolster screening by widening buffers and increasing setbacks thereby effectively reducing net developable area. Evaluate, document, and address neighbor-caused tree damage as early as possible.

Encumbrances

Encumber or be encumbered. For some in the industry, encumbering property as a means of controlling borders or establishing presence can be a sensitive subject to broach. While that's understandable, two facts remain clear. First, if a project owner doesn't take steps necessary to protect his property and investment potential, nobody-else will. Second, more land is restricted from use each year than becomes available for development. The term "shortage of land" is a phrase that we continually hear in real estate, land development, and construction. As a general strategy, it can be wise to legally encumber adjoining property as soon as it's practical to do so. Encumbrances can allow project owners to indirectly control property lines, improve adjoining property, and maintain buildable area. Encumbering land can protect an owner's interest and preserve land value while affording that same owner leverage if, and when, he may have to negotiate with neighbors and adjacent landowners. Here are some thoughts on encumbering land as well as strategies useful for maintaining project control and maximum buildable area:

- Procure easements from neighbors early.
- Establish backbone infrastructure including utility stubs a soon as possible.

- Establish structures to be located on or near property boundaries as early as possible.
- Establish unilateral setback distances early by installing utilities along property boundaries.
- Build water reservoirs and utility structures while they can be hid behind a forested backdrop or screening.
- As a rule, when encumbering property, it's better to work from the project's outer boundary in, rather than from the center of a project out.
- Log and clear as soon as permits allow to counter those who may attempt to challenge subsequent road and utility permits based on the fact that trees are still standing.
- Clear to approved boundary setback distances early if it appears that neighbor pressure or municipal policy may result in enacting new rules to increase setback distances, prohibit logging, or widen buffers.
- The structural nature of retaining walls typically forbids filling against or undermining them. Therefore, whichever party *first* places a retaining wall on his side of the property line controls that line and may encumber his neighbor's property for a short distance.
- Grant easements only after receiving full payment and *before* the buyer secures construction permits within the easement, or, affix a "shelf-life" to the easement that will expire unless full payment is received. Grantees may be less motivated to fulfill their payment obligations after securing easements and permits.
- Fence property lines early that separate your project from adjoining land that is soon to be developed. New residents on adjoining property may attempt to force you to install an elaborate, expensive, or out-of-proportion fence for visual or acoustic reasons. Let *their* developer install specialty fencing and sound barriers.
- Don't count on "grandfathering." Complete work in preserved and sensitive areas, including open and green space, as soon as you receive construction permits to do so. For example, establish storm water outfalls, place culverts or bridges through wetlands, install utilities, remove hazard trees, and build trails before regulations or newly-adopted policy negate being able to begin or complete such types of work.
- Design and layout utility systems and stubs so that they are numerous and large enough in capacity to serve adjacent undeveloped property. Also, design roads that will adequately handle maximum (densest) zoning on adjacent property. Landowners on both sides of the property line benefit. One owner collects fees via a "latecomer's agreement" and the other is more likely to procure quick approval to develop land that is fully serviced by existing infrastructure.
- Views easements are valuable commodities, but granting one across *your* project may unintendedly impact planning and later trigger

architectural and landscape constraints. Grantors should also remember that it might be prudent to delay logging, thinning, or pruning to satisfy a view easement if the value of the easement is apt to appreciate over time. The best time for grantees to pursue view easements is before the grantor realizes either the magnitude of the project that the grantee is developing, or the value that the easement will create. When granting or seeking view easements, describe them in measurable and objective terms that a survey crew can easily calculate and stake in the field.

- Have a qualified engineer inspect pre-existing retaining walls that abut the edge of your property. Perform whatever tests are necessary to evaluate walls that appear structurally unsound or that seem to whisper something darker—like they could collapse at any moment. Research a wall's design, age, and history at the local building department. A structural retaining wall built without permits may be dysfunctional and a catastrophe waiting to happen. Verify wall locations relative to property line regardless of wall condition. The owner of a decrepit wall will have to repair or replace it in accordance with municipal standards. Dealings with property lines, off-site retaining walls, encroachments, and adjacent landowner issues require the services and support of a qualified attorney.

Researching existing features

Site-work planning requires researching a property's history such as past land use, homesteads, utilities, well history, septic and underground storage tank locations, shoreline boundaries, and past mining or landfill operations. Other bits of desired information should include meets and bounds, plat information, public easements, rights of access, and an accurate accounting of water, timber, surface, and subsurface mineral rights. It's particularly valuable to know whether or not a property's subsurface mineral rights have been separated from the oil and gas rights which could make a property a potential window for oil or gas drilling. For some properties, there may be as much as 200 years' of useful research information available dating back to the early 1800s. Any of the following sources may serve as a good starting point when researching a property's history:

- Court records
- ALTA surveys
- Title companies
- Fire department records
- County auditor's offices
- Utility company records
- Historic real estate atlases
- Local and regional libraries
- Historic fire insurance maps

- Phase 1 environmental reports
- Local colleges and universities
- Local aerial photography companies
- Regional branches of the National Archives
- Local building and public works departments
- State and local preservation and historical societies
- Private company or state-run mining and timber departments
- United States Coast Survey (U.S.C.S.) historical coastline surveys
- United States Geological Survey (U.S.G.S.) historical topographical surveys
- National Digital Library (at the Library of Congress) for historic hand-drawn panoramic maps depicting city and street layouts
- General Land Office (GLO) and later, Bureau of Land Management (BLM) Cadastral surveys showing land ownership, lot dimensions, bearings and distances, and corners

As an owner digs deeper into existing records, additional specialized sources for information and possible leads may crop up that either work to clarify past details or, sometimes, further cloud them.

Here's a partial list of items that owner may have to contend with on property destined for development:

- Abandoned fuel tanks may lead to hazardous materials and environmental studies of the area.
- Existing fuel lines, sewer lines, and other utilities that are located may require removal and backfilling, or plugging and abandonment.
- In some localities, poor drainage can create wetlands in less than a year. Inspect recent activity that may have left ground depressions that can fill with water and potentially form a wetland.
- Depending upon condition, run-down structures and handyman specials slated for demolition work well as temporary offices and for storing weather-sensitive materials such as jute mat and grass seed.
- Missing timber can be a real eye-opener. So can having somebody use an owner's land as a borrow pit or convenient dumpsite. Unfortunately, by the time the damage is done, it's usually too late to do much about it, which is all the more reason to gate, to use locks, and to regulate access to vacant property.
- Locate unknown underground septic tanks and cesspools before a piece of heavy equipment crashes through and finds them for you. While it may be permissible to pump out and backfill such structures, beware of tanks that fall into developable area as they'll have to be removed, and the hole will have to be backfilled with structural material and compacted to specification.

- Changes in the size and direction of surface water can lead to poor, out-of-date, or inaccurate flood hazard mapping. While up-to-date mapping might not lead to fewer permitting and environmental regulations, it does help owners by providing them with more accurate flood-zone delineations. Accurate mapping allows owners to better plan and design projects, may yield more buildable area, and can save owners time and money if previously required construction or mitigation measures are shown to be unnecessary. Better mapping also reduces the likelihood of surprise flooding and helps to define where flood insurance is absolutely necessary.

The ALTA survey

By use of an ALTA survey, field verify encroachments and boundary conditions and immediately deal with any outstanding issues. Known as an ALTA survey, the American Land Title Survey can be used to confirm conditions of public record disclosed by a title report and verify items such as:

- Land features;
- boundary line conflicts and error;
- discrepancies in the deed or title report;
- physical encroachments by adjacent landowners; and,
- encumbrances including undocumented and existing easements, right-of-ways, leases, and zoning issues.

Monitoring weather

Owners wanting to bid site-work, to develop an unpredictable or risky property, or to finance a project likely to be negatively impacted by weather should install an on-site weather station. A small, private weather station can gain credibility if it's used as a third-party source for providing local weather data *before* construction begins. This requires tying a new weather station into a network of surrounding stations that constantly share and cross-check data. Credibility is mandatory for owners wanting the option of using on-site station data to support or dispute a weather-related construction issue, or, if actual on-site rain data is required for culvert design, to produce storm-water requirements, or to verify a storm's intensity. For owners, this entails coordinating with other weather reporters in your area and developing a weather-reporting track record. Weather hobbyists and professionals can provide excellent support should later, they be asked to they be asked to make a statement, describe a weather event, or endorse your station's data. If it makes sense to have your own weather station, install one immediately and begin developing a preconstruction baseline of historic weather conditions.

THIS BOOK AND THE PROJECT LOGIC BOOK SERIES

Preparing Projects for Site Construction is the second book in the Project Logic Book Series and the first in a trilogy of books planned for site construction and smart land development. The next book in the Project Logic Book Series will begin with bidding and cover the actual site construction component of land development.

INDEX